Novel Developments in Fish Biology

Novel Developments in Fish Biology

Edited by **Roger Creed**

R CALLISTO
REFERENCE

New York

Published by Callisto Reference,
106 Park Avenue, Suite 200,
New York, NY 10016, USA
www.callistoreference.com

Novel Developments in Fish Biology
Edited by Roger Creed

International Standard Book Number: 978-1-63239-481-1 (Hardback)

The publisher's policy is to use permanent paper from mills that operate a sustainable forestry policy. Furthermore, the publisher ensures that the text paper and cover boards used have met acceptable environmental accreditation standards.

Trademark Notice: Registered trademark of products or corporate names are used only for explanation and identification without intent to infringe.

Printed in the United States of America.

Contents

Permissions

List of Contributors

Preface

This book provides comprehensive and research-focused information regarding novel developments in the field of fish biology. It provides the readers with a comprehensive study on fish biology. Recent advancements on cellular and molecular biology and ecology have led to the collection of advanced scientific knowledge in this area. Prominent research works from across the globe have been compiled in this book to provide a useful source of reference for researchers, teachers as well as students of advanced biological sciences. This book provides the readers with a basic description of the complex biology of immune response. In-depth descriptions have also been included for the intensive understanding of cytokine regulation in teleost immune system. Important details about the environmental stressors on the responses of fish from molecular to population level have also been provided. The book also discusses topics like the roles of the atrium and the ventricular system in teleost species and the tracer methodologies used for measurements of carbohydrate metabolism. It talks about the variables responsible for the feeding action of a predacious freshwater fish species.

This book has been the outcome of endless efforts put in by authors and researchers on various issues and topics within the field. The book is a comprehensive collection of significant researches that are addressed in a variety of chapters. It will surely enhance the knowledge of the field among readers across the globe.

It is indeed an immense pleasure to thank our researchers and authors for their efforts to submit their piece of writing before the deadlines. Finally in the end, I would like to thank my family and colleagues who have been a great source of inspiration and support.

Editor

New Advances and Contributions to Fish Biology

Regulation of Teleost Macrophage and Neutrophil Cell Development by Growth Factors and Transcription Factors

Barbara A. Katzenback, Fumihiko Katakura and Miodrag Belosevic

Additional information is available at the end of the chapter

1. Introduction

Macrophages and neutrophils are the sentinel cells of the innate immune response of verte‐brates, such as bony fish (teleosts). As phagocytic myeloid cells, they are involved in homeo‐static mechanisms, wound healing, and the detection, elimination and clearance of foreign entities including tumors, virus-infected cells and invading pathogens. Furthermore, macro‐phages and neutrophils are responsible for producing hundreds of bioactive molecules that are important in pathogen recognition and destruction, cellular communication and activa‐tion, initiation of an adaptive immune response and later, resolution of an inflammatory re‐sponse and tissue repair. Neutrophils and macrophages, while essential to survival, have a finite lifespan. Therefore, a manufacturing centre, the hematopoietic niche, is needed for the production of myeloid cells. The hematopoietic niche must maintain basal myeloid cell pro‐duction levels during homeostasis, yet retain the flexibility to ramp-up cell production in re‐sponse to physiological demands, such as pathogenic insult. The development of macrophages (monopoiesis) and neutrophils (granulopoiesis) is collectively known as mye‐lopoiesis, and is regulated by the complex interaction of colony-stimulating factors (CSFs), their receptors, and intracellular transcription factor machinery that control lineage fate de‐cisions and terminal differentiation events.

Over the past 50 years, research using the mouse model system has culminated in the identi‐fication of the site(s) of myelopoiesis, the progenitor cell types that give rise to mature mye‐loid cells, the extracellular and intracellular cues required, and a detailed understanding of the complex intracellular and extracellular milieu of factors that drive this tightly controlled process. From these studies we understand hematopoiesis as an exquisitely fine-tuned, highly regulated, process whereby all blood cells develop from a small number of hemato- poietic stem cells (HSCs). HSCs are characterized as long-term repopulating, pluripotent, quiescent cells that undergo symmetrical self-renewal to sustain the population of HSCs within the hematopoietic niche, or asymmetrical division to give rise to hematopoietic pro- genitor cells (HPCs) [1]. HPCs can develop along the lymphoid lineage, termed lymphopoi- esis, to give rise to B-cells, T-cells, natural killer (NK) cells and dendritic cells (DCs). Alternatively, HPCs can develop along an erythroid lineage, termed erythropoiesis, to give rise to erythrocytes and megakaryocytes, or develop along a myeloid lineage to give rise to granulocytes (neutrophils, basophils, eosinophils, mast cells), mononuclear phagocytes (monocytes and macrophages), and DCs. The lymphoid lineage represents the adaptive arm of the immune response, while the myeloid lineage represents the innate arm of the immune response. Regardless of lineage, the decisions made to commit and develop along a given lineage are controlled by extracellular growth factors and intracellular transcription factors that act in concert to regulate gene and protein expression to achieve the desired outcome.

When compared to the mechanisms of myelopoiesis in the mouse, studies using lower vertebrates, such as teleosts, have identified both evolutionary conservation as well as di- vergence in the mechanisms of myelopoiesis. With over 30,000 identified species, teleosts are the most expansive class of vertebrates, and represent an excellent model system to study the evolution of vertebrate myelopoiesis as they are one of the ancient classes of vertebrates to retain the production of myeloid cells. Within the teleost system, much re- search surrounds the characterization of teleost cytokines and receptors involved in in- flammation and their cellular targets (primarily macrophages). In comparison, little is known about the mechanisms that govern myeloid cell production. Research on teleost myelopoiesis is hampered by the lack of reagents, the difficulty in isolating appreciable numbers of relatively pure populations of HSCs/HPCs, and in identifying key growth factors important for myeloid cell development due to evolutionary selection pressures. As such, the focus of this review is to provide an overview of the current knowledge of the fish model systems used and the growth factors, receptors and transcription factors involved in teleost myelopoiesis, using information from the mammalian model systems as a scaffold to put the advances into context.

2. Teleost model systems of myelopoiesis

2.1. Zebrafish model system

The zebrafish model has been instrumental in advancing our knowledge of the sites of hem- atopoiesis/myelopoiesis in teleosts, the development, differentiation and migration of HSCs, and through genetic manipulation, the characterization of the early acting growth factors, receptors and transcription factors involved in hematopoiesis. By far, the major advantage provided by zebrafish is the ease of generating transgenic zebrafish, morphant zebrafish (morpholinos) and knockout zebrafish (zinc finger nucleases), as well as many others, due to genetic manipulation. In conjunction, the rapid generation of embryos, embryonic transparency, and small embryo size allows for mass screening strategies. The advantages of zebrafish as a model system, and their contributions to hematopoiesis have been extensively reviewed elsewhere [2-5], and thus will not be covered in this review. While the zebrafish is an excellent *in vivo* model, it does not lend itself to *ex-vivo* studies due to the small size of the fish and the difficulty of isolating sufficient number of cells for *in vitro* studies.

2.2. Ginbuna crucian carp model system

In vivo transplantation to test the repopulation activity of donor cells has been the gold standard for the characterization of HSCs and HPCs [6-9]. In cyprinid fish, there is a unique transplantation model system for detecting HSCs and HPCs using clonal ginbuna crucian carp (*Carassius auratus langsdorfii*, S3n strain) and ginbuna-goldfish (*Carassius auratus*) hybrids (S4n strain). Ginbuna crucian carp have advantages for transplantation experiments because they are easily maintained, tolerate handling and are large enough to allow for the collection of sufficient hematopoietic cells. Clonal ginbuna are unisexual triploid fish (all female, 3n = 156) that principally reproduce gynogenetically. A unique clone (S3n) can reproduce by not only gynogenesis but also bisexual reproduction. When eggs from the S3n clone are inseminated with UV-irradiated goldfish sperm, triploid clones result. In contrast, when the eggs are inseminated with normal goldfish sperm, tetraploid hybrids (S4n) are obtained [10]. These S4n fish possessed four sets of chromosomes, three from the S3n clone and one from the goldfish. Therefore, when the cells from S3n clones are transferred into S4n recipients, transplants are accepted, whereas the reverse transplants are rejected [11, 12]. Moreover, the donor cells in the recipient tissues are easily distinguished by their difference in DNA content by flow cytometric analysis (ploidy analysis) [13].

The ginbuna crucian carp model system has been instrumental in serving as a close parallel to the mouse model system in terms of hematopoietic reconstitution experiments to demonstrate the existence of HSCs in teleosts. Identification of donor and recipient HSCs/HPCs and characterization of their progeny by ploidy analysis is useful for assessing the multipotency of different progenitor cell populations. Furthermore, the use of this model system has allowed for the determination of the location of HSCs within the hematopoietic organs of cyprinids. Use of the ginbuan crucian carp system will be particularly important for future work as antibodies are developed against markers on the surface of fish HSCs and HPCs to allow for the analysis of the potency of progenitor cell subpopulations.

2.3. Goldfish model system

The goldfish model system represents a unique opportunity to study myelopoiesis *in vitro*. Firstly, teleost monopoiesis can be examined using the previously developed primary kidney macrophage (PKM) culture system [14, 15] and has provided information on the growth factors, receptors, and transcription factors involved. Secondly, large numbers of relatively pure neutrophils can be isolated from the goldfish kidney [16] and represents a starting point for studying granulopoiesis in goldfish and will be discussed in the following sections.

Together, these two model systems will prove instrumental in understanding the factors that regulate teleost myelopoiesis.

In the *in vitro* PKM system, small mononuclear cells isolated from the goldfish kidney proliferated and differentiated over 8-10 days giving rise to three cell sub-populations, R1-, R2- and R3-gated cells [14, 15]. The cytochemical, molecular and functional characterization of these cell sub-populations demonstrated the presence of putative progenitor cells (R1 gate), monocytes (R3 gate), and mature macrophages (R2 gate). These three cell sub-populations in PKM cultures represent distinct junctures of macrophage development simultaneously occurring *in vitro* [14, 15].

The spontaneous proliferation and differentiation of PKMs suggested the production of endogenous growth factors and prompted the examination of the target cell sub-population(s) upon which they acted and their effects on cell proliferation and differentiation. The putative progenitors (R1 cells) and macrophages (R2 cells), but not monocytes, were determined to be responsible for the production of endogenous growth factors that act in an autocrine and paracrine fashion [15]. Addition of cell-conditioned medium (CCM) to sorted cell populations demonstrated the capacity of putative progenitors and monocytes to proliferate and differentiate in response to endogenous growth factors. However, treatment of macrophages (R2 cells) with CCM demonstrated their apparent terminal differentiation, while their capacity to proliferate suggested they were capable of self-renewal [15, 17]. Clearly, different endogenous growth factors present in CCM exert distinct actions on macrophage cell sub-populations.

Two pathways of macrophage development were proposed to occur in the PKM cultures. The predominant pathway was classical macrophage development in which progenitor cells differentiated into monocytes and then macrophages [17]. The second was an alternative pathway of macrophage development in which progenitor cells differentiated into macrophages without a prominent monocytic stage [17]. The possible retention of the alternative pathway of macrophage production in addition to the classical pathway may provide a mechanism for rapid generation of macrophages during injury or infection *in vivo*.

The observed kinetics of the PKM cultures suggested three phases of growth. Initially, there is a lag phase (days 1-4) where many cells die, followed by a proliferative phase (days 5-9) where cell numbers rapidly increase [14], and finally, a senescence phase (days 10-14) characterized by cell clumping and cell apoptosis [17, 18]. Differential cross screening of proliferative versus senescence phase PKMs identified a number of differentially expressed genes including those involved in hematopoiesis, signal transduction, transcription, translation and protein processing [19]. The involvement of the identified transcripts in the regulation of cell development [20-22] will be discussed in the following sections.

These seminal observations from PKM cultures established three important ideas regarding goldfish monopoiesis: (1) kidney leukocytes produce their own endogenous growth factors important for driving proliferation and differentiation [14, 15]. (2) Within the population of small leukocyte R1 cells, a population of macrophage progenitor cells must exist. (3) Unlike mammalian systems, the progenitor cell population gives rise to fully differentiated macrophages *in vitro* in the absence of exogenous growth factors. Thus, the goldfish PKM model system allows for comprehensive analysis of the interactions between developing macrophage subpopulations in vitro.

3. Site of hematopoiesis/myelopoiesis

3.1. Two waves of hematopoiesis in vertebrates

There are two waves of hematopoiesis in vertebrates. The first wave is primitive hematopoiesis and occurs during embryonic development. Definitive hematopoiesis follows primitive hematopoiesis and occurs in the post-natal or adult animal. Primitive and definitive hematopoiesis are different on a temporal scale, a spatial scale, and in the types of cellular progeny generated. With the exception of T-cells, that undergo maturation in the thymus, lymphopoiesis and myelopoiesis occur in the major hematopoietic organs. The major hematopoietic organ of teleosts is the kidney, akin to that of mammalian bone marrow.

3.2. Primitive myelopoiesis in teleosts

The development of myelopoiesis in fish has primarily been studied using the zebrafish model system. Primitive myelopoiesis is predominated by HPCs with primarily erythroid and myeloid development potential. Initially, primitive hematopoiesis is initiated in the anterior lateral mesoderm (ALM), that gives rise to the rostral blood island (RBI), and in the posterior lateral mesoderm (PLM), that gives rise to the intermediate cell mass (ICM). The RBI is the site of primitive myeloid cell development, generating primarily primitive macrophages that undergo rapid differentiation, lacking or having a very short monocytic stage [23] and a few neutrophils [24], while the ICM is the site of primitive erythroid cell development [25]. This stage of primitive hematopoiesis occurs early during development of zebrafish, approximately 11 hours post fertilization (hpf). Following the onset of circulation, at around 24 hpf, the site of hematopoiesis then switches to the posterior blood island (PBI) [26] and produces multi-lineage progenitor cells capable of producing both primitive erythroid and myeloid cells [27]. Primitive macrophages act as phagocytes during tissue remodeling throughout embryonic development and in clearance of bacterial pathogens [23]. While primitive neutrophils also migrate to a site of infection, they were not observed to phagocytose bacteria [24]. The temporal, spatial and transcriptional control of zebrafish primitive hematopoiesis has been reviewed by [28-30]. Differences in the initial site of hematopoiesis occur between fish species, however, the production of erythrocytes and macrophages during primitive hematopoiesis is consistent [31, 32].

3.3. Definitive myelopoiesis in teleosts

The onset of definitive myelopoiesis occurs around 36 hpf in the zebrafish. Here, HSCs seed the aorta-gonad-mesonephros (AGM) and the caudal hematopoietic tissue (CHT) [33, 34]. By 48 hpf, the HSCs seed the kidney [33], the final hematopoietic site equivalent to mammalian bone marrow [35-37].

The existence of teleost kidney HSCs and HPCs capable of generating all hematopoietic lineages was demonstrated using transplantation studies in zebrafish and ginbuna crucian carp. Transplantation of whole kidney marrow from gata1^{eGFP} zebrafish into pre-thymic *vlad tepes* (gata1$^{-/-}$) zebrafish [37] or whole kidney marrow from β-actin^{eGFP} zebrafish into lethally irradiated zebrafish [38], resulted in rescue of the phenotype and produced lymphoid and myeloid cell types suggestive of the presence of HSCs capable of long-term reconstitution. However, these studies were complicated by the use of whole kidney marrow during transplantation. Using ginbuna crucian carp, HSCs, found to be associated with the trunk kidney renal tubules, were identifiable by their ability to efflux Hoechst 33342 using the ATP-binding cassette (ABC) transporter, ABCG2a, and HPCs were identified by their ability to efflux rhodamine 123 by another ABC transporter, P-glycoprotein [39-42]. HSCs, consisting of 0.33% ± 0.15 of the total body kidney cells, were capable of engraftment and long-term production (>9 months) of all hemopoietic progeny, including erythrocytes, granulocytes, monocytes, thrombocytes and lymphocytes [40, 41, 43]. HPCs, while they could also give rise to all hemopoietic progeny, were only capable of short-term reconstitution [42]. However, engraftment of donor HSCs and HPCs only occurred in anemia-induced or gamma irradiated recipients [40, 43, 44] suggesting that space within the hematopoietic niche is required for successful engraftment of HSCs to occur [40, 43]. Experiments using zebrafish and ginbuna crucian carp provide strong evidence that the teleost trunk kidney contains HSCs and HPCs capable of multi-lineage differentiation, including myelopoiesis [45].

4. Commitment to the myeloid lineage

4.1. Progression of cell development

From the mouse model we know that the commitment of a pluripotent, self-renewing HSC to a common myeloid progenitor (CMP) is a progression of lineage fate decisions controlled by extracellular cues, such as growth factors, within the hematopoietic niche [46-48], as well as the modulation of intracellular transcription factors [49-52]. The process of committing to a CMP begins with long-term HSCs (LT-HSCs), capable of self-renewal and multi-lineage differentiation. LT-HSCs give rise to short-term HSCs (ST-HSCs) with limited capacity for self-renewal, which then differentiate into multipotent progenitors (MPPs) with no ability to self-renew, reviewed by [53]. The MPPs can give rise to the CMP or the lymphoid-myeloid primed multipotent progenitors (LMPPs) [54-57]. The CMP can differentiate into megakaryocyte/erythroid progenitor (MEP) or to a granulocyte/macrophage progenitor (GMP) [58] (Figure 1). The LMPPs can differentiate into a common lymphoid precursor (CLP) that gives rise to T- and B-lymphocytes, or can also give rise to GMPs [54-57, 59, 60], and reviewed in [61].

On the other hand, the "myeloid-based model" of hematopoiesis, in which myeloid potential is retained in erythroid, T, and B cell branches even after these lineages have segregated from each other, has been proposed [62]. Notably, there is no CLP in this model [63-65]. According to this model, hematopoiesis can be understood as follows: specification toward erythroid, T, and B cell lineages proceeds on a basis of a prototypical developmental program to construct myeloid cells [66, 67]. Indeed, several findings in teleosts are supportive of the myeloid-based model [68, 69]. In the future, the myeloid-based model may bring a paradigm shift in the concept of blood cell lineage development. In the following sections the key growth factors and transcription factors studied in the teleost system will be discussed.

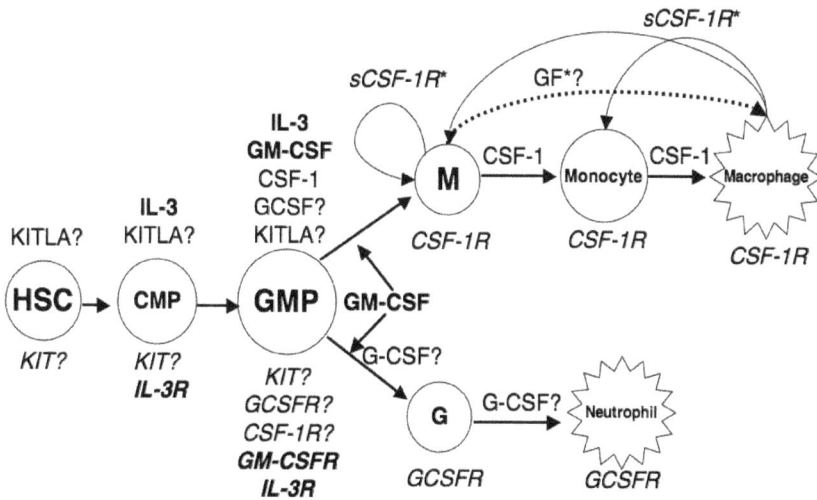

Figure 1. Growth factors and their receptors involved in goldfish myelopoiesis. Goldfish growth factors are shown in uppercase lettering, goldfish growth factor receptors/surface receptors are shown in uppercase italics lettering, and growth factors and their receptors important in mammalian myelopoiesis, but have yet to be identified in teleosts are shown in uppercase **italics**. The dashed arrow denotes the alternative pathway of macrophage development in goldfish, the solid curved arrows denote negative regulation of macrophage development by sCSF-1R. Question marks denote the hypothesized role of growth factors or receptors and further studies are required to test the hypothesis. Asterisks mark differences between teleosts and mammals. Abbreviations used: (1) **Cellular stages**: HSC, hematopoietic stem cell; CMP, common myeloid progenitor; GMP, granulocyte-macrophage progenitor; M, monocytic precursor; G, granulocytic precursor. (2) **Growth factors**: KITLA, kit ligand a; IL-3, interleukin 3; GM-CSF, granulocyte-macrophage colony-stimulating factor; CSF-1, colony-stimulating factor 1 (macrophage colony-stimulating factor); GCSF, granulocyte colony-stimulating factor; GF, growth factor. **Receptors**: IL-3R, interleukin 3 receptor; GM-CSFR, granulocyte-macrophage colony-stimulating factor receptor; CSF-1R, colony-stimulating factor-1 receptor (macrophage colony-stimulating factor receptor); sCSF-1R, soluble colony-stimulating factor-1 receptor; GCSFR, granulocyte colony-stimulating factor receptor.

4.2. Receptors and growth factors

4.2.1. Mammalian stem cell factor and Kit receptor

Stem cell factor (SCF) was identified [70-72] as short-chain four-helix bundle [73] encoded by the *Steel* locus in the mouse [74]. Mutations in the *Steel* locus were associated with defects in stromal cells, and resulted in reduced numbers of HSCs and HPCs [75]. The *SCF* gene produces two alternatively spliced mRNAs that differ in the presence or absence of exon 6 [71]. Although the two *SCF* splice variants can be expressed in the same tissues, they have tissue specific regulation of expression [71, 76]. Both SCF isoforms are produced as extensively glycosylated [77, 78] membrane bound forms (mSCF) that can undergo proteolytic cleavage to produce a soluble form of SCF (sSCF) [79, 80]. In human blood, sSCF is at a concentration of 3.0 ± 1.1 ng/mL [77]. Alternatively, mSCF may provide a means for cell-to-cell contact with the stromal cells in the hematopoietic niche [71], and may act to increase the signal strength provided to the HSC/HPCs, reviewed in [81]. Both mSCF and sSCF are capable of forming dimers [78, 82] and signal through their receptor, c-KIT.

The SCF receptor, c-KIT (CD117), was first identified as the cellular oncogene (*c-onc*) equivalent of the viral oncogene (*v-onc*), *v-Kit*, isolated from the Hardy-Zuckerman 4 feline sarcoma virus [83]. Based on structural analysis, the c-KIT protein was grouped within the Type III tyrosine kinase receptor family that includes colony-stimulating factor-1 receptor (CSF-1R), platelet derived growth factor receptor (PDGFR), and FLT3/FLK2 receptor [84-87]. Studies mapped *c-KIT* to the *White* locus (*W*) in the mouse [74, 83], and demonstrated that mice with mutations in the *White* or *Steel* loci exhibit hypopigmentation, mast cell deficiency, macrocytic anemia, and sterility, while the complete loss of either of these genes was lethal [74, 88].

The c-KIT protein is primarily found on hematopoietic cells and is a marker of long-term reconstituting HSCs in humans [89] and mice [90-92]. c-KIT is expressed on pluripotent and multipotent HSCs and myeloerythroid precursors, but not on differentiating or mature cell types [90-92], with the exception of mast cells [93]. Approximately 2×10^4 c-KIT receptors are found on normal human HPCs [94], and can undergo proteolytic cleavage to release a soluble form of c-KIT [95-97]. The soluble c-KIT receptor is thought to regulate membrane bound c-KIT activity, *in vivo*, by blocking SCF binding [95, 98].

Binding of homodimeric SCF to c-KIT results in receptor homodimerization, conformational changes in the extracellular and intracellular domains and autophosphorylation of the intracellular tyrosines (reviewed extensively in [73, 78, 99-105]) leading to a number of downstream signaling pathways that mediate the action of SCF through c-KIT. These signaling pathways include phosphatidylinositol-3-kinase (PI3K), phospholipase Cγ (PLCγ), members of the Janus family of protein tyrosine kinases (JAK) and signal transducers and activators of transcription (STATs), Src family members, the Ras/Raf/MAP kinase pathway, and others. The signaling pathway initiated depends on the cell type, and the strength and duration of the signal, reviewed in [106-108].

4.2.2. Biological functions of stem cell factor

SCF and its type III tyrosine kinase receptor c-KIT, are involved in hematopoiesis [81, 107, 108], spermatogenesis [109-111], and development of melanocytes [110, 112-114] and mast cells [93, 96, 115-120]. Within the hematopoietic niche, one role of SCF/c-KIT is to mediate HSC and HPC survival, important for the generation of spleen, interleukin-3 (IL-3), granulocyte/macrophage, and macrophage colony-forming units (CFU-S, CFU-IL-3, CFU-GM, and CFU-M) [121]. Further studies have confirmed SCF/c-KIT to mediate the survival of long-term HSCs by blocking cell cycling or by inhibiting apoptosis [122, 123]. Furthermore, SCF can synergizing with other growth factors, such as granulocyte-macrophage colony-stimulating factor (GM-CSF) [124], granulocyte colony-stimulating factor (G-CSF), IL-1, IL-3 [98], IL-6, and IL-7, among others, to promote the proliferation and differentiation of HPCs [125, 126] and reviewed in [101]. Often, the progeny of HPC differentiation depends on the particular growth factor and SCF. Lastly, SCF acts as a homing signal to HPCs, such as CFU-GEMM (granulocyte-erythrocyte-macrophage-megakaryocyte), CFU-GM, CFU-Meg (megakaryocyte) and burst forming units-erythrocyte (BFU-E) [127]

4.2.3. Teleost Kit and Kit ligand

Whole genome duplication has resulted in two orthologues of *c-KIT* and *SCF* in teleosts. Teleost orthologues of *c-KIT*, termed kit a (*kita*) and kit b (*kitb*), were first identified in zebrafish and have subsequently been predicted from genomic analysis of *Takifugu rubripes* and *Tetraodon nigroviridis* [128, 129]. The *kita* orthologue has also been identified and characterized in *Carassius auratus* [130]. The two orthologues of mammalian *SCF* are termed kit ligand a (*kitla*) and kit ligand b (*kitlb*) [128, 131]. The *kitla* and *kitlb* have been identified in zebrafish, and predicted in fugu, medaka, and stickleback genomes [131]. The *kitla* orthologue has been identified and characterized in goldfish [130].

Zebrafish *kita*, located on chromosome 20, and *kitb*, located on chromosome 1, are the orthologues of human and mouse *c-KIT* [128, 129]. Both *kita* and *kitb* genes contain 21 exons, however, their respective proteins only retain 55% identity to each other [129]. The partitioning of gene distribution and function was proposed to explain the retention of *kita* and the duplicated gene, *kitb* [128, 129]. From studies on developing zebrafish, *kita* is expressed in hematopoietic progenitors, melanoblasts and melanocytes derived from the neural crest, along the lateral line, the notochord and pineal gland [128, 129]. The expression of *kitb* occurs by 9 hpf and does not overlap that of *kita*. Instead, *kitb* expression is restricted to the Rohon-Beard neurons, trigeminal ganglia, and otic vesicle [129]. Together, the expression of *kita* and *kitb* approximates that of *c-KIT* in the mouse model system, with the notable exception of *c-KIT* expression in primordial germ cells (PGCs).

The *kitla* gene is located on chromosome 25 and the *kitlb* gene is located on chromosome 4 of the zebrafish genome [132]. *Kitla* has 9 exons while *kitlb* has 8 exons [131]. The nine *kitla* exons correspond to the 9 exons of mammalian SCF isoform 1, including exon 6 which allows for cleavage of membrane bound SCF into a soluble form [131]. However, *kitlb* appears to correspond to SCF isoform 2, in which exon 6 has been spliced out. The expression of *kitla* is first observed at 19 hpf in the zebrafish and is found in the developing tail bud, pineal gland, sensory epithelium of the ear, ventral otic vesicles, and in the somites [131]. Similar to the expression of goldfish *kita*, *kitla* showed constitutive mRNA levels in tissues [130] and this expression pattern was similar to what was observed in adult zebrafish tissues [132]. Goldfish *kitla* showed high levels of mRNA in isolated putative progenitor cells and monocytes compared to macrophages [130]. Zebrafish *kitlb* mRNA expression was observed in the brain ventricles, ear and cardinal vein plexus and at lower levels in the skin as zebrafish development progressed [131].

4.2.4. Biological functions of teleost kit ligands and receptors

Based on the non-overlapping expression of *kita* and *kitb*, the functional roles of c-KIT in mammals may be partitioned between teleost KITA and KITB. The zebrafish mutant *sparse*, shown to map to *kita* [128], or *kit*[w34] mutants [133] show defects in their pigmentation pattern. Zebrafish KITA was shown to be involved in the dispersion and maintenance of melanocytes [128], and may play a transient role in melanocyte differentiation when melanoblast development is perturbed [134]. Furthermore, knock-down of zebrafish *kitla* or *kitlb* using morpholinos supported the involvement of KITLA in the migration and survival of melanocytes [131]. Teleost *kit* expression in melanocytes has been implicated in the pigment pattern formation in a number of fish species [128, 135-137] and suggests that the functions in myelocyte development have been partitioned to the *kita* orthologue.

The role of teleost *kita/kitla and kitb/kitlb* during hematopoiesis is not clear. Examination of hematopoiesis in zebrafish *sparse* mutants revealed no obvious defects in hematopoiesis during development. Although, slight decreases in promyelocyte and neutrophil cell numbers, and slight increases in band cells and monocytes were observed in the kidney [128]. In addition, zebrafish injected with *kitla* morpholinos or *kitlb* morpholinos also did not show defects in hematopoiesis. However, studies in the goldfish model system demonstrated the expression of *kita* mRNA in isolated kidney progenitor cells, and the functional role of goldfish KITLA in progenitor cell chemotaxis, proliferation, and maintenance [130]. Taken together, these data suggest that KITA and KITLA proteins play a central role in myelopoiesis (Figure 1). However, redundancy between the two ligands and receptors may account for the absence of hematopoietic defects in the zebrafish system, or there may be redundancy with another tyrosine kinase receptor. Additionally, the absence of hematopoietic defects in the zebrafish may represent KIT-independent and KIT-dependent stages of hematopoiesis. The function of KITLB and KITB during hematopoiesis in teleosts remains to be determined.

Lastly, c-KIT plays a role in the development of primordial germ cells (PGCs) in mice. Examination of primordial germ cell development in fish revealed that *kita* and *kitb* expression was not detected in PGCs, and suggests teleost KITs do not play a role in the development of PGCs [128, 129]. However, it appears that *kita, kitb, kitla* and *kitlb* play a role in ovarian folliculogenesis in zebrafish and provides evidence of neofunctionalization of these genes [132].

4.2.5. Interleukin-3 and Interleukin-3 receptor

Interleukin-3 (IL-3) is a multi-lineage colony-stimulating factor (multi-CSF) that acts through the IL-3 receptor alpha and common beta chain on multipotent erythro/myeloid HPCs to promote their self renewal, proliferation and differentiation [138-140]. IL-3 can also act on committed myeloid progenitors to promote their proliferation and differentiation [138-142]. Interestingly, *IL-3, IL-4, IL-5* and *GM-CSF* are all found on chromosome 5q in humans. The close proximity of the CSFs on the chromosome, along with their similar structure and function may suggest they arose from a common ancestral gene [143]. However, genes encoding IL-3 and the specific IL-3 receptor alpha (IL-3Rα) have not been identified in any teleosts to date, despite genome sequencing (Figure 1). The lack of IL-3 in teleosts may be due to difficulties in identifying the IL-3 orthologue in teleosts due to the low sequence conservation of IL-3 observed between mammals, or may represent the evolutionary loss of IL-3 in teleosts. As IL-3 and IL-3R have not been identified in teleosts, IL-3 and IL-3R will not be discussed here. The structure, function and regulation of mammalian IL-3 and its receptor have been extensively reviewed elsewhere by [144-146].

4.2.6. Granulocyte-macrophage colony-stimulating factor/Granulocyte-macrophage colony-stimulating factor receptor

GM-CSF shares redundancy with IL-3 in terms of its function. However, GM-CSF acts on a more mature population of HPCs and has been associated with the formation of both granulocyte and macrophage colonies from CFU-GM [147, 148]. GM-CSF is produced by activated T-lymphocytes [147, 149], endothelial cells [150], and lung fibroblasts [151] and suggests the importance of GM-CSF during emergency hematopoiesis. GM-CSF promotes the survival, proliferation and differentiation of GMPs [147, 148, 152]. Furthermore, GM-CSF is chemoattractive to immature and mature neutrophils *in vitro* and *in vivo* [153, 154] and enhances neutrophil anti-microbial functions and neutrophil survival [155]. GM-CSF can also promote monocytes to differentiate into inflammatory dendritic cells [156, 157]. The GM-CSF receptor (GM-CSFR) is composed of heterodimeric alpha and beta chains as described for IL-3. Since the βc chain is common to IL-3, IL-5 and GM-CSF, the βc chain signals through JAK/STAT, MAPK, and PI3K pathways [145, 158].

Similar to that of *IL-3*, *GM-CSF* has not been identified in teleosts (Figure 1). The close proximity of *IL-3* and *GM-CSF* on the same chromosome may suggest that a genomic deletion occurred on this chromosome, subsequent to the divergence of fish and mammals. The hematopoietic CSFs that compensate for the loss of IL-3 and GM-CSF in teleosts are not known.

4.3. Transcription factors

Commitment of LT-HSCs to the myeloid lineage is an intricate regulation of the transcription factors expressed, their relative levels to one another, and their expression on a temporal scale. Transcription factors (TFs) can act antagonistically or co-operatively. Thus, the presence or absence of a TF partner, or the relative levels of a TF to its antagonistic counterpart, determine lineage fate decisions. Furthermore, the expression of a transcription factor in an HSC does not exert the same effect as when it is expressed in a committed progenitor cell. The transcriptional regulation of mammalian hematopoiesis/myelopoiesis has been extensively reviewed elsewhere [159-162], and will only be briefly described here for the purpose of putting advances in the teleost model systems into context. A visual representation of which stages these transcription factors are important is shown in Figure 2.

4.3.1. MafB

MAFB, a bZIP transcription factor family member, is highly expressed in LT-HSCs, but not in MPPs, CMPs, or GMPs and was recently found to be involved in restricting proliferation and myeloid lineage differentiation of LT-HSCs [163]. MAFB[-/-] LT-HSCs showed increased proliferative activity and gave rise to large numbers of primarily myeloid progeny in a mouse repopulation assay [163]. The MAFB[-/-] HSCs had higher proliferative ability and gives rise to greater numbers of myeloid progeny in response to CSF-1 compared to wild type HSCs, *in vitro*. Furthermore, *in vitro* studies demonstrated that treatment of MAFB[-/-] HSCs with CSF-1 led to the rapid activation of *PU.1* transcription that suggested MAFB must be down-regulated to allow expression of *PU.1* in MPPs [163]. It appears that MAFB plays an important role in antagonizing the expression of PU.1 and the commitment of MPPs to CMPs. Furthermore, MAFB has been shown to bind ETS-1 though its zipper-binding domain and can act to repress erythroid lineage commitment in CMPs [164].

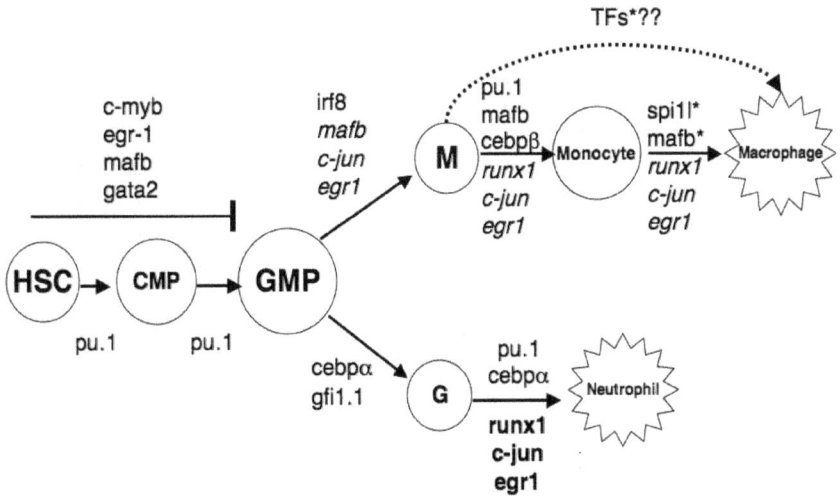

Figure 2. Transcription factors involved in goldfish myelopoiesis. Goldfish transcription factors shown in lower case lettering are up-regulated, goldfish transcription factors shown in bold are down-regulated, transcription factors that are important in cellular differentiation in mammalian systmes but have yet to be studied in the teleost system are shown in italics. The dashed arrow denotes the alternative pathway of macrophage development in teleosts. Question marks denote unknown transcription factors involved in the alternative pathway of macrophage development. Asterisks mark differences between teleosts and mammals. Abbreviations used: (1) Cellular stages: HSC, hematopoietic stem cell; CMP, common myeloid progenitor; GMP, granulocyte-macrophage progenitor; M, monocytic precursor; G, granulocytic precursor. (2) Transcription factors: c-MYB, cellular myelobastosis oncogene; EGR-1, early growth response-1; MAFB, musculoaponeurotic fibrosarcoma oncogene homologue B; GATA2, GATA binding protein 2; IRF8, interferon regulatory factor 8; CEBPα, CCAAT/enhancer-binding protein alpha; GFI1, growth factor independent 1; RUNX1, runt-related transcription factor 1.

In zebrafish, the *mafb* orthologue has been identified and mRNA was found expressed in the blood forming regions of the developing embryo [165]. However, the role of MAFB in zebrafish HSCs has not yet been assessed.

4.3.2. C/EBPs

CCAAT/enhancer binding proteins (C/EBPs) are members of the family of transcription factors that contain a C-terminal basic leucine zipper domain (bZIP) comprised of a basic region involved in DNA binding and a leucine zipper domain involved in protein interactions [166]. Six members of the C/EBP family have been identified in mammals: alpha, beta, gamma, delta, epsilon and zeta [167]. Orthologues of the C/EBP family of transcription factors have been identified in teleosts [168-171], corresponding to C/EBPα, C/EBPβ, C/EBPγ, C/EBPε, and C/EBPδ.

Expressed in HSCs, CMPs and GMPs [172, 173], C/EBPα has been shown to be involved in directing granulocyte cell fate and terminal differentiation of neutrophils, along with C/EBPε. Mice deficient in C/EBPα show diminished numbers of CFU-GM, CFU-M, CFU-G, macrophages and neutrophils [174, 175]. The loss of myeloid cells in C/EBPα deficient mice is reflective of the role that CEBPα plays in determining the fate of a CMP to a GMP lineage versus an MEP lineage [176]. C/EBPα is capable of binding to the *PU.1* promoter [175] and up-regulating *PU.1* expression, to dictate a GMP cell fate [175, 177] (see discussion on PU.1 below). The increase in C/EBPα in GMPs has been shown to inhibit monocyte/macrophage differentiation [178] and initiate differentiation along the granulocyte lineage by regulating *GCSFR*, elastase and myeloperoxidase gene expression [179-181].

The zebrafish CEBPα orthologue showed 66% amino acid identity to human C/EBPα, while the bZIP domains showed 99% amino acid identity [168]. In zebrafish, *cebpa* was expressed in myeloid cells on the surface of the yolk sac during embryogenesis [168]. At 16 hpf, a population of blood cells co-expressed the transcription factors *gata1*, *pu.1* and *cebpa*, and by 22 hpf, the majority of the *cebpa*+ cells co-expressed *pu.1*, however, not all *pu.1*+ cells expressed *cebpa* [182]. Furthermore, *cebpa* was co-expressed with myeloperoxidase (*mpo*), a marker for granulocytes, but *cebpa*+ cells did not always express *mpo* [182]. These three cell sub-populations likely represent distinct junctures in myeloid cell development: erythromyeloid cells, GMPs and committed neutrophils and their precursors, respectively. The expression of *cebpa* with these additional markers mirrors the importance of C/EBPα in the mammalian system in which C/EBPα is important for commitment to a myeloid lineage versus an erythroid lineage, to a granulocyte lineage over a macrophage lineage, and in terminal differentiation of neutrophils. An orthologue of *cebpa* was also identified in Japanese flounder and mRNA was observed in the head and posterior kidney, spleen, liver, gill, heart, brain, skin, intraperitoneal cells, and weakly in the intestine, muscle and PBLs [171]. However, expression of *cebpa* in isolated cells populations was not performed.

Two studies have examined the function of CEBPα in zebrafish primitive myelopoiesis. The injection of a deletion mutant of *cebpa* into zebrafish embryos functioned as a dominant-negative mutation and blocked the production of full-length CEBPα. These embryos exhibited an increase in *gata1*+ expression in the posterior lateral plate mesoderm at 22 hpf and in the intermediate cell mass at 26 hpf, reflective of an erythroid progenitor cell expansion. This expression corresponded to a subsequent increase in circulating erythrocytes based on the increase in *α-hemoglobin* expression, indicative of erythrocytes [182]. However, the expressions of the myeloid specific genes, *mpo* and *l-plastin*, were normal [182]. Based on the pattern of expression, it was suggested that PU.1 acts upstream or in parallel with C/EBPα during zebrafish primitive myelopoiesis [182]. Recently, it has been shown that the sumoylation (a post-translational protein modification) of zebrafish CEBPα inhibited CEBPα transcriptional activity and its ability to interact with and repress GATA1, thus driving lineage commitment of a myelo-erythroid progenitor to that of the erythroid lineage [183]. Taken together, these studies demonstrate the conserved role of CEBPα in the commitment of a CMP to a GMP. However, due to the toxicity of *cebpa* morpholinos to zebrafish embryos, knockdown experiments could not be performed.

Cebpb was identified in rainbow trout as a single intron-less gene and the predicted CEBPβ protein showed 30-34% amino acid identity to mammalian C/EBPβ [169]. The *cebpb* mRNA was detected in the head and posterior kidney, spleen, liver, gill, intestine, muscle and peripheral blood leukocytes (PBLs) [169]. Japanese flounder CEBPβ also showed a low (33-38%) amino acid identity to other vertebrate sequences, but retained 95% amino acid identity in the bZIP domain. The *cebpb* mRNA was expressed in the head and posterior kidney, liver, gill, brain, peritoneal cavity fluid and PBLs, with low mRNA levels in the heart, intestine, mucus, eye and spleen [170]. In zebrafish, CEBPβ showed 49% amino acid identity to human C/EBPβ and *cebpb mRNA* was detected in cells on the surface of the yolk sac, corresponding to the myeloid cells that normally spread over the yolk sac early in embryogenesis [168]. A *cebpb* transcript was also identified in a differential cross-screen of goldfish proliferative phase and senescence phase PKMs, and was up-regulated in goldfish monocytes, and expressed in low levels in progenitors and macrophages [19]. However, the functional role of CEBPβ has not been examined in teleost myelopoiesis.

The orthologues of C/EBPδ, C/EBPγ and C/EBPε exist in teleosts. The *cebpd* and *cebpg* transcripts were identified in zebrafish and show a ubiquitous expression pattern in embryos [168]. CEBPδ and CEBPγ showed 57 and 50% identity to their human counterparts on the amino acid level. However, their bZIP domains showed higher conservation to their human counterparts, with 86% and 76% amino acid identity, respectively [168]. The *cebpe* orthologue was identified in Japanese flounder and its corresponding predicted protein had a 27% overall amino acid identity and a 90% amino acid identity in the bZIP domain compared to the mammalian counterparts, but failed to cluster with other *cebpe* sequences in phylogenetic analysis [170]. The *cebpe* mRNA was detected in the head and posterior kidney, spleen, brain, peritoneal cavity fluid and at low levels in the PBLs. However, the functional role of these C/EBPs in teleost myelopoiesis is unknown.

4.3.3. PU.1

The Ets transcription family member PU.1 is well known as the master transcriptional regulator of mammalian myelopoiesis through an antagonistic relationship with GATA1, recently reviewed by [184]. At the N-terminus, PU.1 comprises of an acidic domain and a glutamine rich domain that are involved in activation of transcription, and a PEST domain important for protein interactions [184]. At the C-terminus, PU.1 has an Ets domain important for binding the DNA consensus sequence AAAG(A/C/G)GGAAG [185]. Mice deficient in PU.1 (*PU.1*[-/-]) have reduced CLPs, and GMPs, increased numbers of MEPs, and lack B-cells, T-cells, monocytes/macrophages as well as severely reduced numbers of granulocytes [186-190]. *PU.1* is expressed in HSCs, CLPs and at varying levels in CMPs, increasing as these progenitors are induced to differentiate into monocytes/macrophages and neutrophils [191]. At the CMP stage, PU.1 antagonizes with GATA1 to determine whether the CMP commits to a GMP or a MEP. PU.1 binds to GATA1 and inhibits GATA1 from binding to and initiating transcriptional activation of a number of erythroid genes that are important for commitment to an erythroid lineage [184, 192, 193]. The reverse is also true; GATA1 can bind to PU.1 and inhibit the binding of PU.1 and transcriptional activation of a number of myeloid genes [184, 192, 193], including to the promoters of *CSF-1R* [194-196] and *GCSFR* genes [181, 196, 197]. Therefore, the lineage fate decision along a GMP or a MEP fate is a balancing act in timing and relative protein levels of PU.1 and GATA1.

PU.1 also plays a role at the GMP stage to regulate commitment to a granulocyte or macrophage lineage. Increased levels of PU.1 at the GMP stage, along with AP-1 association, drives a monocyte cell fate, while lower levels of PU.1 drives granulocyte cell fate [175, 177]. Furthermore, PU.1 induces *EGR-2* and *NAB-2* expression [177]. The EGR-2/NAB-2 transcription factors function to repress neutrophil genes by antagonizing GFI1, an important transcription factor in the initiation of neutrophil differentiation [177], discussed in section 5.3.2.

An orthologue of PU.1 has been identified in teleosts. In the Japanese flounder, *pu.1* mRNA was detected in the head and posterior kidney, spleen, heart, PBLs, intraperitoneal cells, and weakly in the intestine and gill, but was absent from the liver, skin, muscle and brain [171]. In zebrafish, *pu.1* was identified as a single gene copy and analysis of the predicted protein sequence showed the conserved transactivation, PEST, and DNA-binding domains. Although the overall amino acid identity to other PU.1 proteins was 48-53%, the DNA-binding domain of zebrafish PU.1 showed 83% amino acid identity to mammalian PU.1 [198]. Examination of the zebrafish *pu.1* promoter region predicted potential binding sites for PU.1 and CEBPα [199]. The expression of *pu.1* is first detected at 12 hpf in blood cells from the PLM, later in the ICM, and finally in the kidney, and these *pu.1+* blood cells give rise to myeloid cells [198-200]. The population of *pu.1+* cells represents myeloid HPCs, myeloid precursors, monocytes/macrophages and neutrophils during both primitive and definitive myelopoiesis in the zebrafish [24, 200].

Knockdown of *pu.1* in zebrafish using morpholinos showed a large reduction in the number of cells positive for *mpo* and *l-plastin* mRNA, markers of granulocytes and monocytes/macrophages [201, 202]. In addition to the loss of myeloid cells, an increase in *gata1* expression was observed, and these *gata1+* cells gave rise to mature erythrocytes [201]. Conversely, *gata1* morphants failed to develop mature erythrocytes and showed an increase in the number of *pu.1+*, *mpo+* and *l-plastin+* cells [201, 202]. Ectopic expression of *pu.1* or *gata1* was observed in *gata1* or *pu.1* morphants, respectively, suggesting the conversion of progenitors to an alternate lineage [201, 202]. Microarray analysis of genes regulated by PU.1 revealed the regulation of ~250 genes, including *cebpa*, *csf-1r* and myeloid-specific peroxidase *(mpx)*, among others [203]. Taken together, PU.1 has a conserved role in dictating a myeloid lineage, opposing GATA1 and the transcriptional activation of erythroid genes.

A pu.1-like gene (spi-1 like, *spi-1l*) was also identified in zebrafish. The predicted amino acid sequence of SPI-1l showed 45% amino acid identity to zebrafish PU.1, and retained all three domains [204]. *In situ* hybridization revealed a population of blood cells positive for *pu.1* and *spi-1l*, in addition to a population of single positive *pu.1* cells [204]. However, only a few single-positive *spi-1l* cells were observed. *Spi-1l* morphants showed a loss of *mpx* and *l-plastin* positive cells, indicative of a loss in granulocytes and monocytes/macrophages [204]. Unlike *pu.1* morphants, no change in *gata1* expression was observed, suggesting that SPI-1l acts downstream of PU.1, and plays an important role in myeloid cell differentiation [204].

5. Commitment of bi-potent myeloid progenitors to the macrophage or neutrophil lineage

5.1. Macrophage development

5.1.1. Progression of cell development

In mammalian systems, the progression of macrophage development proceeds from a committed macrophage progenitor, monoblast, promonocyte, monocyte and then to a mature tissue macrophage, reviewed by [205-207] (Figure 1). While the presence of a unipotent committed macrophage progenitor has yet to be unequivocally demonstrated in the teleost systems, progenitor/precursor cells that give rise to monocytes and macrophages have been demonstrated. *In vitro*, a spontaneous proliferating trout RTS-11 cell line has two predominant cell types; a round non-adherent cell type that appears to be a pre-monocyte or myeloid precursor and an adherent macrophage-like cell, arising from the non-adherent cell type [208]. The cultivation of trout kidney progenitor-like cells developed a trout primary kidney monocyte culture that contained progenitor cells, promonocyte-like cells, and monocytes [209]. Furthermore, the generation of goldfish primary kidney macrophage cultures demonstrated that small mononuclear cells became monocytes and mature macrophages, *in vitro*. In the zebrafish model system, whole kidney marrow was added to a kidney fibroblast layer and was shown to maintain HPCs and precursor cells that then differentiated into myeloid and lymphoid cells [210]. Recently, the development of a zebrafish methylcellulose colony forming unit assay suggested the presence of a common erythro-myeloid HPC [211]. *In vivo* studies, primarily in the zebrafish, have demonstrated that monocytes/ macrophages arise from the hematopoietic organ [45, 212-215], migrate to various tissues [216], and both primitive and definitive macrophages are motile, migrate to the site of insult, and readily phagocytose particles or pathogens [23, 45, 217-219]. The identification of progenitor cells that are capable of differentiating into monocytes and macrophages suggests a conserved macrophage differentiation pathway in vertebrates.

5.1.2. Receptors and growth factors

5.1.2.1. Colony-stimulating factor-1

The central growth factor that regulates the survival, proliferation, and differentiation of macrophages and their precursors is colony-stimulating factor-1 (CSF-1) [220-223]. Alternative splicing of *CSF-1* transcripts leads to production of a secreted glycoprotein, a secreted proteoglycan, or a membrane-bound glycoprotein that can be proteolytically cleaved from the surface, reviewed by [144, 221]. However, only the first 149-150 aa of the N-terminal portion of the CSF-1 core protein has shown to be important for biological function [224, 225]. CSF-1 homodimers, are covalently linked by an interchain disulphide bond to form a dimer [226] that then binds the CSF-1 receptor. CSF-1 is produced by an array of cell types including fibroblasts, endothelial cells, and bone marrow stromal cells, reviewed by [144]. In addition, activated T-cells [227-229], monocytes, macrophages [230, 231], fibroblasts, and endothelial cells [144] can produce CSF-1. CSF-1 production by activated cell types suggests a role for CSF-1 at the site of inflammation, which may be necessary for the rapid recruitment, differentiation and activation of macrophages and their precursors.

5.1.2.2. Interleukin-34

Recently, IL-34 was identified as another growth factor involved in mediating macrophage development in mammals, in addition to CSF-1 [232-234]. The IL-34 protein does not show homology to any other human protein and or contain any known conserved structural motifs [232]. Homodimeric IL-34 binds to CSF-1R, although with a different affinity than that of CSF-1, and to different sites on the receptor [232, 233] [235]. The hierarchy in binding of the CSF-1R ligands may provide a mechanism for differential signaling depending on the bound ligand. To date, IL-34 has not been identified in teleosts.

5.1.2.3. Colony-stimulating factor-1 receptor

The *CSF-1R* gene, shown to map to the proto-oncogene *c-fms*, is a member of the type III tyrosine kinase family of receptors [236, 237], reviewed by [238]. The binding of homodimeric CSF-1 to CSF-1R, triggers receptor homodimerization and activation [239]. Receptor activation triggers autophosphorylation of the intracellular tyrosine residues and activation of JAK/STAT, PI3K/Akt, and MAPK pathways, as well as pathways for receptor-mediated internalization and destruction, reviewed by [162, 238, 240]. Within the hematopoietic system, CSF-1R protein is primarily found on macrophages and their precursors and has been used as a marker of cells along the macrophage lineage in mammalian systems [222, 237]. CSF-1R progressively increased with macrophage differentiation [144].

5.1.2.4. Biological functions of colony stimulating factor-1

In addition to the regulation of survival, proliferation, and differentiation of macrophages and their precursors [220-223], CSF-1 has been shown to exert pro-inflammatory effects on monocytes and macrophages. These effects include the enhancement of macrophage chemotaxis, phagocytosis of pathogens, and the production of antimicrobial agents, reviewed by [162, 238]. CSF-1 is a pleiotropic cytokine and functions in a number of other biological systems such as regulation of macrophage and osteoclast numbers, bone remodeling, tooth production and fertility and breast development [241-245]

5.1.2.5. Teleost colony stimulating factor-1

Teleost csf-1 (mcsf) was first identified in the goldfish as a 600 bp mRNA transcript that was present at high levels in spleen tissue, monocytes, and phorbol ester-activated monocytes [246]. The csf-1 transcript encoded for a 199 aa precursor protein, with the mature CSF-1 protein predicted to have a molecular weight of 22 kDa. The goldfish CSF-1 has 27% aa identity to human CSF-1 [246]. Alignment of goldfish CSF-1 with mammalian CSF-1s showed conservation of four cysteine residues required for protein folding, similar to that of mammalian CSF-1 [246]. Ligand-receptor binding studies demonstrated that homodimeric CSF-1 could bind to soluble CSF-1R (see teleost CSF-1R section below, Figure 1). Functional characterization of a recombinant goldfish CSF-1 was shown to induce monocyte proliferation and differentiation (Figure 1), which was abrogated in the presence of sCSF-1R or in monocytes transfected with csf-1r RNAi oligos [246, 247]. Recombinant goldfish CSF-1 also aided in the long-term survival of mature macrophages in vitro [247]. The recombinant CSF-1 protein was chemoattractive to PKMs, and promoted their ability to perform phagocytosis and produce antimicrobial compounds [248], suggesting a pro-inflammatory role for CSF-1 in goldfish.

Two csf-1 genes were later identified in trout and zebrafish, termed mcsf-1 and mcsf-2, and a second goldfish mcsf transcript was identified [249]. The trout and zebrafish mcsf-1 genes encoded for proteins of 593 and 526 aa, the trout and zebrafish mcsf-2 genes encoded for proteins of 276 and 284 aa, respectively, while the goldfish mcsf gene encoded for a 544 aa protein [249]. All of the identified transcripts possessed a signal peptide, a CSF-1 domain, a transmembrane domain, and a short cytoplasmic domain [249]. However, the N-terminal region of all teleost CSF-1 proteins showed high homology (46-88%), consistent with the important role of the CSF-1 N-terminal portion for biological function.

The genomic structure of the identified mcsfs also differed. The zebrafish mcsf-1, found on chromosome 11, possessed seven exons and mcsf-2, found on chromosome 8, possessed nine exons. Based on syntenic analysis, the two mcsf genes appeared to have arose through a chromosomal or genome duplication [249]. Examination of the intron-exon structure of trout mcsfs showed mcsf-1 to possess 10 exons and 9 introns, and mcsf-2 to have 9 exons and 8 introns [249].

Along with differing genomic organizations, trout mcsf-1/-2 are differentially expressed in tissues. The mcsf-1 transcript was predominantly expressed in the spleen, intestine and brain, while mcsf-2 was predominantly expressed in the head kidney, gills, muscle and liver [249]. While a recombinant trout MCSF-1 protein was produced and demonstrated to induce the proliferation of head kidney macrophages, a recombinant trout MCSF-2 protein was not produced to examine whether there was differential regulation of macrophage function by the MCSFs [249]. Whether MCSF-1 and MCSF-2 are functionally redundant or functionally partitioned (sub-functionalization), remains to be determined.

5.1.2.6. Teleost colony-stimulating factor-1 receptor

The *csf-1r* sequences have been identified in a number of teleost species including puffer fish [250, 251], zebrafish [252], rainbow trout [253], gilthead seabream [254] and goldfish [20]. CSF-1R protein appears to be a marker of monocytes and macrophages in teleosts [20, 254, 255] (Figure 1). Analysis of the puffer fish *csf-1r* gene shows a 21 exon gene structure in fish, same as in mammals. However, the puffer fish *csf-1r* gene only spans 10.5 kbp versus the mammalian 55 kbp, due to the decrease in the size of the intronic sequences [250]. The *csf-1r* mRNA open reading frame encodes for a 975 aa protein, with a signal peptide, an extracellular domain with 10 conserved cysteine residues characteristic of immunoglobulin domains, transmembrane domain, and an intracellular kinase domain with an interruption of 70 bp [250]. While CSF-1R of puffer fish is only 39% similar to human CSF-1R, the kinase domain is considerably more conserved, particularly in the motifs associated with signaling. The fish *csf-1r* gene was linked with *pdgfrb-1* [250].

A second *csf-1r* gene (*csf-1r-2*) was also identified in puffer fish, and linked with a second *pdgfrb* (*pdgfrb-2*). The *csf-1r-2* gene was comprised of 22 exons and had a different intron-exon organization than *csf-1r-1* [251]. Despite the similar protein structure of the two CSF-1Rs, the amino acid sequences were only 39% identical. The *csf-1r* mRNAs were differentially expressed in tissues. The *csf-1r-1* was expressed in blood, brain, eye, gill, heart, kidney, ovary, skin, and spleen, while *csf-1r-2* was expressed in the blood, brain, eye, gill, heart, kidney, liver, muscle, skin, spleen and testis. [251].

The duplication of *csf-1r* genes was also observed in cichlids, the green-spotted pufferfish, medaka, and *Tetraodon* (found on chromosomes 1 and 7), with the *csf-1r-2* duplicated genes appearing to have undergone evolutionary selection or diversification while the *csf-1r-1* gene appeared to resemble that of the ancestral gene [256]. It was proposed that the fish specific whole genome duplication generated the two paralogues of *csf-1r* in fish, as well as two *pdgfrb* and *kit* genes, and that *kit* and *csf-1r-2* may have been retained to play a role in the survival, migration and differentiation of melanocytes and xanthophores, important pigment cells involved in fish coloration patterns [256].

The *panther* (*fms*) mutant zebrafish have a defect in the *csf-1r* gene, and mutant fish fail to develop their characteristic pigment pattern of black and yellow stripes. The CSF-1R was found to be important in the survival, migration and differentiation of precursors to yellow xanthophores in zebrafish [257, 258]. However, unlike that of the CSF-1R[-/-] mice, there were no reports of hematopoietic defects in *panther* zebrafish. The lack of hematopoietic defects may be due to the presence of another *csf-1r* gene, a low level of *csf-1r* expression, or a differential requirement for CSF-1R during embryonic macrophage development versus adult macrophage development in teleosts. However, CSF-1R was shown to be important in the migration of primitive macrophages to tissues, such as the brain, retina and epidermis upon comparing primitive macrophage distribution and migration in wild-type and *panther* zebrafish [252]. Furthermore, *csf-1r* mRNA was detected in inflammatory macrophages from 3 dpf zebrafish embryos [219]. Taken together, these results support a role for CSF-1R in teleost macrophage biology.

A full-length *csf-1r* cDNA sequence was identified in trout, with an open reading frame of 2904 bp encoding for a 967 aa protein, predicted to be ~109 kDa. Trout CSF-1R had 40% aa identity to that of human and mouse, and 54% and 52% identity to that of puffer fish and zebrafish CSF-1R [253]. The trout *csf-1r* gene was similar to that of the ancestral gene, and mRNA was found in the head-kidney, spleen, blood, ovary, and showed lower mRNA levels in the liver, brain, heart, muscle, gill, and skin [253]. Southern blotting revealed two bands in each lane, suggestive of a second *csf-1r* gene in trout. However, a second *csf-1r* gene was never identified.

CSF-1R was also identified in goldfish as a 975 aa integral membrane bound protein (mCSF-1R) that possessed the five Ig extracellular domains with multiple N-linked glycosylation sites, a transmembrane domain, and an intracellular tyrosine kinase domain [20]. The mRNA of mCSF-1R could be detected in progenitor, monocyte and macrophage subpopulations, and an antibody produced against the first two Ig domains of CSF-1R was able to recognize monocytes and macrophages [20]. However, unlike mammalian neutrophils, zebrafish and goldfish neutrophils do not appear to express mRNA for *csf-1r* [16, 219]. Additionally, alternative splicing of the *csf-1r* transcript encoded for a soluble form of the CSF-1R (sCSF-1R), possessing only the D1 and D2 Ig domains, important for binding of CSF-1. The s*csf-1r* mRNA was expressed by leukocytes within the progenitor and macrophage populations, but not in the monocyte subpopulation [20]. Furthermore, addition of a recombinant purified sCSF-1R dampened the proliferation of spontaneously growing and differentiating PKMs [20]. The increased production of the sCSF-1R by PKMs during senescence phase suggested that sCSF-1R was involved in the negative regulation of CSF-1 signaling through mCSF-1R [20, 246] (Figure 1).

5.2. Neutrophil development

5.2.1. Progression of cell development

Following the commitment of the CFU-GM to a committed granulocyte progenitor cell, terminal differentiation through a promyelocyte, myelocyte, and metamyelocyte stages occur to give rise to a mature neutrophil, and are regulated through growth factor and transcription factor signaling, reviewed by [259] (Figure 1). Similar to that of mammals, the differentiation of fish neutrophils appears to occur through various stages, based on morphological and cytochemical characteristics, and include the promyelocyte, myelocyte, metamyelocyte and the mature neutrophil, which sometimes had a segmented nucleus [45, 212, 213, 215, 260]. These neutrophils were shown to migrate from the hematopoietic organ to the site of wounding, pathogen injection, or transformed cell injection [24, 45, 261], in response to a hydrogen peroxide attractant produced by cells at the site of damage [217]. However, the responding neutrophils had low phagocytic activity [24], or engulfed small fragments of the pathogen [217]. *In vitro*, treatment of zebrafish kidney marrow cells with G-CSF gave rise to CFU-GM in a methylcellulose assay [211]. However, there is a lack of *in vitro* culture sys-

tems for studying progenitor cell to neutrophil differentiation. The identification of functional neutrophils and their precursors suggests the presence of a committed granulocyte
progenitor cell in teleosts.

5.2.2. Receptors and growth factors

5.2.2.1. Granulocyte colony-stimulating factor

Neutrophils contribute to both innate and adaptive immune responses. They are capable of
chemotaxis, phagocytosis, antimicrobial molecule production, and formation of extracellular
traps [262-267]. Upon activation, neutrophils produce a number of chemokines, pro-inflammatory and anti-inflammatory cytokines, as well as the colony-stimulating factors G-CSF,
CSF-1, GM-CSF, IL3 and SCF, reviewed by [268, 269]. However, neutrophils are short lived,
6-90 hrs, and need to be continuously replaced.

GCSF, a member of the class I cytokine family, is the primary CSF that mediates the proliferation, differentiation, survival and activation of neutrophils and their progenitors, and has
been reviewed extensively by [144, 270]. The transcription of GCSF is controlled by an upstream promoter region with a tumor necrosis factor alpha response region that is bound by
NF-kB p65 and NF-IL6, reviewed elsewhere by [144, 271]. As such, GCSF can be produced
by activated monocytes/macrophages, neutrophils, fibroblasts and endothelial cells in response to a number of pro-inflammatory stimuli, reviewed elsewhere by [144, 270, 271]. In
humans, the normal GCSF concentration in blood ranges from 30-162 pg/mL, and can be
massively up-regulated during infection up to 3200 pg/mL [272-274].

5.2.2.2. Granulocyte colony-stimulating factor receptor

The protein structure of GCSFR is comprised of a signal peptide, an immunoglobulin-like
domain, a cytokine receptor homology (CRH) domain containing the class I cytokine receptor superfamily motif W-S-X-W-S, three fibronectin domains, a transmembrane domain, and
an intracellular cytoplasmic signaling domain containing three motifs termed Box 1, Box 2,
and Box 3, important for signal transduction [270, 275]. Based on their protein structure and
conserved motifs, the human and mouse integral membrane GCSFR proteins were placed in
the type I cytokine receptor family.

While there are reports of GCSFR on other hematopoietic cells such as monocytes [276] and
lymphocytes, as well as some non-hematopoietic cells, GCSFR is primarily found on neutrophils and their precursors [270, 277]. Neutrophils up-regulate their levels of GCSFR as they
differentiate from progenitor cell to mature neutrophil, with 50-500 GCSF receptors per cell
[278]. Structural analysis showed GCSF forms a homodimer, binds two GCSFRs, and leads
to receptor homodimerization in a 2:2 complex [279-281]. Binding of a homodimeric GCSF
to two GCSF receptors triggers intracellular signaling through the JAK/STAT, Ras/Raf/Erk,
or PI3K pathways [275, 277, 282]. These signaling pathways ultimately lead to the migration,
survival, proliferation, and differentiation of neutrophils. Control of GCSFR signaling in
neutrophils is modulated through (1) transcriptional activation of the GCSFR by AP-1, AP-2,

C/EBPα, NF-IL6, GATA-1, and PU.1/SPI1 transcription factors [181, 197], (2) the production of a soluble receptor through alternative splicing [275], and (3) cleavage of surface GCSFR by elastase [283].

5.2.2.3. Biological activity of granulocyte colony stimulating factor

The targeted gene disruption of *GCSF* and *GCSFR* has demonstrated the important functional roles of GCSF *in vivo*. GCSF and GCSFR deficient mice display severe neutropenia (70%-88% reduction in circulating neutrophils), reduction in monocyte and macrophage numbers, and ~50% reduction in the numbers of neutrophil precursors present in the bone marrow [284, 285] [282, 286] and are unable to control *Listeria monocytogenes* infections [284, 285]. GCSF treatment of bone marrow cells, *in vitro*, induced CFU activity to produce mainly neutrophil colonies [287] and promoted the proliferation of neutrophil precursors [270]. The release of mature neutrophils, their terminal differentiation, survival, and activation, is also mediated by GCSF *in vitro* and *in vivo*, reviewed by [270]. Lastly, GCSF has been used in the clinical setting to increase peripheral blood neutrophil numbers for treatment of disease and for stem cell mobilization from the bone marrow into the peripheral blood, reviewed by [288, 289].

5.2.2.4. Teleost granulocyte colony-stimulating factor

The teleost *gcsf* gene was first identified in Japanese flounder, fugu, and the green-spotted pufferfish [290]. Both the fugu and green-spotted pufferfish have two *gcsf* genes, termed *gcsf-1* and *gcsf-2*, while only an orthologue of *gcsf-2* was identified in flounder [290]. Phylogenetic analysis of vertebrate *gcsf*s predicted fish *gcsf-1* to be the ancestral gene, while *gcsf-2* was predicted to be the duplicated gene. Alignment of the fish GCSFs with human and mouse GCSF showed low identity, ranging from no significant identity to 34% amino acid identity [290]. Despite the low amino acid identity of fish to mammalian GCSF, all fish *gcsf* genes retained a 5 exon/ 4 intron structure with a conserved tumor necrosis factor alpha response element in the promoter region. Furthermore, the predicted transcripts have an open reading frame of 561-636 bp, corresponding to a predicted protein of 20-23 kDa, and 4-5 AU rich sequences in their 3' UTRs shown to be involved in mRNA instability and degradation [290]. Determination of the ratio of synonymous to asynonymous nucleotide substitutions (Ks/Ka) in fish *gcsf* genes ranged from 0.467 to 0.961 with an average of 0.793, demonstrating that positive selection was occurring in GCSFs of fish (and chicken) [290]. Two *gcsf* genes were also identified in the black rockfish (*Sebastes schlegelii*) [291] and in zebrafish [292] (O. Svoboda and P. Bartunek, personal communication), while only one *gcsf* gene has been identified in trout (NM_001195184).

Flounder *gcsf-2* mRNA levels were highest in the spleen, kidney, and gill. However, *gcsf-2* mRNA was still detected in the brain, eyes, heart, peripheral blood leukocytes, ovary, skin, and stomach, but was not detected in intestine, liver, or muscle tissue [290]. As expected, *gcsf-2* mRNA levels were up-regulated in kidney and peripheral blood leukocytes following treatment with lipopolysaccharide (LPS) or a mixture of concanavalin A and phorbol esters (ConA/PMA) [290]. The black rockfish *gcsf-1* showed expression in the peripheral blood leukocytes, spleen, gill, intestine and muscle [291]. However, black rockfish *gcsf-2* was ubiquitously expressed in the peripheral blood leukocytes, head and trunk kidney, spleen, gill, intestine, muscle, liver and brain [291]. Although both *gcsf-1* and *gcsf-2* black rockfish mRNA levels were upregulated in PBLs treated with LPS or ConA/PMA, differential kinetics and levels of expression were observed between the two *gcsfs* [291]. It appears that *gcsf-1* may be rapidly induced with sustained levels following stimulation, whereas *gcsf-2* is only slightly upregulated and showed a drastic increase in mRNA levels after ConA/PMA treatment for 24 hrs [291]. Taken together, these data suggest that GCSF-1 may play an important role during inflammation, although functional studies are required to determine the roles of GCSF-1 and GCSF-2 in teleost granulopoiesis and inflammation.

Functional studies on fish GCSF-1 are limited. Only two manuscripts report on the function of GCSF-1 and both utilize the zebrafish model system. *In vitro*, precursor cells from whole kidney marrow were sorted, plated in a methylcellulose colony forming unit assay and treated with either GCSF or a combination of GCSF and erythropoietin (EPO). While both treatments led to CFUs containing granulocytes and macrophages, the combination of GCSF and EPO also supported the formation of erythroid CFUs [211]. *In vivo*, morpholino mediated knockdown of *gcsfr* in zebrafish showed a decrease in numbers and migration of cells expressing both neutrophil and macrophage specific transcripts, during both primitive and definitive hematopoiesis in the zebrafish embryo. However, a population of myeloid cells remained, despite morpholino mediated knockdown of *gcsfr*, suggesting the presence of a GCSFR-independent pathway of myeloid cell development and migration [292]. Injection of wild-type zebrafish with *gcsf* mRNA increased the number of myeloid and *gcsfr⁺* cells, while injection of *gcsf* mRNA into *gcsfr* morpholino zebrafish did not result in an increase in myeloid cell numbers [292]. These studies suggested GCSF-1 participates in myeloid cell development, similar to that observed in mammalian systems (Figure 1). No functional studies have been performed using GCSF-2, and the role(s) of GCSF-2 in myelopoiesis remain to be elucidated.

5.2.2.5. Teleost granulocyte colony-stimulating factor receptor

The *gcsfr* has been identified in zebrafish [292], goldfish [293], and trout (AJ616901). Only one gene copy has been identified, although Southern blotting for goldfish *gcsfr* suggested the presence of more than one gene [293]. Analysis of the upstream promoter region of the 16 exon zebrafish *gcsfr* gene showed conserved putative sites for binding of the transcription factors HOXA5, PU.1 and CEBP family members [292], similar to the human *gcsfr* promoter region. These data suggest the conserved regulation of *gcsfr* gene expression in teleosts.

The predicted protein structure of zebrafish and goldfish GCSFRs is conserved across vertebrates. The teleost GCSFR extracellular domain is comprised of a signal peptide, an Ig-like domain, a cytokine homology domain containing the WSXWS motif and four cysteine residues, and three fibronectin domains. Following the transmembrane region, the intracellular region contains predicted Box1, Box2, and Box 3 signaling motifs and 6 tyrosine residues [292, 293], shown to be involved in receptor activation and internalization in higher vertebrates.

In zebrafish, *gcsfr* mRNA is expressed by 14 hpf in the RBI, followed by the yolk sac, the ICM, and finally in the kidney by 96 hpf, consistent with the production of neutrophils during primitive and definitive hematopoiesis. In adult goldfish, *gcsfr* mRNA levels were highest in kidney and spleen, followed by the gill, intestine, heart, brain and blood [293]. The *gcsfr* mRNA was highly expressed in goldfish neutrophils and was up-regulated in response to mitogens or pathogens [293] (Figure 1).

5.3. Transcription factors

In addition to the transcription factors described in section 4.3, there are a number of transcription factors downstream that participate in determining GMP fate decisions and that play a role in macrophage and neutrophil cell development, reviewed by [51, 294]. A visual representation of the stage(s) in which these transcription factors are important are shown in Figure 2.

5.3.1. Early growth response (Egr)

The four Egr proteins, EGR1 [295, 296], EGR2 [297], EGR3 [298] and EGR4 [299], are members of the zinc finger transcription factor family and have an N-terminus activation domain, a repressor domain capable of binding to NAB1/2, and a DNA binding domain comprised of three zinc fingers that bind to the GC rich sequence, 5'-GCGGGGGC'3' [300]. EGR1 promotes commitment to the macrophage lineage at the expense of granulocytic lineage [301, 302] and has been shown to be essential for myeloblast differentiation into monocytes/macrophages [303, 304]. Treatment of mouse bone marrow cells with CSF-1 has been shown to induce *EGR1* mRNA levels by 6-7 fold three hours post treatment, as well as *EGR2* and *EGR3* mRNA levels by 2-4 fold [305]. Although *EGR*[-/-] mice display normal macrophage development [306], it is thought that there is redundancy amongst the Egr transcription factors. Consistent with this idea, EGR2 is also abundant in monoblasts and monocytes [307], and may be involved in monocyte differentiation. Although a zebrafish orthologue of *EGR1* has been identified [308], the role of *egr1* in teleost macrophage development has not been examined.

5.3.2. Growth factor independence 1 (Gfi1)

Growth factor independence 1 (GFI1) is a zinc finger transcription factor comprised of an N-terminal Snail/Gfi1 (SNAG) domain that is involved in recruiting proteins to modify histones, and a C-terminal domain containing six zinc fingers involved in DNA recognition [309]. *GFI1* is expressed in T-cells, B-cells, mature granulocytes and activated macrophages [310, 311]. GFI1[-/-] mice showed slight defects in lymphocyte development, increased monocyte and monocyte precursor numbers, an absence of granulocytes and were highly susceptible to infections [310, 311]. Furthermore, myeloid progenitors from GFI1[-/-] mice did not differentiate into mature granulocytes in the presence of GCSF *in vitro* [310] or *in vivo* [311]. C/EBPα can up-regulate *GFI1* expression, promoting a neutrophil cell fate, and GFI1 also acts as a negative regulator on *PU.1* to decrease its expression [177, 180]. This lower level of *PU.1* drives granulocyte cell fate [175, 177]. GFI1 is important for neutrophil differentiation [177,

310, 312] and acts by activating Ras guanine nucleotide releasing protein 1 (RasGRP1) which is necessary for activating Ras in the Ras/MEK/Erk pathway that is initiated during GCSF signaling [313]. The expression of GFI1 is sustained during differentiation and the transcription factor functions by blocking the expression of *EGR-2/NAB-2*, effectively antagonizing the EGR1/2 transcription factor and preventing initiation of a monocytic differentiation pathway, thereby promoting neutrophil differentiation [177, 312]. Like that of PU.1 and GATA1, GFI1 and EGR-1/EGR-2 act as an antagonistic pair to regulate neutrophil versus macrophage lineage fate.

In zebrafish, two *gfi1* genes have been identified, termed *gfi1* and *gfi1.1*. *gfi1* is primarily expressed in neural tissues, and not in the hematopoietic system [314], suggesting that this is not the functional orthologue of mammalian *GFI1*. However, *gfi1.1* was expressed in the different hematopoietic organs of the developing zebrafish embryo, suggesting that *gfi1.1* is expressed in hematopoietic cells [315]. Zebrafish *gfi1.1* morphants displayed a three-fold increase in the number of *pu.1*+ cells, along with an increase in *l-plastin* expression and a decrease in *mpo* expression [315]. These data are consistent with the known functional role of mammalian GFI1, suggesting that zebrafish GFI1.1, and not zebrafish GFI1, is the functional orthologue of mammalian GFI.

5.3.3. Interferon response factor-8 (IRF-8)

Interferon response factor-8 (IRF-8, also known as ICSBP) is one out of nine members of the IRF transcription factor family and is characterized by an N-terminal DNA binding domain and a C-terminus IRF association domain that can associate with other IRF or Ets transcription family members [316, 317]. IRF8-/- mice and BXH-2 mice with a mutation in their IRF association domain show a drastic expansion of granulocytes at the expense of macrophages [318, 319]. Enforced expression of *IRF8* in myeloid progenitor cells *in vitro* led to the induced expression of a number of macrophage lineage differentiation transcripts including *CSF-1R* and *EGR1*. Additionally, enforced expression of *IRF8* in myeloid cell lines prevented their differentiation into granulocytes when treated with GCSF [320]. It is clear from the *in vivo* and *in vitro* studies that IRF8 promotes the commitment of myeloid progenitors along the macrophage lineage at the expense of the granulocyte lineage.

The homologue of *irf-8* was identified in rainbow trout [321] and zebrafish [322] with 53-55% amino acid identity to human IRF8 [321, 322]. In trout tissues, *irf8* mRNA was detected in the spleen, head kidney, gill, brain, intestine, skin, muscle, and liver [321] and mRNA levels could be up-regulated in splenocytes upon treatment with Poly I:C, PMA, PHA and recombinant IL-15. However, the role of IRF8 in GMP fate decisions or during macrophage development was not assessed. In zebrafish developing embryos, *irf8* mRNA was first detected in the rostral blood island, the site of primitive myelopoiesis, and was co-expressed with *csf-1r* mRNA, but not in cells positive for *mpx*, suggesting that *irf8* is expressed in cells committed to the macrophage lineage [322]. In zebrafish *irf8* morphants, *csf-1r*+ cells were absent, while *mpx*+ cells and mature neutrophils were increased by approximately three-fold, suggesting IRF8 is required for macrophage development. This phenotype could be rescued by injecting embryos with *irf8* mRNA.

Conversely, the over-expression of *irf8* mRNA in zebrafish resulted in an increase in macrophages by approximately 50% and a decrease in neutrophil numbers by about 40% [322]. These data are similar to those of the mammalian system and suggest a conserved role for IRF8 in determining macrophage over neutrophil cell lineage during primitive myelopoiesis. However, whether IRF8 plays the same role during definitive myelopoiesis in teleosts remains to be determined.

5.3.4. MafB

In addition to the previously described role of MAFB in HSCs and CMPs (see section 4.3.1), *MAFB* is highly expressed in monocytes and macrophages [164, 323] and has been shown to induce differentiation of myeloblasts into monocytes and macrophages [163, 307, 324, 325]. Furthermore, MAFB and c-MAF double knockout mice displayed differentiated macrophages that were capable of proliferating in response to CSF-1 in semi-solid and liquid culture [325]. Therefore, it appears that *MAFB* expression is sustained in monocytes and macrophages in order to prevent proliferation in these terminally differentiated cell populations.

Studies examining the role of MAFB in teleost myelopoiesis are limited. In the goldfish PKM system, a *mafb* transcript was identified and showed increasing mRNA levels with macrophage development [19]. The increasing mRNA levels of *mafb* during macrophage differentiation are similar to what has been observed in mammalian systems and suggest that MAFB may play a role in teleost macrophage differentiation.

6. Conclusion

Myelopoiesis is an orchestration of a multitude of growth factors and transcription factors that control cell fate decisions and differentiation along a chosen cell lineage. It is evident that there exists some functional redundancy in the action of myelopoietic growth factors, most likely put in place to ensure the production of these critical innate immune cells. Studies have focused on examining the regulation of myelopoiesis in the mouse model system, and have only just begun in the teleost model system.

The divergence of teleosts and mammals occurred approximately 400-450 Mya, thus teleosts represent one of the most basal groups of vertebrates [326]. Comparison of soluble factors and their receptors in teleosts and mammals show retention of many of important hematopoietic growth factors and receptors, including PDGFR [250, 251, 327], c-KIT [128-130], FLT3 (accession number DQ317446), CSF-1R [20, 250-254], GCSFR [292, 293], and their ligands PDGF [328], KIT ligand [128, 130, 131], CSF-1 [246, 249], and GCSF [290-292], although FLK2, the ligand to FLT3, has not yet been reported. However, it appears that teleosts do not possess the key myeloid growth factors IL-3 and GM-CSF, and their cognate receptors. In addition, teleosts possess all of the TF families required for hematopoiesis in higher vertebrates, reviewed in [28, 30]. Based on studies performed to date, the regulation of hematopoiesis is largely similar between mammals and teleosts. However, teleosts often possess a number of gene duplications for many of the soluble factors, receptors, and to some extent, transcription factors as a result of a teleost-specific whole genome duplication predicted to have occurred approximately 350 Mya, and is believed to be responsible for the radiation of teleosts [329, 330]. Many of these teleost genes are rapidly evolving, often undergoing sub-functionalization or neo-functionalization making the identification of teleost orthologues difficult. By developing an understanding of the soluble mediators, receptors, and the intra-cellular machinery that govern teleost myelopoiesis, we may be better equipped to develop strategies to promote host defense against pathogens, particularly in aquaculture in which fish are predisposed to infection.

Acknowledgements

This work was supported by a grant from Natural Sciences and Engineering Research Council of Canada (NSERC) to MB. BAK was supported by NSERC and Alberta Ingenuity Fund (AIF) doctoral scholarships, FK was supported by a Japan Society for the Promotion of Science (JSPS) research fellowship.

List of abbreviations:

ABC ATP-binding cassette; **AGM** aorta-gonad-mesonephros; **ALM** anterior lateral meso-derm; **ATP** adenosine triphosphate; **BFU-E** burst-forming unit-erythrocyte; **bZIP** basic leu-cine zipper domain; **CCM** cell-conditioned medium; **C/EBPs** CCAAT/enhancer binding proteins; **CFU** colony forming unit; **CFU-G** colony-forming unit-granulocyte; **CFU-GEMM** colony-forming unit granulocyte/erythrocyte/macrophage/megakaryocyte; **CFU-GM** colo-ny-forming unit-granulocyte/macrophage; **CFU-M** colony-forming unit-macrophage; **CFU-Meg** colony-forming unit-megakaryocyte; **CFU-S** colony-forming unit-spleen; **CHT** caudal hematopoietic tissue; **CLP** common lymphoid progenitor; **CMP** common myeloid progeni-tor; **ConA** concanavalin A; **CRH** cytokine receptor homology; **CSFs** colony-stimulating fac-tors; **CSF-1** colony-stimulating factor 1; **CSF-1R** colony-stimulating factor-1 receptor; **DCs** dendritic cells; **EGR/egr** early growth response; **EPO** erythropoietin; **GCSF** granulocyte col-ony-stimulating factor; **GFI1** growth factor independence 1; **GM-CSF** granulocyte/macro-phage colony-stimulating factor; **GM-CSFR** granulocyte/macrophage colony-stimulating factor receptor; **GM-CSFRα** granulocyte/macrophage colony-stimulating factor receptor al-pha chain; **GMP** granulocyte/macrophage progenitor; **HPCs** hematopoietic progenitor cells; **hpf** hours post fertilization; **HSCs** hematopoietic stem cells; **IL** interleukin; **IL-3Rα** interleu-kin-3 receptor alpha; **IRF** interferon response factor; **JAK** Janus family of protein tyrosine kinases; **Ks/Ka** ratio of synonymous to asynonymous nucleotide substitutions; **LMPPs** lym-phoid-myeloid primed multipotent progenitors; **LPS** lipopolysaccharide; **LT-HSCs** long-term hematopoietic stem cells; **MCSF** macrophage colony-stimulating factor; **mCSF-1R** membrane-bound colony-stimulating factor-1 receptor; **MEP** megakaryocyte/erythroid pro-genitor; **MMP** matrix metalloprotease; **mpo** myeloperoxidase; **MPPs** multipotent progenitors; **mpx** myeloid-specific peroxidase; **mRNA** messenger ribonucleic acid; **mSCF** membrane-bound stem cell factor; **multi-CSF** multi-lineage colony-stimulating factor; **NF-kB** nuclear factor kappa B; **NK** natural killer; **PBI** posterior blood island; **PBL** peripheral blood leukocytes; **PDGFR** platelet-derived growth factor receptor; **PGCs** primordial germ cells; **PKM** primary kidney macrophages; **PI3K** phosphatidylinositol-3-kinase; **PLCγ** phospholipase Cγ; **PLM** posterior lateral mesoderm; **PMA** phorbol esters; **Poly I:C** polyinosinic:polycytidylic; **RasGRP1** Ras guanine nucleotide releasing protein 1; **RBI** rostral blood island; **SCF** stem cell factor; **sCSF-1R** soluble colony-stimulating factor-1 receptor; **SOCS** suppressor of cytokine signaling; **sSCF** solube stem cell factor; **STATs** signal transducers and activators of transcription; **ST-HSCs** short-term hematopoietic stem cells; **TFs** transcription factors; **UTR** untranslated region;

Author details

Barbara A. Katzenback[1], Fumihiko Katakura[1] and Miodrag Belosevic[1,2*]

*Address all correspondence to: mike.belosevic@ualberta.ca

1 Department of Biological Sciences, University of Alberta, Edmonton, Alberta, Canada

2 Department of Biological Sciences, School of Public Health, University of Alberta, Edmonton, Alberta, Canada

References

[1] Martinez-Agosto JA, Mikkola HK, Hartenstein V, Banerjee U. The hematopoietic stem cell and its niche: a comparative view. Genes and Development 2007;21(23) 3044-60.

[2] Yoder JA, Nielsen ME, Amemiya CT, Litman GW. Zebrafish as an immunological model system. Microbes and Infection 2002;4(14) 1469-78.

[3] Zhang C, Patient R, Liu F. Hematopoietic stem cell development and regulatory signaling in zebrafish. Biochimica et Biophysica Acta 2012.

[4] Renshaw SA, Trede NS. A model 450 million years in the making: zebrafish and vertebrate immunity. Disease Models and Mechanisms 2012;5(1) 38-47.

[5] Jing L, Zon LI. Zebrafish as a model for normal and malignant hematopoiesis. Disease Models and Mechanisms 2011;4(4) 433-8.

[6] Zhong R, Astle CM, Harrison DE. Distinct developmental patterns of short-term and long-term functioning lymphoid and myeloid precursors defined by competitive limiting dilution analysis *in vivo*. Journal of Immunology 1996;157(1) 138-45.

[7] Osawa M, Hanada K, Hamada H, Nakauchi H. Long-term lymphohematopoietic reconstitution by a single CD34-low/negative hematopoietic stem cell. Science
1996;273(5272) 242-5.

[8] Wognum AW, Eaves AC, Thomas TE. Identification and isolation of hematopoietic
stem cells. Archives of medical research 2003;34(6) 461-75.

[9] Szilvassy SJ. The biology of hematopoietic stem cells. Archives of medical research
2003;34(6) 446-60.

[10] Nakanishi T. Histocompatibility analyses in tetraploids induced from clonal triploid
crucian carp and in gynogenetic diploid goldfish. Journal of Fish Biology
1987;3135-40.

[11] Nakanishi T, Ototake M. The graft-versus-host reaction (GVHR) in the ginbuna crucian carp, *Carassius auratus langsdorfii*. Developmental and Comparative Immunology
1999;23(1) 15-26.

[12] Nakanishi T, Okamoto N. Cell-mediated cytotoxicity in isogeneic ginbuna crucian
carp. Fish and Shellfish Immunology 1999;9(4) 259-67.

[13] Fischer U, Ototake M, Nakanishi T. Life span of circulating blood cells in ginbuna
crucian carp (*Carassius auratus langsdorfii*). Fish and Shellfish Immunology 1998;8(5)
339-49.

[14] Neumann NF, Barreda D, Belosevic M. Production of a macrophage growth factor(s)
by a goldfish macrophage cell line and macrophages derived from goldfish kidney
leukocytes. Developmental and Comparative Immunology 1998;22(4) 417-32.

[15] Neumann NF, Barreda DR, Belosevic M. Generation and functional analysis of distinct macrophage sub-populations from goldfish (*Carassius auratus* L.) kidney leukocyte cultures. Fish and Shellfish Immunology 2000;10(1) 1-20.

[16] Katzenback BA, Belosevic M. Isolation and functional characterization of neutrophil-
like cells, from goldfish (*Carassius auratus* L.) kidney. Developmental and Comparative Immunology 2009;33(4) 601-11.

[17] Barreda DR, Neumann NF, Belosevic M. Flow cytometric analysis of PKH26-labeled
goldfish kidney-derived macrophages. Developmental and Comparative Immunology 2000;24(4) 395-406.

[18] Barreda DR, Belosevic M. Characterisation of growth enhancing factor production in
different phases of *in vitro* fish macrophage development. Fish and Shellfish Immunology 2001;11(2) 169-85.

[19] Barreda DR, Hanington PC, Walsh CK, Wong P, Belosevic M. Differentially expressed genes that encode potential markers of goldfish macrophage development *in
vitro*. Developmental and Comparative Immunology 2004;28(7-8) 727-46.

[20] Barreda DR, Hanington PC, Stafford JL, Belosevic M. A novel soluble form of the CSF-1 receptor inhibits proliferation of self-renewing macrophages of goldfish (*Carassius auratus* L.). Developmental and Comparative Immunology 2005;29(10) 879-94.

[21] Walsh JG, Barreda DR, Belosevic M. Cloning and expression analysis of goldfish (*Carassius auratus* L.) prominin. Fish and Shellfish Immunology 2007;22(4) 308-17.

[22] Hanington PC, Barreda DR, Belosevic M. A novel hematopoietic granulin induces proliferation of goldfish (*Carassius auratus* L.) macrophages. Journal of Biological Chemistry 2006;281(15) 9963-70.

[23] Herbomel P, Thisse B, Thisse C. Ontogeny and behaviour of early macrophages in the zebrafish embryo. Development 1999;126(17) 3735-45.

[24] Le Guyader D, Redd MJ, Colucci-Guyon E, Murayama E, Kissa K, Briolat V, et al. Origins and unconventional behavior of neutrophils in developing zebrafish. Blood 2008;111(1) 132-41.

[25] Detrich HW, 3rd, Kieran MW, Chan FY, Barone LM, Yee K, Rundstadler JA, et al. Intraembryonic hematopoietic cell migration during vertebrate development. Proceedings of the National Academy of Science U S A 1995;92(23) 10713-7.

[26] Jin H, Xu J, Wen Z. Migratory path of definitive hematopoietic stem/progenitor cells during zebrafish development. Blood 2007;109(12) 5208-14.

[27] Bertrand JY, Kim AD, Violette EP, Stachura DL, Cisson JL, Traver D. Definitive hematopoiesis initiates through a committed erythromyeloid progenitor in the zebrafish embryo. Development 2007;134(23) 4147-56.

[28] Davidson AJ, Zon LI. The 'definitive' (and 'primitive') guide to zebrafish hematopoiesis. Oncogene 2004;23(43) 7233-46.

[29] Chen AT, Zon LI. Zebrafish blood stem cells. J Cell Biochem 2009;108(1) 35-42.

[30] de Jong JL, Zon LI. Use of the zebrafish system to study primitive and definitive hematopoiesis. Annual Review of Genetics 2005;39481-501.

[31] Zapata A, Diez B, Cejalvo T, Gutierrez-de Frias C, Cortes A. Ontogeny of the immune system of fish. Fish and Shellfish Immunology 2006;20(2) 126-36.

[32] Rombout JH, Huttenhuis HB, Picchietti S, Scapigliati G. Phylogeny and ontogeny of fish leucocytes. Fish and Shellfish Immunology 2005;19(5) 441-55.

[33] Bertrand JY, Kim AD, Teng S, Traver D. CD41+ cmyb+ precursors colonize the zebrafish pronephros by a novel migration route to initiate adult hematopoiesis. Development 2008;135(10) 1853-62.

[34] Murayama E, Kissa K, Zapata A, Mordelet E, Briolat V, Lin HF, et al. Tracing hematopoietic precursor migration to successive hematopoietic organs during zebrafish development. Immunity 2006;25(6) 963-75.

[35] Willett CE, Cortes A, Zuasti A, Zapata AG. Early hematopoiesis and developing lymphoid organs in the zebrafish. Dev Dyn 1999;214(4) 323-36.

[36] Zapata A. Ultrastructural study of the teleost fish kidney. Developmental and Comparative Immunology 1979;3(1) 55-65.

[37] Traver D, Paw BH, Poss KD, Penberthy WT, Lin S, Zon LI. Transplantation and in vivo imaging of multilineage engraftment in zebrafish bloodless mutants. Nature Immunology 2003;4(12) 1238-46.

[38] Traver D, Winzeler A, Stern HM, Mayhall EA, Langenau DM, Kutok JL, et al. Effects of lethal irradiation in zebrafish and rescue by hematopoietic cell transplantation. Blood 2004;104(5) 1298-305.

[39] Kobayashi I, Saito K, Moritomo T, Araki K, Takizawa F, Nakanishi T. Characterization and localization of side population (SP) cells in zebrafish kidney hematopoietic tissue. Blood 2008;111(3) 1131-7.

[40] Kobayashi I, Sekiya M, Moritomo T, Ototake M, Nakanishi T. Demonstration of hematopoietic stem cells in ginbuna carp (*Carassius auratus langsdorfii*) kidney. Developmental and Comparative Immunology 2006;30(11) 1034-46.

[41] Kobayashi I, Moritomo T, Ototake M, Nakanishi T. Isolation of side population cells from ginbuna carp (*Carassius auratus langsdorfii*) kidney hematopoietic tissues. Developmental and Comparative Immunology 2007;31(7) 696-707.

[42] Kobayashi I, Kusakabe H, Toda H, Moritomo T, Takahashi T, Nakanishi T. *In vivo* characterization of primitive hematopoietic cells in clonal ginbuna crucian carp (*Carassius auratus langsdorfii*). Veterinary Immunology and Immunopathology 2008;126(1-2) 74-82.

[43] Kobayashi I, Kuniyoshi S, Saito K, Moritomo T, Takahashi T, Nakanishi T. Long-term hematopoietic reconstitution by transplantation of kidney hematopoietic stem cells in lethally irradiated clonal ginbuna crucian carp (*Carassius auratus langsdorfii*). Developmental and Comparative Immunology 2008;32(8) 957-65.

[44] Moritomo T, Asakura N, Sekiya M, Ototake M, Inoue Y, Nakanishi T. Cell culture of clonal ginbuna crucian carp hematopoietic cells: differentiation of cultured cells into erythrocytes *in vivo*. Developmental and Comparative Immunology 2004;28(9) 863-9.

[45] Lieschke GJ, Oates AC, Crowhurst MO, Ward AC, Layton JE. Morphologic and functional characterization of granulocytes and macrophages in embryonic and adult zebrafish. Blood 2001;98(10) 3087-96.

[46] Rieger MA, Hoppe PS, Smejkal BM, Eitelhuber AC, Schroeder T. Hematopoietic cytokines can instruct lineage choice. Science 2009;325(5937) 217-8.

[47] Metcalf D. Lineage commitment and maturation in hematopoietic cells: the case for extrinsic regulation. Blood 1998;92(2) 345-7; discussion 52.

[48] Metcalf D. Hematopoietic cytokines. Blood 2008;111(2) 485-91.

[49] Zhu J, Emerson SG. Hematopoietic cytokines, transcription factors and lineage com-
      mitment. Oncogene 2002;21(21) 3295-313.

[50] Cantor AB, Orkin SH. Transcriptional regulation of erythropoiesis: an affair involv-
      ing multiple partners. Oncogene 2002;21(21) 3368-76.

[51] Friedman AD. Transcriptional regulation of granulocyte and monocyte develop-
      ment. Oncogene 2002;21(21) 3377-90.

[52] Ye M, Graf T. Early decisions in lymphoid development. Current Opinion in Immu-
      nology 2007;19(2) 123-8.

[53] Wang LD, Wagers AJ. Dynamic niches in the origination and differentiation of hae-
      matopoietic stem cells. Nature Reviews Molecular Cell Biology 2011;12(10) 643-55.

[54] Lai AY, Kondo M. Asymmetrical lymphoid and myeloid lineage commitment in
      multipotent hematopoietic progenitors. Journal of Experimental Medicine
      2006;203(8) 1867-73.

[55] Adolfsson J, Mansson R, Buza-Vidas N, Hultquist A, Liuba K, Jensen CT, et al. Iden-
      tification of Flt3+ lympho-myeloid stem cells lacking erythro-megakaryocytic poten-
      tial a revised road map for adult blood lineage commitment. Cell 2005;121(2) 295-306.

[56] Chi AW, Chavez A, Xu L, Weber BN, Shestova O, Schaffer A, et al. Identification of
      Flt3CD150 myeloid progenitors in adult mouse bone marrow that harbor T lymphoid
      developmental potential. Blood 2011;118(10) 2723-32.

[57] Ng SY, Yoshida T, Zhang J, Georgopoulos K. Genome-wide lineage-specific tran-
      scriptional networks underscore Ikaros-dependent lymphoid priming in hemato-
      poietic stem cells. Immunity 2009;30(4) 493-507.

[58] Pronk CJ, Rossi DJ, Mansson R, Attema JL, Norddahl GL, Chan CK, et al. Elucidation
      of the phenotypic, functional, and molecular topography of a myeloerythroid pro-
      genitor cell hierarchy. Cell Stem Cell 2007;1(4) 428-42.

[59] Montecino-Rodriguez E, Leathers H, Dorshkind K. Bipotential B-macrophage pro-
      genitors are present in adult bone marrow. Nature Immunology 2001;2(1) 83-8.

[60] Kondo M, Weissman IL, Akashi K. Identification of clonogenic common lymphoid
      progenitors in mouse bone marrow. Cell 1997;91(5) 661-72.

[61] Luc S, Buza-Vidas N, Jacobsen SE. Biological and molecular evidence for existence of
      lymphoid-primed multipotent progenitors. Annals of the New York Academy of Sci-
      ences 2007;110689-94.

[62] Katsura Y, Kawamoto H. Stepwise lineage restriction of progenitors in lympho-mye-
      lopoiesis. International Reviews of Immunology 2001;20(1) 1-20.

[63] Kawamoto H, Ohmura K, Katsura Y. Direct evidence for the commitment of hemato-
      poietic stem cells to T, B and myeloid lineages in murine fetal liver. International Im-
      munology 1997;9(7) 1011-9.

[64] Lu M, Kawamoto H, Katsube Y, Ikawa T, Katsura Y. The common myelolymphoid progenitor: a key intermediate stage in hemopoiesis generating T and B cells. Journal of Immunology 2002;169(7) 3519-25.

[65] Wada H, Masuda K, Satoh R, Kakugawa K, Ikawa T, Katsura Y, et al. Adult T-cell progenitors retain myeloid potential. Nature 2008;452(7188) 768-72.

[66] Kawamoto H, Katsura Y. A new paradigm for hematopoietic cell lineages: revision of the classical concept of the myeloid-lymphoid dichotomy. Trends in Immunology 2009;30(5) 193-200.

[67] Kawamoto H, Ikawa T, Masuda K, Wada H, Katsura Y. A map for lineage restriction of progenitors during hematopoiesis: the essence of the myeloid-based model. Immunological reviews 2010;238(1) 23-36.

[68] Li J, Barreda DR, Zhang YA, Boshra H, Gelman AE, LaPatra S, et al. B lymphocytes from early vertebrates have potent phagocytic and microbicidal abilities. Nature Immunology 2006;7(10) 1116-24.

[69] Katakura F, Yamaguchi T, Yoshida M, Moritomo T, Nakanishi T. Demonstration of T cell and macrophage progenitors in carp (*Cyprinus carpio*) kidney hematopoietic tissues. Development of clonal assay system for carp hematopoietic cells. Developmental and Comparative Immunology 2010;34(6) 685-9.

[70] Huang E, Nocka K, Beier DR, Chu TY, Buck J, Lahm HW, et al. The hematopoietic growth factor KL is encoded by the Sl locus and is the ligand of the c-kit receptor, the gene product of the W locus. Cell 1990;63(1) 225-33.

[71] Copeland NG, Gilbert DJ, Cho BC, Donovan PJ, Jenkins NA, Cosman D, et al. Mast cell growth factor maps near the steel locus on mouse chromosome 10 and is deleted in a number of steel alleles. Cell 1990;63(1) 175-83.

[72] Witte ON. Steel locus defines new multipotent growth factor. Cell 1990;63(1) 5-6.

[73] Zhang Z, Zhang R, Joachimiak A, Schlessinger J, Kong XP. Crystal structure of human stem cell factor: implication for stem cell factor receptor dimerization and activation. Proceedings of the National Academy of Science U S A 2000;97(14) 7732-7.

[74] Russell ES. Hereditary anemias of the mouse: a review for geneticists. Advances in Genetics 1979;20357-459.

[75] Dexter TM, Moore MA. *In vitro* duplication and "cure" of haemopoietic defects in genetically anaemic mice. Nature 1977;269(5627) 412-4.

[76] Anderson DM, Lyman SD, Baird A, Wignall JM, Eisenman J, Rauch C, et al. Molecular cloning of mast cell growth factor, a hematopoietin that is active in both membrane bound and soluble forms. Cell 1990;63(1) 235-43.

[77] Langley KE, Bennett LG, Wypych J, Yancik SA, Liu XD, Westcott KR, et al. Soluble stem cell factor in human serum. Blood 1993;81(3) 656-60.

[78] Arakawa T, Yphantis DA, Lary JW, Narhi LO, Lu HS, Prestrelski SJ, et al. Glycosylated and unglycosylated recombinant-derived human stem cell factors are dimeric and have extensive regular secondary structure. Journal of Biological Chemistry 1991;266(28) 18942-8.

[79] Pandiella A, Bosenberg MW, Huang EJ, Besmer P, Massague J. Cleavage of membrane-anchored growth factors involves distinct protease activities regulated through common mechanisms. Journal of Biological Chemistry 1992;267(33) 24028-33.

[80] Tajima Y, Moore MA, Soares V, Ono M, Kissel H, Besmer P. Consequences of exclusive expression in vivo of Kit-ligand lacking the major proteolytic cleavage site. Proceedings of the National Academy of Science U S A 1998;95(20) 11903-8.

[81] Ashman LK. The biology of stem cell factor and its receptor C-kit. International Journal of Biochemistry and Cell Biology 1999;31(10) 1037-51.

[82] Tajima Y, Huang EJ, Vosseller K, Ono M, Moore MA, Besmer P. Role of dimerization of the membrane-associated growth factor kit ligand in juxtacrine signaling: the Sl17H mutation affects dimerization and stability-phenotypes in hematopoiesis. Journal of Experimental Medicine 1998;187(9) 1451-61.

[83] Besmer P, Murphy JE, George PC, Qiu FH, Bergold PJ, Lederman L, et al. A new acute transforming feline retrovirus and relationship of its oncogene v-kit with the protein kinase gene family. Nature 1986;320(6061) 415-21.

[84] Yarden Y, Kuang WJ, Yang-Feng T, Coussens L, Munemitsu S, Dull TJ, et al. Human proto-oncogene c-kit: a new cell surface receptor tyrosine kinase for an unidentified ligand. EMBO Journal 1987;6(11) 3341-51.

[85] Qiu FH, Ray P, Brown K, Barker PE, Jhanwar S, Ruddle FH, et al. Primary structure of c-kit: relationship with the CSF-1/PDGF receptor kinase family--oncogenic activation of v-kit involves deletion of extracellular domain and C terminus. EMBO Journal 1988;7(4) 1003-11.

[86] Ullrich A, Schlessinger J. Signal transduction by receptors with tyrosine kinase activity. Cell 1990;61(2) 203-12.

[87] Rosnet O, Schiff C, Pebusque MJ, Marchetto S, Tonnelle C, Toiron Y, et al. Human FLT3/FLK2 gene: cDNA cloning and expression in hematopoietic cells. Blood 1993;82(4) 1110-9.

[88] Galli SJ, Zsebo KM, Geissler EN. The kit ligand, stem cell factor. Advances in Immunology 1994;551-96.

[89] Ratajczak J, Machalinski B, Majka M, Kijowski J, Marlicz W, Rozmyslowicz T, et al. Evidence that human haematopoietic stem cells (HSC) do not reside within the CD34+KIT- cell population. Annnals of Transplantation 1999;4(1) 22-30.

[90] Cairns LA, Moroni E, Levantini E, Giorgetti A, Klinger FG, Ronzoni S, et al. Kit regu-
      latory elements required for expression in developing hematopoietic and germ cell
      lineages. Blood 2003;102(12) 3954-62.

[91] Ikuta K, Weissman IL. Evidence that hematopoietic stem cells express mouse c-kit
      but do not depend on steel factor for their generation. Proceedings of the National
      Academy of Science U S A 1992;89(4) 1502-6.

[92] Cerisoli F, Cassinelli L, Lamorte G, Citterio S, Bertolotti F, Magli MC, et al. Green flu-
      orescent protein transgene driven by Kit regulatory sequences is expressed in hema-
      topoietic stem cells. Haematologica 2009;94(3) 318-25.

[93] Okayama Y, Kawakami T. Development, migration, and survival of mast cells. Im-
      munologic Research 2006;34(2) 97-115.

[94] Cole SR, Aylett GW, Harvey NL, Cambareri AC, Ashman LK. Increased expression
      of c-Kit or its ligand Steel Factor is not a common feature of adult acute myeloid leu-
      kaemia. Leukemia 1996;10(2) 288-96.

[95] Turner AM, Bennett LG, Lin NL, Wypych J, Bartley TD, Hunt RW, et al. Identifica-
      tion and characterization of a soluble c-kit receptor produced by human hemato-
      poietic cell lines. Blood 1995;85(8) 2052-8.

[96] Yee NS, Langen H, Besmer P. Mechanism of kit ligand, phorbol ester, and calcium-
      induced down-regulation of c-kit receptors in mast cells. Journal of Biological Chem-
      istry 1993;268(19) 14189-201.

[97] Broudy VC, Kovach NL, Bennett LG, Lin N, Jacobsen FW, Kidd PG. Human umbili-
      cal vein endothelial cells display high-affinity c-kit receptors and produce a soluble
      form of the c-kit receptor. Blood 1994;83(8) 2145-52.

[98] Dahlen DD, Lin NL, Liu YC, Broudy VC. Soluble Kit receptor blocks stem cell factor
      bioactivity in vitro. Leukemia Research 2001;25(5) 413-21.

[99] Jiang X, Gurel O, Mendiaz EA, Stearns GW, Clogston CL, Lu HS, et al. Structure of
      the active core of human stem cell factor and analysis of binding to its receptor kit.
      EMBO Journal 2000;19(13) 3192-203.

[100] Philo JS, Wen J, Wypych J, Schwartz MG, Mendiaz EA, Langley KE. Human stem cell
      factor dimer forms a complex with two molecules of the extracellular domain of its
      receptor, Kit. Journal of Biological Chemistry 1996;271(12) 6895-902.

[101] Broudy VC. Stem cell factor and hematopoiesis. Blood 1997;90(4) 1345-64.

[102] Blechman JM, Lev S, Brizzi MF, Leitner O, Pegoraro L, Givol D, et al. Soluble c-kit
      proteins and antireceptor monoclonal antibodies confine the binding site of the stem
      cell factor. Journal of Biological Chemistry 1993;268(6) 4399-406.

[103] Lev S, Blechman J, Nishikawa S, Givol D, Yarden Y. Interspecies molecular chimeras
      of kit help define the binding site of the stem cell factor. Molecular and Cellular Biol-
      ogy 1993;13(4) 2224-34.

[104] Yuzawa S, Opatowsky Y, Zhang Z, Mandiyan V, Lax I, Schlessinger J. Structural basis for activation of the receptor tyrosine kinase KIT by stem cell factor. Cell 2007;130(2) 323-34.

[105] Narhi LO, Wypych J, Li T, Langley KE, Arakawa T. Changes in conformation and stability upon SCF/sKit complex formation. Journal of Protein Chemistry 1998;17(5) 387-96.

[106] Linnekin D. Early signaling pathways activated by c-Kit in hematopoietic cells. International Journal of Biochemistry and Cell Biology 1999;31(10) 1053-74.

[107] Roskoski R, Jr. Signaling by Kit protein-tyrosine kinase--the stem cell factor receptor. Biochemical and Biophysical Research Communications 2005;337(1) 1-13.

[108] Kent D, Copley M, Benz C, Dykstra B, Bowie M, Eaves C. Regulation of hematopoietic stem cells by the steel factor/KIT signaling pathway. Clinical Cancer Research 2008;14(7) 1926-30.

[109] Mauduit C, Hamamah S, Benahmed M. Stem cell factor/c-kit system in spermatogenesis. Humum Reproduction Update 1999;5(5) 535-45.

[110] Besmer P, Manova K, Duttlinger R, Huang EJ, Packer A, Gyssler C, et al. The kit-ligand (steel factor) and its receptor c-kit/W: pleiotropic roles in gametogenesis and melanogenesis. Journal of Reproduction and Development Supplement 1993125-37.

[111] Bedell MA, Mahakali Zama A. Genetic analysis of Kit ligand functions during mouse spermatogenesis. Journal of Adrology 2004;25(2) 188-99.

[112] Reid K, Nishikawa S, Bartlett PF, Murphy M. Steel factor directs melanocyte development in vitro through selective regulation of the number of c-kit+ progenitors. Developmental Biology 1995;169(2) 568-79.

[113] Wehrle-Haller B, Weston JA. Soluble and cell-bound forms of steel factor activity play distinct roles in melanocyte precursor dispersal and survival on the lateral neural crest migration pathway. Development 1995;121(3) 731-42.

[114] Wehrle-Haller B. The role of Kit-ligand in melanocyte development and epidermal homeostasis. Pigment Cell Research 2003;16(3) 287-96.

[115] Yee NS, Paek I, Besmer P. Role of kit-ligand in proliferation and suppression of apoptosis in mast cells: basis for radiosensitivity of white spotting and steel mutant mice. Journal of Experimental Medicine 1994;179(6) 1777-87.

[116] Serve H, Yee NS, Stella G, Sepp-Lorenzino L, Tan JC, Besmer P. Differential roles of PI3-kinase and Kit tyrosine 821 in Kit receptor-mediated proliferation, survival and cell adhesion in mast cells. EMBO Journal 1995;14(3) 473-83.

[117] Wershil BK, Tsai M, Geissler EN, Zsebo KM, Galli SJ. The rat c-kit ligand, stem cell factor, induces c-kit receptor-dependent mouse mast cell activation in vivo. Evidence that signaling through the c-kit receptor can induce expression of cellular function. Journal of Experimental Medicine 1992;175(1) 245-55.

[118] Tanaka A, Arai K, Kitamura Y, Matsuda H. Matrix metalloproteinase-9 production, a newly identified function of mast cell progenitors, is downregulated by c-kit receptor activation. Blood 1999;94(7) 2390-5.

[119] Lukacs NW, Kunkel SL, Strieter RM, Evanoff HL, Kunkel RG, Key ML, et al. The role of stem cell factor (c-kit ligand) and inflammatory cytokines in pulmonary mast cell activation. Blood 1996;87(6) 2262-8.

[120] Reber L, Da Silva CA, Frossard N. Stem cell factor and its receptor c-Kit as targets for inflammatory diseases. European Journal of Pharmacology 2006;533(1-3) 327-40.

[121] Ogawa M, Matsuzaki Y, Nishikawa S, Hayashi S, Kunisada T, Sudo T, et al. Expression and function of c-kit in hemopoietic progenitor cells. Journal of Experimental Medicine 1991;174(1) 63-71.

[122] Keller JR, Ortiz M, Ruscetti FW. Steel factor (c-kit ligand) promotes the survival of hematopoietic stem/progenitor cells in the absence of cell division. Blood 1995;86(5) 1757-64.

[123] Thoren LA, Liuba K, Bryder D, Nygren JM, Jensen CT, Qian H, et al. Kit regulates maintenance of quiescent hematopoietic stem cells. J Immunol 2008;180(4) 2045-53.

[124] Mantel C, Hendrie P, Broxmeyer HE. Steel factor regulates cell cycle asymmetry. Stem Cells 2001;19(6) 483-91.

[125] Zsebo KM, Wypych J, McNiece IK, Lu HS, Smith KA, Karkare SB, et al. Identification, purification, and biological characterization of hematopoietic stem cell factor from buffalo rat liver--conditioned medium. Cell 1990;63(1) 195-201.

[126] Bernstein ID, Andrews RG, Zsebo KM. Recombinant human stem cell factor enhances the formation of colonies by CD34+ and CD34+lin- cells, and the generation of colony-forming cell progeny from CD34+lin- cells cultured with interleukin-3, granulocyte colony-stimulating factor, or granulocyte-macrophage colony-stimulating factor. Blood 1991;77(11) 2316-21.

[127] Okumura N, Tsuji K, Ebihara Y, Tanaka I, Sawai N, Koike K, et al. Chemotactic and chemokinetic activities of stem cell factor on murine hematopoietic progenitor cells. Blood 1996;87(10) 4100-8.

[128] Parichy DM, Rawls JF, Pratt SJ, Whitfield TT, Johnson SL. Zebrafish sparse corresponds to an orthologue of c-kit and is required for the morphogenesis of a subpopulation of melanocytes, but is not essential for hematopoiesis or primordial germ cell development. Development 1999;126(15) 3425-36.

[129] Mellgren EM, Johnson SL. kitb, a second zebrafish ortholog of mouse Kit. Developmental Genes and Evolution 2005;215(9) 470-77.

[130] Katzenback BA, Belosevic M. Molecular and functional characterization of kita and kitla of the goldfish (Carassius auratus L.). Developmental and Comparative Immunology 2009;33(11) 1165-75.

[131] Hultman KA, Bahary N, Zon LI, Johnson SL. Gene Duplication of the zebrafish kit ligand and partitioning of melanocyte development functions to kit ligand a. PLoS Genetics 2007;3(1) e17.

[132] Yao K, Ge W. Kit system in the zebrafish ovary: evidence for functional divergence of two isoforms of kit (kita and kitb) and kit ligand (kitlga and kitlgb) during folliculogenesis. Biology of Reproduction 2010;82(6) 1216-26.

[133] Cooper CD, Linbo TH, Raible DW. Kit and foxd3 genetically interact to regulate melanophore survival in zebrafish. Developmental Dynamics 2009;238(4) 875-86.

[134] Mellgren EM, Johnson SL. A requirement for kit in embryonic zebrafish melanocyte differentiation is revealed by melanoblast delay. Developmental Genes and Evolution 2004;214(10) 493-502.

[135] Rawls JF, Johnson SL. Requirements for the kit receptor tyrosine kinase during regeneration of zebrafish fin melanocytes. Development 2001;128(11) 1943-9.

[136] Yamada T, Okauchi M, Araki K. Origin of adult-type pigment cells forming the asymmetric pigment pattern, in Japanese flounder (Paralichthys olivaceus). Developmental Dynamics 2010;239(12) 3147-62.

[137] Mills MG, Nuckels RJ, Parichy DM. Deconstructing evolution of adult phenotypes: genetic analyses of kit reveal homology and evolutionary novelty during adult pigment pattern development of Danio fishes. Development 2007;134(6) 1081-90.

[138] Ihle JN, Keller J, Oroszlan S, Henderson LE, Copeland TD, Fitch F, et al. Biologic properties of homogeneous interleukin 3. I. Demonstration of WEHI-3 growth factor activity, mast cell growth factor activity, p cell-stimulating factor activity, colony-stimulating factor activity, and histamine-producing cell-stimulating factor activity. Journal of Immunology 1983;131(1) 282-7.

[139] Yang YC, Ciarletta AB, Temple PA, Chung MP, Kovacic S, Witek-Giannotti JS, et al. Human IL-3 (multi-CSF): identification by expression cloning of a novel hematopoietic growth factor related to murine IL-3. Cell 1986;47(1) 3-10.

[140] Rennick DM, Lee FD, Yokota T, Arai KI, Cantor H, Nabel GJ. A cloned MCGF cDNA encodes a multilineage hematopoietic growth factor: multiple activities of interleukin 3. Journal of Immunology 1985;134(2) 910-4.

[141] Spivak JL, Smith RR, Ihle JN. Interleukin 3 promotes the in vitro proliferation of murine pluripotent hematopoietic stem cells. Journal of Clinical Investigation 1985;76(4) 1613-21.

[142] Quesenberry PJ, Ihle JN, McGrath E. The effect of interleukin 3 and GM-CSA-2 on megakaryocyte and myeloid clonal colony formation. Blood 1985;65(1) 214-7.

[143] van Leeuwen BH, Martinson ME, Webb GC, Young IG. Molecular organization of the cytokine gene cluster, involving the human IL-3, IL-4, IL-5, and GM-CSF genes, on human chromosome 5. Blood 1989;73(5) 1142-8.

[144] Barreda DR, Hanington PC, Belosevic M. Regulation of myeloid development and function by colony stimulating factors. Developmental and Comparative Immunology 2004;28(5) 509-54.

[145] Martinez-Moczygemba M, Huston DP. Biology of common beta receptor-signaling cytokines: IL-3, IL-5, and GM-CSF. Journal of Allergy and Clinical Immunology 2003;112(4) 653-65; quiz 66.

[146] Scott CL, Begley CG. The beta common chain (beta c) of the granulocyte macrophage-colony stimulating factor, interleukin-3 and interleukin-5 receptors. International Journal of Biochemistry and Cell Biology 1999;31(10) 1011-5.

[147] Gasson JC, Weisbart RH, Kaufman SE, Clark SC, Hewick RM, Wong GG, et al. Purified human granulocyte-macrophage colony-stimulating factor: direct action on neutrophils. Science 1984;226(4680) 1339-42.

[148] Cantrell MA, Anderson D, Cerretti DP, Price V, McKereghan K, Tushinski RJ, et al. Cloning, sequence, and expression of a human granulocyte/macrophage colony-stimulating factor. Proceedings of the National Academy of Science U S A 1985;82(18) 6250-4.

[149] Otsuka T, Miyajima A, Brown N, Otsu K, Abrams J, Saeland S, et al. Isolation and characterization of an expressible cDNA encoding human IL-3. Induction of IL-3 mRNA in human T cell clones. Journal of Immunology 1988;140(7) 2288-95.

[150] Broudy VC, Kaushansky K, Segal GM, Harlan JM, Adamson JW. Tumor necrosis factor type alpha stimulates human endothelial cells to produce granulocyte/macrophage colony-stimulating factor. Proceedings of the National Academy of Science U S A 1986;83(19) 7467-71.

[151] Munker R, Gasson J, Ogawa M, Koeffler HP. Recombinant human TNF induces production of granulocyte-monocyte colony-stimulating factor. Nature 1986;323(6083) 79-82.

[152] Kimura A, Rieger MA, Simone JM, Chen W, Wickre MC, Zhu BM, et al. The transcription factors STAT5A/B regulate GM-CSF-mediated granulopoiesis. Blood 2009;114(21) 4721-8.

[153] Gomez-Cambronero J, Horn J, Paul CC, Baumann MA. Granulocyte-macrophage colony-stimulating factor is a chemoattractant cytokine for human neutrophils: involvement of the ribosomal p70 S6 kinase signaling pathway. Journal of Immunology 2003;171(12) 6846-55.

[154] Khajah M, Millen B, Cara DC, Waterhouse C, McCafferty DM. Granulocyte-macrophage colony stimulating factor (GM-CSF): a chemoattractive agent for murine leukocytes in vivo. Journal of Leukocyte Biology 2011;89(6) 945-53.

[155] Fossati G, Mazzucchelli I, Gritti D, Ricevuti G, Edwards SW, Moulding DA, et al. In vitro effects of GM-CSF on mature peripheral blood neutrophils. International Journal of Molecular Medicine 1998;1(6) 943-51.

[156] Watowich SS, Liu YJ. Mechanisms regulating dendritic cell specification and development. Immunological Reviews 2010;238(1) 76-92.

[157] Schmid MA, Kingston D, Boddupalli S, Manz MG. Instructive cytokine signals in dendritic cell lineage commitment. Immunological Reviews 2010;234(1) 32-44.

[158] Hercus TR, Thomas D, Guthridge MA, Ekert PG, King-Scott J, Parker MW, et al. The granulocyte-macrophage colony-stimulating factor receptor: linking its structure to cell signaling and its role in disease. Blood 2009;114(7) 1289-98.

[159] Laiosa CV, Stadtfeld M, Graf T. Determinants of lymphoid-myeloid lineage diversification. Annual Review of Immunology 2006;24705-38.

[160] Iwasaki H, Akashi K. Myeloid lineage commitment from the hematopoietic stem cell. Immunity 2007;26(6) 726-40.

[161] Rosenbauer F, Tenen DG. Transcription factors in myeloid development: balancing differentiation with transformation. Nature Reviews Immunology 2007;7(2) 105-17.

[162] Hanington PC, Tam J, Katzenback BA, Hitchen SJ, Barreda DR, Belosevic M. Development of macrophages of cyprinid fish. Developmental and Comparative Immunology 2009;33(4) 411-29.

[163] Sarrazin S, Mossadegh-Keller N, Fukao T, Aziz A, Mourcin F, Vanhille L, et al. MafB restricts M-CSF-dependent myeloid commitment divisions of hematopoietic stem cells. Cell 2009;138(2) 300-13.

[164] Sieweke MH, Tekotte H, Frampton J, Graf T. MafB is an interaction partner and repressor of Ets-1 that inhibits erythroid differentiation. Cell 1996;85(1) 49-60.

[165] Kajihara M, Kawauchi S, Kobayashi M, Ogino H, Takahashi S, Yasuda K. Isolation, characterization, and expression analysis of zebrafish large Mafs. Journal of Biochemistry 2001;129(1) 139-46.

[166] Landschulz WH, Johnson PF, McKnight SL. The leucine zipper: a hypothetical structure common to a new class of DNA binding proteins. Science 1988;240(4860) 1759-64.

[167] Lekstrom-Himes J, Xanthopoulos KG. Biological role of the CCAAT/enhancer-binding protein family of transcription factors. Journal of Biological Chemistry 1998;273(44) 28545-8.

[168] Lyons SE, Shue BC, Lei L, Oates AC, Zon LI, Liu PP. Molecular cloning, genetic mapping, and expression analysis of four zebrafish c/ebp genes. Gene 2001;281(1-2) 43-51.

[169] Fujiki K, Gerwick L, Bayne CJ, Mitchell L, Gauley J, Bols N, et al. Molecular cloning and characterization of rainbow trout (*Oncorhynchus mykiss*) CCAAT/enhancer binding protein beta. Immunogenetics 2003;55(4) 253-61.

[170] Tucker CS, Hirono I, Aoki T. Molecular cloning and expression of CCAAT/enhancer binding proteins in Japanese flounder *Paralichthys olivaceus*. Developmental and Comparative Immunology 2002;26(3) 271-82.

[171] Hikima J, Ohtani M, Kondo H, Hirono I, Jung TS, Aoki T. Characterization and gene expression of transcription factors, PU.1 and C/EBPalpha driving transcription from the tumor necrosis factor alpha promoter in Japanese flounder, *Paralichthys olivaceus*. Developmental and Comparative Immunology 2011;35(3) 304-13.

[172] Traver D, Miyamoto T, Christensen J, Iwasaki-Arai J, Akashi K, Weissman IL. Fetal liver myelopoiesis occurs through distinct, prospectively isolatable progenitor subsets. Blood 2001;98(3) 627-35.

[173] Hohaus S, Petrovick MS, Voso MT, Sun Z, Zhang DE, Tenen DG. PU.1 (Spi-1) and C/EBP alpha regulate expression of the granulocyte-macrophage colony-stimulating factor receptor alpha gene. Molecular and Cellular Biology 1995;15(10) 5830-45.

[174] Heath V, Suh HC, Holman M, Renn K, Gooya JM, Parkin S, et al. C/EBPalpha deficiency results in hyperproliferation of hematopoietic progenitor cells and disrupts macrophage development *in vitro* and *in vivo*. Blood 2004;104(6) 1639-47.

[175] Friedman AD. C/EBPalpha induces PU.1 and interacts with AP-1 and NF-kappaB to regulate myeloid development. Blood Cells, Molecules and Diseases 2007;39(3) 340-3.

[176] Zhang P, Iwasaki-Arai J, Iwasaki H, Fenyus ML, Dayaram T, Owens BM, et al. Enhancement of hematopoietic stem cell repopulating capacity and self-renewal in the absence of the transcription factor C/EBP alpha. Immunity 2004;21(6) 853-63.

[177] Laslo P, Spooner CJ, Warmflash A, Lancki DW, Lee HJ, Sciammas R, et al. Multilineage transcriptional priming and determination of alternate hematopoietic cell fates. Cell 2006;126(4) 755-66.

[178] Radomska HS, Huettner CS, Zhang P, Cheng T, Scadden DT, Tenen DG. CCAAT/enhancer binding protein alpha is a regulatory switch sufficient for induction of granulocytic development from bipotential myeloid progenitors. Molecular and Cellular Biology 1998;18(7) 4301-14.

[179] Ford AM, Bennett CA, Healy LE, Towatari M, Greaves MF, Enver T. Regulation of the myeloperoxidase enhancer binding proteins Pu1, C-EBP alpha, -beta, and -delta during granulocyte-lineage specification. Proceedings of the National Academy of Science U S A 1996;93(20) 10838-43.

[180] Oelgeschlager M, Nuchprayoon I, Luscher B, Friedman AD. C/EBP, c-Myb, and PU.1 cooperate to regulate the neutrophil elastase promoter. Molecular and Cellular Biology 1996;16(9) 4717-25.

[181] Smith LT, Hohaus S, Gonzalez DA, Dziennis SE, Tenen DG. PU.1 (Spi-1) and C/EBP alpha regulate the granulocyte colony-stimulating factor receptor promoter in myeloid cells. Blood 1996;88(4) 1234-47.

[182] Liu TX, Rhodes J, Deng M, Hsu K, Radomska HS, Kanki JP, et al. Dominant-interfering C/EBPalpha stimulates primitive erythropoiesis in zebrafish. Experimental Hematology 2007;35(2) 230-9.

[183] Yuan H, Zhou J, Deng M, Zhang Y, Chen Y, Jin Y, et al. Sumoylation of CCAAT/enhancer-binding protein alpha promotes the biased primitive hematopoiesis of zebrafish. Blood 2011;117(26) 7014-20.

[184] Burda P, Laslo P, Stopka T. The role of PU.1 and GATA-1 transcription factors during normal and leukemogenic hematopoiesis. Leukemia 2010;24(7) 1249-57.

[185] Klemsz MJ, McKercher SR, Celada A, Van Beveren C, Maki RA. The macrophage and B cell-specific transcription factor PU.1 is related to the ets oncogene. Cell 1990;61(1) 113-24.

[186] Kim HG, de Guzman CG, Swindle CS, Cotta CV, Gartland L, Scott EW, et al. The ETS family transcription factor PU.1 is necessary for the maintenance of fetal liver hematopoietic stem cells. Blood 2004;104(13) 3894-900.

[187] Iwasaki H, Somoza C, Shigematsu H, Duprez EA, Iwasaki-Arai J, Mizuno S, et al. Distinctive and indispensable roles of PU.1 in maintenance of hematopoietic stem cells and their differentiation. Blood 2005;106(5) 1590-600.

[188] Dakic A, Metcalf D, Di Rago L, Mifsud S, Wu L, Nutt SL. PU.1 regulates the commitment of adult hematopoietic progenitors and restricts granulopoiesis. Journal of Experimental Medicine 2005;201(9) 1487-502.

[189] McKercher SR, Torbett BE, Anderson KL, Henkel GW, Vestal DJ, Baribault H, et al. Targeted disruption of the PU.1 gene results in multiple hematopoietic abnormalities. EMBO Journal 1996;15(20) 5647-58.

[190] Scott EW, Simon MC, Anastasi J, Singh H. Requirement of transcription factor PU.1 in the development of multiple hematopoietic lineages. Science 1994;265(5178) 1573-7.

[191] Nutt SL, Metcalf D, D'Amico A, Polli M, Wu L. Dynamic regulation of PU.1 expression in multipotent hematopoietic progenitors. Journal of Experimental Medicine 2005;201(2) 221-31.

[192] Zhang P, Zhang X, Iwama A, Yu C, Smith KA, Mueller BU, et al. PU.1 inhibits GATA-1 function and erythroid differentiation by blocking GATA-1 DNA binding. Blood 2000;96(8) 2641-8.

[193] Rekhtman N, Radparvar F, Evans T, Skoultchi AI. Direct interaction of hematopoietic transcription factors PU.1 and GATA-1: functional antagonism in erythroid cells. Genes and Development 1999;13(11) 1398-411.

[194] Xie Y, Chen C, Hume DA. Transcriptional regulation of c-fms gene expression. Cell Biochemistry and Biophysics 2001;34(1) 1-16.

[195] Tagoh H, Himes R, Clarke D, Leenen PJ, Riggs AD, Hume D, et al. Transcription fac-
       tor complex formation and chromatin fine structure alterations at the murine c-fms
       (CSF-1 receptor) locus during maturation of myeloid precursor cells. Genes and De-
       velopment 2002;16(13) 1721-37.

[196] Gangenahalli GU, Gupta P, Saluja D, Verma YK, Kishore V, Chandra R, et al. Stem
       cell fate specification: role of master regulatory switch transcription factor PU.1 in
       differential hematopoiesis. Stem Cells and Devevelopment 2005;14(2) 140-52.

[197] Ito Y, Seto Y, Brannan CI, Copeland NG, Jenkins NA, Fukunaga R, et al. Structural
       analysis of the functional gene and pseudogene encoding the murine granulocyte
       colony-stimulating-factor receptor. European Journal of Biochemistry 1994;220(3)
       881-91.

[198] Lieschke GJ, Oates AC, Paw BH, Thompson MA, Hall NE, Ward AC, et al. Zebrafish
       SPI-1 (PU.1) marks a site of myeloid development independent of primitive erythro-
       poiesis: implications for axial patterning. Developmental Bioliology 2002;246(2)
       274-95.

[199] Ward AC, McPhee DO, Condron MM, Varma S, Cody SH, Onnebo SM, et al. The ze-
       brafish spi1 promoter drives myeloid-specific expression in stable transgenic fish.
       Blood 2003;102(9) 3238-40.

[200] Hsu K, Traver D, Kutok JL, Hagen A, Liu TX, Paw BH, et al. The pu.1 promoter
       drives myeloid gene expression in zebrafish. Blood 2004;104(5) 1291-7.

[201] Rhodes J, Hagen A, Hsu K, Deng M, Liu TX, Look AT, et al. Interplay of pu.1 and
       gata1 determines myelo-erythroid progenitor cell fate in zebrafish. Developmental
       Cell 2005;8(1) 97-108.

[202] Galloway JL, Wingert RA, Thisse C, Thisse B, Zon LI. Loss of gata1 but not gata2 con-
       verts erythropoiesis to myelopoiesis in zebrafish embryos. Developmental Cell
       2005;8(1) 109-16.

[203] Zakrzewska A, Cui C, Stockhammer OW, Benard EL, Spaink HP, Meijer AH. Macro-
       phage-specific gene functions in Spi1-directed innate immunity. Blood 2010;116(3)
       e1-11.

[204] Bukrinsky A, Griffin KJ, Zhao Y, Lin S, Banerjee U. Essential role of spi-1-like (spi-1l)
       in zebrafish myeloid cell differentiation. Blood 2009;113(9) 2038-46.

[205] Takahashi K. Development and differentiation of macrophages and related cells: his-
       torical review and current concepts. Journal of Clinical and Experimental Hematopa-
       thology 2001;411-33.

[206] Auffray C, Sieweke MH, Geissmann F. Blood monocytes: development, heterogenei-
       ty, and relationship with dendritic cells. Annual Review of Immunology
       2009;27669-92.

[207] Geissmann F, Manz MG, Jung S, Sieweke MH, Merad M, Ley K. Development of monocytes, macrophages, and dendritic cells. Science 2010;327(5966) 656-61.

[208] Ganassin RC, Bols NC. Development of a monocyte/macrophage-like cell line, RTS11, from rainbow trout spleen. Fish and Shellfish Immunology 1998;8457-76.

[209] Stafford JL, McLauchlan PE, Secombes CJ, Ellis AE, Belosevic M. Generation of primary monocyte-like cultures from rainbow trout head kidney leukocytes. Developmental and Comparative Immunology 2001;25(5-6) 447-59.

[210] Stachura DL, Reyes JR, Bartunek P, Paw BH, Zon LI, Traver D. Zebrafish kidney stromal cell lines support multilineage hematopoiesis. Blood 2009;114(2) 279-89.

[211] Stachura DL, Svoboda O, Lau RP, Balla KM, Zon LI, Bartunek P, et al. Clonal analysis of hematopoietic progenitor cells in the zebrafish. Blood 2011;118(5) 1274-82.

[212] Bennett CM, Kanki JP, Rhodes J, Liu TX, Paw BH, Kieran MW, et al. Myelopoiesis in the zebrafish, *Danio rerio*. Blood 2001;98(3) 643-51.

[213] Crowhurst MO, Layton JE, Lieschke GJ. Developmental biology of zebrafish myeloid cells. International Journal of Developmental Biology 2002;46(4) 483-92.

[214] Zuasti A, Ferrer C. Haemopoiesis in the head kidney of *Sparus auratus*. Archives of Histology and Cytology 1989;52(3) 249-55.

[215] Abdel-Aziz el SH, Abdu SB, Ali Tel S, Fouad HF. Haemopoiesis in the head kidney of tilapia, *Oreochromis niloticus* (Teleostei: Cichlidae): a morphological (optical and ultrastructural) study. Fish Physiology and Biochemistry 2010;36(3) 323-36.

[216] Wittamer V, Bertrand JY, Gutschow PW, Traver D. Characterization of the mononuclear phagocyte system in zebrafish. Blood 2011;117(26) 7126-35.

[217] Feng Y, Santoriello C, Mione M, Hurlstone A, Martin P. Live imaging of innate immune cell sensing of transformed cells in zebrafish larvae: parallels between tumor initiation and wound inflammation. PLoS Biology 2010;8(12) e1000562.

[218] Grabher C, Cliffe A, Miura K, Hayflick J, Pepperkock R, Rorth P, et al. Birth and life of tissue macrophages and their migration in embryogenesis and inflammation in medaka. Journal of Leukocyte Biology 2007;81263-71.

[219] Mathias JR, Dodd ME, Walters KB, Yoo SK, Ranheim EA, Huttenlocher A. Characterization of zebrafish larval inflammatory macrophages. Developmental and Comparative Immunology 2009;33(11) 1212-7.

[220] Fixe P, Praloran V. Macrophage colony-stimulating-factor (M-CSF or CSF-1) and its receptor: structure-function relationships. European Cytokine Network 1997;8(2) 125-36.

[221] Stanley ER, Berg KL, Einstein DB, Lee PS, Pixley FJ, Wang Y, et al. Biology and action of colony--stimulating factor-1. Molecular Reproduction and Development 1997;46(1) 4-10.

[222] Guilbert LJ, Stanley ER. Specific interaction of murine colony-stimulating factor with mononuclear phagocytic cells. Journal of Cell Biology 1980;85(1) 153-9.

[223] Tushinski RJ, Oliver IT, Guilbert LJ, Tynan PW, Warner JR, Stanley ER. Survival of mononuclear phagocytes depends on a lineage-specific growth factor that the differentiated cells selectively destroy. Cell 1982;28(1) 71-81.

[224] Ladner MB, Martin GA, Noble JA, Nikoloff DM, Tal R, Kawasaki ES, et al. Human CSF-1: gene structure and alternative splicing of mRNA precursors. EMBO Journal 1987;6(9) 2693-8.

[225] Kawasaki ES, Ladner MB. Molecular biology of macrophage colony-stimulating factor. Immunology Series 1990;49155-76.

[226] Pandit J, Bohm A, Jancarik J, Halenbeck R, Koths K, Kim SH. Three-dimensional structure of dimeric human recombinant macrophage colony-stimulating factor. Science 1992;258(5086) 1358-62.

[227] Praloran V, Gascan H, Papin S, Chevalier S, Trossaert M, Boursier MC. Inducible production of macrophage colony-stimulating factor (CSF-1) by malignant and normal human T cells. Leukemia 1990;4(6) 411-4.

[228] Fretier S, Besse A, Delwail A, Garcia M, Morel F, Leprivey-Lorgeot V, et al. Cyclosporin A inhibition of macrophage colony-stimulating factor (M-CSF) production by activated human T lymphocytes. Journal of Leukocyte Biology 2002;71(2) 289-94.

[229] Hallet MM, Praloran V, Vie H, Peyrat MA, Wong G, Witek-Giannotti J, et al. Macrophage colony-stimulating factor (CSF-1) gene expression in human T-lymphocyte clones. Blood 1991;77(4) 780-6.

[230] Horiguchi J, Warren MK, Kufe D. Expression of the macrophage-specific colony-stimulating factor in human monocytes treated with granulocyte-macrophage colony-stimulating factor. Blood 1987;69(4) 1259-61.

[231] Oster W, Lindemann A, Horn S, Mertelsmann R, Herrmann F. Tumor necrosis factor (TNF)-alpha but not TNF-beta induces secretion of colony stimulating factor for macrophages (CSF-1) by human monocytes. Blood 1987;70(5) 1700-3.

[232] Lin H, Lee E, Hestir K, Leo C, Huang M, Bosch E, et al. Discovery of a cytokine and its receptor by functional screening of the extracellular proteome. Science 2008;320(5877) 807-11.

[233] Wei S, Nandi S, Chitu V, Yeung YG, Yu W, Huang M, et al. Functional overlap but differential expression of CSF-1 and IL-34 in their CSF-1 receptor-mediated regulation of myeloid cells. Journal of Leukocyte Biology 2010;88(3) 495-505.

[234] Chihara T, Suzu S, Hassan R, Chutiwitoonchai N, Hiyoshi M, Motoyoshi K, et al. IL-34 and M-CSF share the receptor Fms but are not identical in biological activity and signal activation. Cell Death and Differentiation 2010;17(12) 1917-27.

[235] Garceau V, Smith J, Paton IR, Davey M, Fares MA, Sester DP, et al. Pivotal Advance: Avian colony-stimulating factor 1 (CSF-1), interleukin-34 (IL-34), and CSF-1 receptor genes and gene products. Journal of Leukocyte Biology 2010;87(5) 753-64.

[236] Rettenmier CW, Chen JH, Roussel MF, Sherr CJ. The product of the c-fms proto-oncogene: a glycoprotein with associated tyrosine kinase activity. Science 1985;228(4697) 320-2.

[237] Sherr CJ, Rettenmier CW, Sacca R, Roussel MF, Look AT, Stanley ER. The c-fms proto-oncogene product is related to the receptor for the mononuclear phagocyte growth factor, CSF-1. Cell 1985;41(3) 665-76.

[238] Pixley FJ, Stanley ER. CSF-1 regulation of the wandering macrophage: complexity in action. Trends in Cell Biology 2004;14(11) 628-38.

[239] Chen X, Liu H, Focia PJ, Shim AH, He X. Structure of macrophage colony stimulating factor bound to FMS: diverse signaling assemblies of class III receptor tyrosine kinases. Proceedings of the National Academy of Science U S A 2008;105(47) 18267-72.

[240] Hamilton JA. CSF-1 signal transduction. Journal of Leukocyte Biology 1997;62(2) 145-55.

[241] Yoshida H, Hayashi S, Kunisada T, Ogawa M, Nishikawa S, Okamura H, et al. The murine mutation osteopetrosis is in the coding region of the macrophage colony stimulating factor gene. Letters to Nature 1990;345442-4.

[242] Marks SC, Jr., Lane PW. Osteopetrosis, a new recessive skeletal mutation on chromosome 12 of the mouse. Journal of Heredity 1976;67(1) 11-8.

[243] Wiktor-Jedrzejczak W, Bartocci A, Ferrante AW, Jr., Ahmed-Ansari A, Sell KW, Pollard JW, et al. Total absence of colony-stimulating factor 1 in the macrophage-deficient osteopetrotic (op/op) mouse. Proceedings of the National Academy of Science U S A 1990;87(12) 4828-32.

[244] Pollard JW, Stanley ER. Pleiotropic roles for CSF-1 in development defined by the mouse mutation osteopetrotic. Advances in Developmental Biochemistry 1996;4153-93.

[245] Dai XM, Ryan GR, Hapel AJ, Dominguez MG, Russell RG, Kapp S, et al. Targeted disruption of the mouse colony-stimulating factor 1 receptor gene results in osteopetrosis, mononuclear phagocyte deficiency, increased primitive progenitor cell frequencies, and reproductive defects. Blood 2002;99(1) 111-20.

[246] Hanington PC, Wang T, Secombes CJ, Belosevic M. Growth factors of lower vertebrates: characterization of goldfish (Carassius auratus L.) macrophage colony-stimulating factor-1. Journal of Biological Chemistry 2007;282(44) 31865-72.

[247] Hanington PC, Hitchen SJ, Beamish LA, Belosevic M. Macrophage colony stimulating factor (CSF-1) is a central growth factor of goldfish macrophages. Fish and Shellfish Immunology 2009;26(1) 1-9.

[248] Grayfer L, Hanington PC, Belosevic M. Macrophage colony-stimulating factor (CSF-1) induces pro-inflammatory gene expression and enhances antimicrobial responses of goldfish (*Carassius auratus* L.) macrophages. Fish and Shellfish Immunology 2009;26(3) 406-13.

[249] Wang T, Hanington PC, Belosevic M, Secombes CJ. Two macrophage colony-stimulating factor genes exist in fish that differ in gene organization and are differentially expressed. Journal of Immunology 2008;181(5) 3310-22.

[250] How GF, Venkatesh B, Brenner S. Conserved linkage between the puffer fish (*Fugu rubripes*) and human genes for platelet-derived growth factor receptor and macrophage colony-stimulating factor receptor. Genome Research 1996;6(12) 1185-91.

[251] Williams H, Brenner S, Venkatesh B. Identification and analysis of additional copies of the platelet-derived growth factor receptor and colony stimulating factor 1 receptor genes in fugu. Gene 2002;295(2) 255-64.

[252] Herbomel P, Thisse B, Thisse C. Zebrafish early macrophages colonize cephalic mesenchyme and developing brain, retina, and epidermis through a M-CSF receptor-dependent invasive process. Developmental Biology 2001;238(2) 274-88.

[253] Honda T, Nishizawa T, Uenobe M, Kohchi C, Kuroda A, Ototake M, et al. Molecular cloning and expression analysis of a macrophage-colony stimulating factor receptor-like gene from rainbow trout, *Oncorhynchus mykiss*. Molecular Immunology 2005;42(1) 1-8.

[254] Roca FJ, Sepulcre MA, Lopez-Castejon G, Meseguer J, Mulero V. The colony-stimulating factor-1 receptor is a specific marker of macrophages from the bony fish gilthead seabream. Molecular Immunology 2006;43(9) 1418-23.

[255] Mulero I, Pilar Sepulcre M, Roca FJ, Meseguer J, Garcia-Ayala A, Mulero V. Characterization of macrophages from the bony fish gilthead seabream using an antibody against the macrophage colony-stimulating factor receptor. Developmental and Comparative Immunology 2008;32(10) 1151-9.

[256] Braasch I, Salzburger W, Meyer A. Asymmetric evolution in two fish-specifically duplicated receptor tyrosine kinase paralogons involved in teleost coloration. Molecular Biology and Evolution 2006;23(6) 1192-202.

[257] Parichy DM, Ransom DG, Paw B, Zon LI, Johnson SL. An orthologue of the kit-related gene fms is required for development of neural crest-derived xanthophores and a subpopulation of adult melanocytes in the zebrafish, *Danio rerio*. Development 2000;127(14) 3031-44.

[258] Parichy DM, Turner JM. Temporal and cellular requirements for Fms signaling during zebrafish adult pigment pattern development. Development 2003;130(5) 817-33.

[259] Theilgaard-Monch K, Jacobsen LC, Borup R, Rasmussen T, Bjerregaard MD, Nielsen FC, et al. The transcriptional program of terminal granulocytic differentiation. Blood 2005;105(4) 1785-96.

[260] Meseguer J, Esteban MA, Garcia Ayala A, Lopez Ruiz A, Agulleiro B. Granulopoiesis in the head-kidney of the sea bass (*Dicentrarchus labrax* L.): an ultrastructural study. Archives of Histology and Cytology 1990;53(3) 287-96.

[261] Yoo SK, Huttenlocher A. Spatiotemporal photolabeling of neutrophil trafficking during inflammation in live zebrafish. Journal of Leukocyte Biology 2011;89(5) 661-7.

[262] Harun NO, Zou J, Zhang YA, Nie P, Secombes CJ. The biological effects of rainbow trout (*Oncorhynchus mykiss*) recombinant interleukin-8. Developmental and Comparative Immunology 2008;32(6) 673-81.

[263] Overland HS, Pettersen EF, Ronneseth A, Wergeland HI. Phagocytosis by B-cells and neutrophils in Atlantic salmon (*Salmo salar* L.) and Atlantic cod (*Gadus morhua* L.). Fish and Shellfish Immunology 2010;28(1) 193-204.

[264] Mayumi M, Takeda Y, Hoshiko M, Serada K, Murata M, Moritomo T, et al. Characterization of teleost phagocyte NADPH oxidase: molecular cloning and expression analysis of carp (*Cyprinus carpio*) phagocyte NADPH oxidase. Molecular Immunology 2008;45(6) 1720-31.

[265] Palic D, Andreasen CB, Menzel BW, Roth JA. A rapid, direct assay to measure degranulation of primary granules in neutrophils from kidney of fathead minnow (*Pimephales promelas* Rafinesque, 1820). Fish and Shellfish Immunology 2005;19(3) 217-27.

[266] Palic D, Ostojic J, Andreasen CB, Roth JA. Fish cast NETs: neutrophil extracellular traps are released from fish neutrophils. Developmental and Comparative Immunology 2007;31(8) 805-16.

[267] Palic D, Andreasen CB, Ostojic J, Tell RM, Roth JA. Zebrafish (*Danio rerio*) whole kidney assays to measure neutrophil extracellular trap release and degranulation of primary granules. Journal of Immunological Methods 2007;319(1-2) 87-97.

[268] Mantovani A, Cassatella MA, Costantini C, Jaillon S. Neutrophils in the activation and regulation of innate and adaptive immunity. Nature Reviews Immunology 2011;11(8) 519-31.

[269] Kumar V, Sharma A. Neutrophils: Cinderella of innate immune system. International Immunopharmacology 2010;10(11) 1325-34.

[270] Panopoulos AD, Watowich SS. Granulocyte colony-stimulating factor: molecular mechanisms of action during steady state and 'emergency' hematopoiesis. Cytokine 2008;42(3) 277-88.

[271] Demetri GD, Griffin JD. Granulocyte colony-stimulating factor and its receptor. Blood 1991;78(11) 2791-808.

[272] Watari K, Asano S, Shirafuji N, Kodo H, Ozawa K, Takaku F, et al. Serum granulocyte colony-stimulating factor levels in healthy volunteers and patients with various
disorders as estimated by enzyme immunoassay. Blood 1989;73(1) 117-22.

[273] Kawakami M, Tsutsumi H, Kumakawa T, Abe H, Hirai M, Kurosawa S, et al. Levels
of serum granulocyte colony-stimulating factor in patients with infections. Blood
1990;76(10) 1962-4.

[274] Cheers C, Haigh AM, Kelso A, Metcalf D, Stanley ER, Young AM. Production of colony-stimulating factors (CSFs) during infection: separate determinations of macro-
phage-, granulocyte-, granulocyte-macrophage-, and multi-CSFs. Infection and
Immunity 1988;56(1) 247-51.

[275] Avalos BR. Molecular analysis of the granulocyte colony-stimulating factor receptor.
Blood 1996;88(3) 761-77.

[276] Shimoda K, Okamura S, Harada N, Niho Y. Detection of the granulocyte colony-
stimulating factor receptor using biotinylated granulocyte colony-stimulating factor:
presence of granulocyte colony-stimulating factor receptor on CD34-positive hematopoietic progenitor cells. Research in Experimental Medicine (Berl) 1992;192(4) 245-55.

[277] Liongue C, Wright C, Russell AP, Ward AC. Granulocyte colony-stimulating factor
receptor: stimulating granulopoiesis and much more. International Journal of Biochemistry and Cell Biology 2009;41(12) 2372-5.

[278] Nicola NA, Metcalf D. Binding of 125I-labeled granulocyte colony-stimulating factor
to normal murine hemopoietic cells. Journal of Cellular Physiology 1985;124(2)
313-21.

[279] Aritomi M, Kunishima N, Okamoto T, Kuroki R, Ota Y, Morikawa K. Atomic structure of the GCSF-receptor complex showing a new cytokine-receptor recognition
scheme. Nature 1999;401(6754) 713-7.

[280] Tamada T, Honjo E, Maeda Y, Okamoto T, Ishibashi M, Tokunaga M, et al. Homodi-
meric cross-over structure of the human granulocyte colony-stimulating factor
(GCSF) receptor signaling complex. Proceedings of the National Academy of Science
U S A 2006;103(9) 3135-40.

[281] Honjo E, Tamada T, Maeda Y, Koshiba T, Matsukura Y, Okamoto T, et al. Crystallization of a 2:2 complex of granulocyte-colony stimulating factor (GCSF) with the li-
gand-binding region of the GCSF receptor. Acta Crystallographica Section F:
Structural Biology and Crystallization Communications 2005;61(Pt 8) 788-90.

[282] Richards MK, Liu F, Iwasaki H, Akashi K, Link DC. Pivotal role of granulocyte colony-stimulating factor in the development of progenitors in the common myeloid
pathway. Blood 2003;102(10) 3562-8.

[283] Piper MG, Massullo PR, Loveland M, Druhan LJ, Kindwall-Keller TL, Ai J, et al. Neutrophil elastase downmodulates native G-CSFR expression and granulocyte-macrophage colony formation. Journal of Inflammation (Lond) 2010;7(1) 5.

[284] Lieschke GJ, Grail D, Hodgson G, Metcalf D, Stanley E, Cheers C, et al. Mice lacking granulocyte colony-stimulating factor have chronic neutropenia, granulocyte and macrophage progenitor cell deficiency, and impaired neutrophil mobilization. Blood 1994;84(6) 1737-46.

[285] Zhan Y, Lieschke GJ, Grail D, Dunn AR, Cheers C. Essential roles for granulocyte-macrophage colony-stimulating factor (GM-CSF) and G-CSF in the sustained hematopoietic response of Listeria monocytogenes-infected mice. Blood 1998;91(3) 863-9.

[286] Liu F, Wu HY, Wesselschmidt R, Kornaga T, Link DC. Impaired production and increased apoptosis of neutrophils in granulocyte colony-stimulating factor receptor-deficient mice. Immunity 1996;5(5) 491-501.

[287] Tsuchiya M, Asano S, Kaziro Y, Nagata S. Isolation and characterization of the cDNA for murine granulocyte colony-stimulating factor. Proceedings of the National Academy of Science U S A 1986;83(20) 7633-7.

[288] Hubel K, Dale DC, Liles WC. Therapeutic use of cytokines to modulate phagocyte function for the treatment of infectious diseases: current status of granulocyte colony-stimulating factor, granulocyte-macrophage colony-stimulating factor, macrophage colony-stimulating factor, and interferon-gamma. Journal of Infectious Diseases 2002;185(10) 1490-501.

[289] Greenbaum AM, Link DC. Mechanisms of G-CSF-mediated hematopoietic stem and progenitor mobilization. Leukemia 2011;25(2) 211-7.

[290] Santos MD, Yasuike M, Hirono I, Aoki T. The granulocyte colony-stimulating factors (CSF3s) of fish and chicken. Immunogenetics 2006;58(5-6) 422-32.

[291] Nam BH, An GH, Baeck GW, Kim MC, Kim JW, Park HJ, et al. Molecular cloning and expression of cDNAs for two distinct granulocyte colony stimulating factor genes from black rockfish Sebastes schlegelii. Fish and Shellfish Immunology 2009;27(2) 360-4.

[292] Liongue C, Hall CJ, O'Connell BA, Crosier P, Ward AC. Zebrafish granulocyte colony-stimulating factor receptor signaling promotes myelopoiesis and myeloid cell migration. Blood 2009;113(11) 2535-46.

[293] Katzenback BA, Belosevic M. Characterization of granulocyte colony stimulating factor receptor of the goldfish (Carassius auratus L.). Developmental and Comparative Immunology 2012;36(1) 199-207.

[294] Tenen DG, Hromas R, Licht JD, Zhang DE. Transcription factors, normal myeloid development, and leukemia. Blood 1997;90(2) 489-519.

[295] Milbrandt J. A nerve growth factor-induced gene encodes a possible transcriptional regulatory factor. Science 1987;238(4828) 797-9.

[296] Sukhatme VP, Kartha S, Toback FG, Taub R, Hoover RG, Tsai-Morris CH. A novel early growth response gene rapidly induced by fibroblast, epithelial cell and lymphocyte mitogens. Oncogene Research 1987;1(4) 343-55.

[297] Chavrier P, Zerial M, Lemaire P, Almendral J, Bravo R, Charnay P. A gene encoding a protein with zinc fingers is activated during G0/G1 transition in cultured cells. EMBO Journal 1988;7(1) 29-35.

[298] Patwardhan S, Gashler A, Siegel MG, Chang LC, Joseph LJ, Shows TB, et al. EGR3, a novel member of the Egr family of genes encoding immediate-early transcription factors. Oncogene 1991;6(6) 917-28.

[299] Crosby SD, Puetz JJ, Simburger KS, Fahrner TJ, Milbrandt J. The early response gene NGFI-C encodes a zinc finger transcriptional activator and is a member of the GCGGGGGCG (GSG) element-binding protein family. Molecular and Cellular Biology 1991;11(8) 3835-41.

[300] Thiel G, Cibelli G. Regulation of life and death by the zinc finger transcription factor Egr-1. Journal of Cellular Physiology 2002;193(3) 287-92.

[301] Krishnaraju K, Nguyen HQ, Liebermann DA, Hoffman B. The zinc finger transcription factor Egr-1 potentiates macrophage differentiation of hematopoietic cells. Molecular and Cellular Biology 1995;15(10) 5499-507.

[302] Krishnaraju K, Hoffman B, Liebermann DA. Early growth response gene 1 stimulates development of hematopoietic progenitor cells along the macrophage lineage at the expense of the granulocyte and erythroid lineages. Blood 2001;97(5) 1298-305.

[303] Nguyen HQ, Hoffman-Liebermann B, Liebermann DA. The zinc finger transcription factor Egr-1 is essential for and restricts differentiation along the macrophage lineage. Cell 1993;72(2) 197-209.

[304] Krishnaraju K, Hoffman B, Liebermann DA. The zinc finger transcription factor Egr-1 activates macrophage differentiation in M1 myeloblastic leukemia cells. Blood 1998;92(6) 1957-66.

[305] Carter JH, Tourtellotte WG. Early growth response transcriptional regulators are dispensable for macrophage differentiation. Journal of Immunology 2007;178(5) 3038-47.

[306] Lee SL, Tourtellotte LC, Wesselschmidt RL, Milbrandt J. Growth and differentiation proceeds normally in cells deficient in the immediate early gene NGFI-A. Journal of Biological Chemistry 1995;270(17) 9971-7.

[307] Gemelli C, Montanari M, Tenedini E, Zanocco Marani T, Vignudelli T, Siena M, et al. Virally mediated MafB transduction induces the monocyte commitment of human CD34+ hematopoietic stem/progenitor cells. Cell Death and Differentiation 2006;13(10) 1686-96.

[308] Drummond IA, Rohwer-Nutter P, Sukhatme VP. The zebrafish egr1 gene encodes a highly conserved, zinc-finger transcriptional regulator. DNA and Cell Biology 1994;13(10) 1047-55.

[309] van der Meer LT, Jansen JH, van der Reijden BA. Gfi1 and Gfi1b: key regulators of hematopoiesis. Leukemia 2010;24(11) 1834-43.

[310] Karsunky H, Zeng H, Schmidt T, Zevnik B, Kluge R, Schmid KW, et al. Inflammatory reactions and severe neutropenia in mice lacking the transcriptional repressor Gfi1. Nature Genetics 2002;30(3) 295-300.

[311] Hock H, Hamblen MJ, Rooke HM, Traver D, Bronson RT, Cameron S, et al. Intrinsic requirement for zinc finger transcription factor Gfi-1 in neutrophil differentiation. Immunity 2003;18(1) 109-20.

[312] Zarebski A, Velu CS, Baktula AM, Bourdeau T, Horman SR, Basu S, et al. Mutations in growth factor independent-1 associated with human neutropenia block murine granulopoiesis through colony stimulating factor-1. Immunity 2008;28(3) 370-80.

[313] de la Luz Sierra M, Sakakibara S, Gasperini P, Salvucci O, Jiang K, McCormick PJ, et al. The transcription factor Gfi1 regulates G-CSF signaling and neutrophil development through the Ras activator RasGRP1. Blood 2010;115(19) 3970-9.

[314] Dufourcq P, Rastegar S, Strahle U, Blader P. Parapineal specific expression of gfi1 in the zebrafish epithalamus. Gene Expression Patterns 2004;4(1) 53-7.

[315] Wei W, Wen L, Huang P, Zhang Z, Chen Y, Xiao A, et al. Gfi1.1 regulates hematopoietic lineage differentiation during zebrafish embryogenesis. Cell Research 2008;18(6) 677-85.

[316] Paun A, Pitha PM. The IRF family, revisited. Biochimie 2007;89(6-7) 744-53.

[317] Wang H, Morse HC, 3rd. IRF8 regulates myeloid and B lymphoid lineage diversification. Immunology Research 2009;43(1-3) 109-17.

[318] Holtschke T, Lohler J, Kanno Y, Fehr T, Giese N, Rosenbauer F, et al. Immunodeficiency and chronic myelogenous leukemia-like syndrome in mice with a targeted mutation of the ICSBP gene. Cell 1996;87(2) 307-17.

[319] Turcotte K, Gauthier S, Tuite A, Mullick A, Malo D, Gros P. A mutation in the Icsbp1 gene causes susceptibility to infection and a chronic myeloid leukemia-like syndrome in BXH-2 mice. Journal of Experimental Medicine 2005;201(6) 881-90.

[320] Tamura T, Nagamura-Inoue T, Shmeltzer Z, Kuwata T, Ozato K. ICSBP directs bipotential myeloid progenitor cells to differentiate into mature macrophages. Immunity 2000;13(2) 155-65.

[321] Holland JW, Karim A, Wang T, Alnabulsi A, Scott J, Collet B, et al. Molecular cloning and characterization of interferon regulatory factors 4 and 8 (IRF-4 and IRF-8) in rainbow trout, *Oncorhynchus mykiss*. Fish and Shellfish Immunology 2010;29(1) 157-66.

[322] Li L, Jin H, Xu J, Shi Y, Wen Z. Irf8 regulates macrophage versus neutrophil fate during zebrafish primitive myelopoiesis. Blood 2011;117(4) 1359-69.

[323] Eichmann A, Grapin-Botton A, Kelly L, Graf T, Le Douarin NM, Sieweke M. The expression pattern of the mafB/kr gene in birds and mice reveals that the kreisler phenotype does not represent a null mutant. Mechanisms of Development 1997;65(1-2) 111-22.

[324] Kelly LM, Englmeier U, Lafon I, Sieweke MH, Graf T. MafB is an inducer of monocytic differentiation. EMBO Journal 2000;19(9) 1987-97.

[325] Aziz A, Soucie E, Sarrazin S, Sieweke MH. MafB/c-Maf deficiency enables self-renewal of differentiated functional macrophages. Science 2009;326(5954) 867-71.

[326] Nobrega MA, Pennacchio LA. Comparative genomic analysis as a tool for biological discovery. Journal of Physiology 2004;554(Pt 1) 31-9.

[327] Williams H, Brenner S, Venkatesh B. Characterization of the platelet-derived growth factor receptor alpha and c-kit genes in the pufferfish *Fugu rubripes*. DNA Sequence 2002;13(5) 263-70.

[328] Liu L, Korzh V, Balasubramaniyan NV, Ekker M, Ge R. Platelet-derived growth factor A (pdgf-a) expression during zebrafish embryonic development. Developmental Genes and Evolution 2002;212(6) 298-301.

[329] Taylor JS, Van de Peer Y, Braasch I, Meyer A. Comparative genomics provides evidence for an ancient genome duplication event in fish. Philosophical Transactions of the Royal Society of London B: Biological Sciences 2001;356(1414) 1661-79.

[330] Ravi V, Venkatesh B. Rapidly evolving fish genomes and teleost diversity. Current Opinion in Genetics and Development 2008;18(6) 544-50.

Chapter 2

Fish Cytokines and Immune Response

Sebastián Reyes-Cerpa, Kevin Maisey,
Felipe Reyes-López, Daniela Toro-Ascuy,
Ana María Sandino and Mónica Imarai

Additional information is available at the end of the chapter

1. Introduction

The immune system can be defined as a complex system that protects the organism against organisms or substances that might cause infection or disease. One of the most fascinating characteristics of the immune system is its capability to recognize and respond to pathogens with significant specificity. Innate and adaptive immune responses are able to recognize foreign structures and trigger different molecular and cellular mechanisms for antigen elimination. The immune response is critical to all individuals; therefore numerous changes have taken place during evolution to generate variability and specialization, although the immune system has conserved some important features over millions of years of evolution that are common for all species. The emergence of new taxonomic categories coincided with the diversification of the immune response. Most notably, the emergence of vertebrates coincided with the development of a novel type of immune response. Apparently, vertebrates inherited innate immunity from their invertebrate ancestors [1].

In higher vertebrates, the immune system consists of primary and secondary lymphoid organs with distinct compartments and morphology located in anatomically distinct sites. The thymus and bone marrow constitute the primary lymphoid organs, while the spleen, lymph nodes, and mucosal associated lymphoid tissue (MALT) comprise the secondary lymphoid organs [2].

Fish are a heterogeneous group divided into three classes: Agnatha (jawless fish such as the hagfish and lampreys), Chrondrichthyes (cartilaginous fish such as sharks, rays and skates) and Osteichthyes (bony fish) [3]. As in all vertebrates, fish have cellular and humoral immune responses and organs, the main function of which is immune defence. Most generative and secondary lymphoid organs in mammals are also found in fish, except for lymphatic nodules and bone marrow [3].

The head kidney or pronephros has hematopoietic functions [3, 4], and unlike in higher vertebrates, it is the immune organ involved in phagocytosis [5], antigen processing, production of IgM [6, 7] and immune memory through melanomacrophagic centres [8, 9]. The thymus, another lymphoid organ situated near the opercular cavity in teleosts, produces T lymphocytes involved in allograft rejection, stimulation of phagocytosis and antibody production by B cells [10, 11]. The spleen is a large, blood-filtering organ that undergoes increasing structural complexity in order to augment its efficiency in trapping and processing antigens [12-15]. Melanomacrophage centres are present for clearance of ingested material and can be surrounded by immunoglobulin-positive cells, especially after immunization [8]. Proliferation of granular cells has also been observed in association with ellipsoids and melanomacrophage centres after immunization [16].

1.1. Innate and adaptive immune response

The development of an immune system is essential for the survival of living organisms. In vertebrates, immunity can be divided into two components, the innate immune response and the adaptive immune response. The innate immune response is the initial line of defence against infection, which includes physical barriers and cellular response. The adaptive immune response is capable of specific antigen recognition and is responsible for the secondary immune response.

The innate immune system recognizes conserved molecular structures common to pathogenic microorganisms such as polysaccharides, lipopolysaccharides (LPS), peptidoglycans, bacterial DNA, and double-strand viral RNA, among others, through their interaction with specific receptors like toll receptors (TLRs). These mechanisms of recognition may lead directly to successful removal of pathogens, for instance by phagocytosis, or may trigger additional protective responses through induction of adaptive immune responses [17]. Cells of the innate immune system have a diverse array of functions. Some cells are phagocytic, allowing them to engulf and degrade pathogenic particles. Other cells produce and secrete cytokines and chemokines that can stimulate and help guide the migration of cells and further direct the immune response [18].

The adaptive system recognizes foreign structures by means of two cellular receptors, the B cell receptor (BCR) and the T cell receptor (TCR). Adaptive immunity is highly regulated by several mechanisms. It increases with antigen exposure and produces immunological memory, which is the basis of vaccine development and the preventive function of vaccines [19, 20]. The adaptive response generally starts days after infection and is capable of recognizing specific protein motifs of peptides, which leads to a response that increases in both speed and magnitude with each successive exposure [21]. The main effector cells of the adaptive immune response are the lymphocytes, specifically B cells and T cells. When B cells are activated, they are capable of differentiating into plasma cells that can secrete antibodies. Upon activation T cells differentiate into either helper T cells or cytotoxic T cells. Helper T cells are capable of activating other cells of the adaptive immune response such as B cells and macrophages, while cytotoxic T cells upon activiation are able to kill cells that have been infected [22].

1.2. Fish immune response

Immune responses in fish have not been as well characterized as they have in higher vertebrates. Consequently, there is not enough information about the components of the fish immune system and its function and regulation. Key immune mammalian homologous genes have been identified in several fish species, suggesting that the fish immune system shares many features with the mammalian system. For example, the identification of α and β T cell receptor genes (TCR) [23], key T cell markers such as CD3, CD4, CD8, CD28, CD40L, and a great number of cytokines and chemokines [24-26] suggest that T helper (Th)1, Th2 and Th17 and the regulatory subset Treg are present in fish. Some cell subsets have been better studied mainly because their activity can be easily differentiated and measured, as in the case of cytotoxic cells [27] and macrophages [28, 29]. Finally, B cells have been much more studied due to the availability of monoclonal antibodies that have been isolated and identified by a number of techniques [30, 31]. Phenotypic characterization of leukocytes has been hampered mainly by the lack of membrane cell markers [32, 33]. Researchers anticipate developing antibodies for cell lineage markers of fish immunocompetent cells that can be used to isolate and characterize immune cells to obtain insights into their regulation and role in immune response [34-36].

Antibodies in teleosts play a key role in the immune response. In general, IgM is the main immunoglobulin in teleosts that can elicit effective specific humoral responses against various antigens. For IgM, one gene alone can generate as many as six structural isoforms. Therefore, diversity is the result of structural organization rather than genetic variability [37]. Recently, several reports have provided evidence for the existence of IgD/IgZ/IgT in fish [38-41]. Interestingly, B cells from rainbow trout and salmon have high phagocytic capacity, suggesting a transition in B lymphocyte during evolution in which a key cell type of the innate immunity and phagocytosis evolved into a highly specialized component of the adaptive immune response in higher vertebrates [42, 43].

1.3. Fish cytokines

Cytokines are secreted proteins with growth, differentiation, and activation functions that regulate the nature of immune responses. Cytokines are involved in several steps of the immune response, from induction of the innate response to the generation of cytotoxic T cells and the production of antibodies. In higher vertebrates, the combination of cytokines that are secreted in response to an immune stimulation induces the expression of immune-related genes through multiple signalling pathways, which contributes to the initiation of the immune response. Cytokines can modulate immune responses through an autocrine or paracrine manner upon binding to their corresponding receptors [44].

Cytokines have overlapping and sometimes contradictory pleiotropic functions that make their classification difficult. Cytokines are produced by macrophages, lymphocytes, granulocytes, DCs, mast cells, and epithelial cells, and can be divided into interferons (IFNs), interleukins (ILs), tumor necrosis factors (TNFs), colony stimulating factors, and chemokines [45]. They are secreted by activated immune-related cells upon induction by various pathogens, such as parasitic, bacterial, or viral components [46]. Macrophages can secret IL-1, IL-6, IL-12, TNFα, and chemokines such as IL-8 and MCP-1, all of which are indispensable for macrophage, neutrophil, and lymphocyte recruitment to the infected tissues and their activation as pathogen eliminators [47]. Meanwhile, cytokines released by phagocytes in tissues can also induce acute phase proteins, including mannose-binding lectin (MBL) and C-reactive protein (CRP), and promote migration of DCs [48].

Fish appear to possess a repertoire of cytokines similar to those of mammals. To date several cytokine homologues and suppressors have been cloned in fish species [24, 25, 49]. Some cytokines described in fish are TNFα, IL-1β, IL-6 or IFN.

Current knowledge of fish cytokines is based on mammal models of the cytokines network and their complex interactions. In this review we included the pro-inflammatory cytokines associated with innate and adaptive immunity, regulatory cytokines and anti-inflammatory cytokines.

1.4. Pro-inflammatory fish cytokines

1.4.1. Tumour necrosis factor α (TNFα)

TNFα (tumour necrosis factor alpha) is a pro-inflammatory cytokine that plays an important role in diverse host responses, including cell proliferation, differentiation, necrosis, apoptosis, and the induction of other cytokines. TNFα can induce either NF-kB mediated survival or apoptosis, depending on the cellular context [50]. TNFα mediates powerful anti-microbial responses, including inducing apoptosis, killing infected cells, inhibiting intracellular pathogen replication, and up-regulating diverse host response genes. Many viruses have evolved strategies to neutralize TNFα by direct binding and inhibition of the ligand or its receptor or modulation of various downstream signalling events [51].

TNFα has been identified, cloned, and characterized in several bony fish, including Japanese flounder [52], rainbow trout [53, 54], gilthead seabream [55], carp [56] catfish [57], tilapia [58], turbot [59] and goldfish [60]. These studies have revealed the existence of some obvious differences from their mammalian counterpart, such as the presence of multiple isoforms of TNFα in some teleost species [54, 56] the high constitutive expression of this gene in different tissues of healthy fish and its relatively poor up-regulation by immune challenge *in vitro* and *in vivo* [53, 55, 57]. However, the most unexpected and interesting difference between fish and mammal TNFα concerns the weak *in vitro* effects of TNFα on phagocyte activation in goldfish [60], rainbow trout [57], turbot [59] and gilthead seabram [61]. This weak *in vitro* activity of fish TNFα sharply contrasts with the powerful actions exerted by the i.p. injection of recombinant TNFα in gilthead seabream, which includes the recruitment of phagocytes to the injection site, with a concomitant strong increase in their respiratory burst [61]. Apparently endothelial cells are the main target cells of fish TNFα, suggesting that TNFα is mainly involved in the recruitment of leukocytes to the inflammatory foci rather than in their activation [62]. Despite the above, differential expression has been observed in studies with rainbow trout leucocytes, which have shown increased response to different pro-inflammatory stimuli, as human recombinant TNFα [63], LPS [53, 64], zimosan and muramyl dipeptide as a peptidoglycan constituent of both gram-positive and gram-negative bacteria [64]. Moreover, it is known that Infectious Pancreatic Necrosis Virus (IPNV)-mediated up-regulation of TNFα regulates both the Bad/Bid-mediated apoptotic pathway and the RIP1 (receptor-interacting protein-1)/ROS-mediated secondary necrosis pathway [65].

1.4.2. Interleukin 1 family

In mammals, the 11 members of the Interleukin-1 family include IL-1α (IL-1F1), IL-1β (IL-1F2), IL-1 receptor antagonist (IL-1ra/IL-1F3), IL-18 (IL-1F4), IL-1F5-10 and IL-33 (IL-1F11). These molecules tend to be either pro-inflammatory or act as antagonists that inhibit the activities of particular family members [66]. Despite these semantic issues, to date only two clear homologues of these molecules have been discovered in fish, IL-1β and IL-18 [24].

1.4.2.1. Interleukin 1β

IL-1β is one of the earliest expressed pro-inflammatory cytokines and enables organisms to respond promptly to infection by inducing a cascade of reactions leading to inflammation. Many of the effector roles of IL-1β are mediated through the up- or down-regulation of expression of other cytokines and chemokines [67]. Mammalian IL-1β is produced by a wide variety of cells, but mainly by blood monocytes and tissue macrophages. IL-1β was the first interleukin to be characterized in fish and has since been identified in a number of fish species, such as rainbow trout [68], carp [69], sea bass [70], gilt head seabram [71], haddock [72], tilapia [73]. A second IL-1β gene (IL-1beta2) has been identified in trout [74]

In mammals pro-IL-1β remains cytosolic and requires cellular proteases to release the mature peptide. It is known that the peptide is cleaved by the IL-1β converting enzyme (ICE) [75]. However, the aspartic acid residue for which this enzyme has specificity is not present in all fish genes sequenced to date. Nevertheless, using a combination of multiple alignments and analysis of the N-terminal sequences of known mature peptides, it is possible to predict fish gene cutting sites. In trout, this gives a mature peptide of 166 and 165 aminoacids for IL-1β1 and IL-1β2 [76].

Like its mammalian counterpart, teleost IL-1β has been found to be regulated in response to various stimuli, such as LPS or poly I:C [68, 70-74, 77-81]. The biological activity of recombinant IL-1β (rIL-1β) has been studied in several fish species, indicating that fish IL-1β is involved in the regulation of immune relevant genes, lymphocyte activation, migration of leucocytes, phagocytosis and bactericidal activities [77, 81-84].

1.4.2.2. Interleukin 18

In mammals, IL-18 is mainly produced by activated macrophages. It is an important cytokine with multiple functions in innate and acquired immunity [85-87]. One of its primary biological properties is to induce interferon gamma (IFNγ) synthesis in Th1 and NK cells in synergy with IL-12 [88, 89]. It promotes T and NK cell maturation, activates neutrophils and enhances Fas ligand-mediated cytotoxicity [90-92]. Like IL-1β, it is synthesized as an inactive precursor of approximately 24 kDa and is stored intracellularly. Activation and secretion of IL-18 is mainly effected through specific cleavage of the precursor after D35 by caspase 1, also termed the IL-1β-converting enzyme (ICE), which is believed to be one of the key processes regulating IL-18 bioactivity [93, 94]. Some other enzymes, including caspase 3 and neutrophil proteinase 3, also cleave the IL-18 precursor to generate active or inactive mature molecules [95, 96].

IL-18 was discovered in fish by analysis of sequenced fish genomes (fugu) and EST databases (medaka) [97, 98]. An alternative splicing form of the IL-18 mRNA was discovered in trout that may have an important role in regulating IL-18 expression and processing in this species. This form shows a lower constitutive expression relative to the full length transcript, but unlike the full length transcript, it increases in response to LPS and polyI:C stimulation in the RTG-2 fibroblast cell line [98]. The expression level of the full length transcript can increase in response to LPS plus IL- 1b in head kidney leucocyte cultures, and by IFNγ in RTS-11 cells [99].

1.4.3. Other pro-inflammatory cytokines

1.4.3.1. Interleukin 6

A number of other interleukins are considered pro-inflammatory, some of which are released during the cytokine cascade that follows bacterial infection. Of these IL-6 is one of the best known, and is itself a member of the IL-6 family of cytokines that includes IL-11 and IL-31, as well as cytokines such as mammalian CNTF, LIF, OSM, CT-1 and CT-2 [24]. Whilst the homology of known fish molecules with many of these IL-6 family members is not conclusive [100], true homologues appear to be present in at least in the cases of IL-6 and IL-11 [24]. IL-6 is produced by a diverse group of cells including T lymphocytes, macrophages, fibroblasts, neurons, endothelial and glial cells. The pleiotropic effects of IL-6 are mediated by a 2-subunit receptor [101] and include the regulation of diverse immune and neuro-endocrine processes. IL-6 has been implicated in the control of immunoglobulin production, lymphocyte and monocyte differentiation, chemokine secretion and migration of leukocytes to inflammation sites [102-104].

IL-6 was first discovered in fugu by analysis of the genome sequence [105] and subsequently in other species as part of EST analysis of immune gene-enriched cDNA libraries [106-108]. However, little is known about the function and signalling pathways of IL-6 in fish. Interestingly, trout IL-6 expression in macrophages is reported to be induced by LPS, poly I:C and IL-1β in the macrophage cell line RTS-11, as well as in head kidney macrophages [109]. Moreover, IL-6 induces the expression of itself, so it can act in an autocrine and paracrine fashion to increase its expression, with the potential to both amplify and exacerbate the inflammatory response. However, IL-6 can significantly down-regulate the expression of trout TNFα1, TNFα2, and IL-1β, suggesting a potential role of trout IL-6 in limiting host damage during inflammation [109].

1.4.3.2. Interleukin 11

In mammals, IL-11 is produced by many cell types throughout the body. Basal and inducible IL-11 mRNA expression can be detected in fibroblasts, epithelial cells, chondrocytes, synoviocytes, keratinocytes, endothelial cells, osteoblasts and certain tumour cells and cell lines [110]. Viral [111] and bacterial infection [112] and cytokine stimulation (IL-1, TNFα and TGF-β1) induce IL-11 expression. IL-11 acts on multiple cell types, including hemotopoietic cells, hepatocytes, adipocytes, intestinal epithelial cells, tumour cells, macrophages, and both osteoblasts and osteoclasts. In the hematopoietic compartment IL-11 supports multilineage and committed progenitors, contributing to myeloid, erythroid, megakaryocyte and lymphoid lineages [113]. IL-11 is also an anti-inflammatory cytokine that inhibits the production of pro-inflammatory cytokines from lipopolysaccharde (LPS)-stimulated macrophages [114]. In combination with its trophic effects on the gastrointestinal epithelium, IL-11 plays an important role in the protection and restoration of gastrointestinal mucosa [115, 116].

The teleostean IL-11 orthologue has been found to consist of duplicate IL-11 genes, named IL-11a and IL-11b [117], with expression patterns indicating that both divergent forms of teleostean IL-11 play roles in antibacterial and antiviral defence mechanisms of fish [117-119]. In trout, IL-11 molecule is grouped with IL-11a and is constitutively expressed in intestine and gills and is highly up-regulated at other immune sites (spleen, head kidney, liver) following bacterial infection. *In vitro*, the macrophage-like RTS-11 cell line has shown enhanced IL-11 expression in response to LPS, bacteria, poly I:C and rIL-1β [118]. In carp, IL-11a is modulated by LPS, ConA and peptidoglycan in head kidney macrophages [117, 120] and cortisol has been found to inhibit IL-11 expression on its own and in combination with LPS [117]. In contrast to carp IL-11a, which shows low levels of constitutive expression in blood leucocytes, IL-11b in Japanese flounder shows higher expression at this site, and strong up-regulation was found in response to rhabdovirus infection in kidney cells [119]. This suggests that these paralogues have some complementarity of function related to their differential expression, although study of both forms in a single experiment is still required [24].

1.5. Chemokines

Chemokines are a superfamily of approximately 40 different small secreted cytokines that direct the migration of immune cells to infection sites. Their activity is coordinated by binding to G-protein-linked receptors with seven transmembrane domains. Four distinct subgroups make up the chemokine superfamily. These are designated as CXC (or a), CC (or b), C (or g) and CX_3C (or d), which are defined by the arrangement of the first two cysteine residues within their peptide structure. The CC subfamily can be further subdivided according to the total number of cysteine residues, as some members of this group contain four cysteines whilst the remainder possesses six (and are known as the C6-b group). Similarly, the

CXC subfamily contains two subgroups based on whether or not the first two cysteines are preceded by a Glu-Leu-Arg (ELR) motif associated with specificity to neutrophils [76, 121].

1.5.1. Interleukin 8

An important chemokine related to the pro-inflammatory process is CXCL-8, also called interleukin 8, this chemokine is a member of the CXC chemokine subfamily and attracts neutrophils, T lymphocytes and basophils *in vitro*, but not macrophages or monocytes [122]. Many cell-types, including macrophages, produce IL-8 in response to a variety of stimuli (LPS, cytokines and viruses). The neutrophil-attracting ability of IL-8 can be attributed to the presence of the ELR motif adjacent to the CXC motifs at its N-terminus, presumably by affecting its binding to specific receptors [123, 124]. In contrast, CXC chemokines lack an ELR motif and specifically attract lymphocytes but not neutrophils. The biological effects of IL-8 on neutrophils include increased cytosolic calcium levels, respiratory burst, a change in neutrophil shape and chemotaxis[125].

The fish IL-8 has been found in flounder [126], trout [125, 127], catfish [128], and lamprey [129]. *In vitro* stimulation of a trout macrophage cell line (RTS-11) [125] or *in vivo* intraperitoneal challenge [78] with either LPS or poly I:C did result in clear up-regulation of IL-8 expression. Moreover, induction of IL-8 expression in primary cultures of rainbow trout leukocytes stimulated for 24 hours with LPS and TNFα confirms that this fish chemokine is associated with inflammatory response, as has been suggested in mammals [127]. Interestingly, the ELR motif associated with the neutrophil-attracting ability is absent from the lamprey molecule and it is similar in flounder, where CXCL8 also lacks the ELR motif and appears to be regulated by a bacterial mechanism, since its transcript has only been detected in the major immune organs (spleen and head kidney) of an LPS stimulated flounder. The case of the trout is different, although there is also no ELR preceding the CXC motif, it has a very similar motif (DLR) in this position [130]. The human CXCL8 molecule, where the ELR motif has been mutated to DLR, retains neutrophil-attracting ability, albeit at lower potency [123]. Consequently, it is possible that the trout molecule has similar chemotactic activity to that of mammalian CXCL8 [130].

1.6. The interleukin 2 family

The IL-2 subfamily of cytokines signals via the common gamma chain (gC or CD132), a member of the type I cytokine receptor family expressed in most leucocytes.These cytokines in mammals include IL-2, IL-4, IL-7, IL-9, IL-15 and IL-21. IL-2, IL-4, IL-9 and IL-21 are all cytokines released from Th cells, which affect their responses [24], whilst IL-7 and IL-15 are particularly important for the maintenance of T cell memory [131]. To date molecules with homology to all of these have been found in fish, except IL-9 [24].

1.6.1. Interleukin 2

Interleukin-2 (IL-2 is an important immunomodulatory cytokine that primarily promotes proliferation, activation and differentiation of T cells [132]. IL-2, initially known as T-cell growth factor (TCGF), is synthesized and secreted mainly by Th1 cells that have been activated by stimulation by certain mitogens or by interaction of the T-cell receptor with the antigen/MHC complex on the surface of antigen-presenting cells [133-135]. Although CD4 T cells are the major source of IL-2 production in response to TCR stimulation, transient induction of IL-2 mRNA and production of the protein has been detected in murine dendritic cells activated by gram-negative bacteria [136]. IL-2 can also be produced by B cells in certain situations [137, 138]. The produced IL-2 promotes the expansion and survival of activated T cells and is also required for the activation of natural killer (NK) cells [139] and for immunoglobulin (Ig) synthesis by B cells [140].

The IL-2 gene has been detected only recently in fish by analysis of the fugu genome sequence, which also identified IL-21 as a neighbouring gene, as in mammals, providing the first direct evidence for the existence of a true IL-2 homologue in bony fish [141]. The gene has a 4 exon/3 intron organisation, as in mammals, and showed no constitutive expression in a range of tissues examined. However, injection of Fugu with poly I:C induced expression of IL-2 in the gut and gills [141]. Moreover, IL-2 could be induced in head kidney cell cultures stimulated with PHA, and in T-cell enriched cultures isolated from PBL when stimulated with B7-H3 or B7- H4 Ig fusions proteins in the presence of PHA [24, 142]. IL-2 has since been cloned in rainbow trout [143, 144]. The trout IL-2 was significantly up-regulated in head kidney leucocytes by the T cell mitogen PHA and in classical mixed leucocyte reactions and *in vivo* following infection with bacteria (*Y. ruckeri*) or the parasite *Tetracapsuloides bryosalmonae*. More importantly, the recombinant trout IL-2 produced in *Escherichia coli* was shown to induce expression of two transcription factors (STAT5 and Blimp-1) known to be involved in IL-2 signalling in mammals [143], as well as interferon-g (IFNγ) and IL-2 itself, and a CXC chemokine known to be induced by IFNγ, termed a IFNγ-inducible protein (γIP) [145].

1.6.2. Interleukin 4

Interleukin-4 IL-4 is a pleiotropic cytokine produced by T cells, mast cells, and basophils and is known to regulate an array of functions in B cells, T cells, macrophages, hematopoietic and non-hematopoietic cells [146, 147]. IL-4 serves as a key cytokine in driving Th2 differentiation and mediating humoral immunity, allergic responses and certain autoimmune diseases [148]. The IL-4 gene is conserved evolutionally in the animal kingdom and has been isolated from various animals including humans [149], mice [150, 151] and bovines [152], in which the IL-4 locus has been mapped in a region adjacent to those of IL-5 and IL-13 on the same chromosome [153, 154].

Teleost fish have two genes of the IL-4/13 family, IL-4/13A and IL-4/13B, which are situated on separate chromosomes in regions that duplicated during the fish-specific whole genome duplication (FS-WGD) around 350 million years ago [155, 156]. A few IL-4-like genes have been found in fish to date. The first was discovered by searching the *Tetraodon nigroviridis* genome [157]. In this work, IL-4 was constitutively expressed in head kidney, spleen, liver, brain, gill, muscle and heart. The ubiquitous expression of IL-4 is consistent with a postulated role in immune cytokines regulation. Stimulating the fish with a mixed stimulant containing ConA, PHA and PMA significantly up-regulated the expression of IL-4, which suggests that IL-4 is involved in the immune inflammatory responses triggered by mitogens [157], as in mammals, where it has been observed that this mitogen increases IL-4 expression [158]. However, the homology (amino acid identity) of this molecule was very low [12–15%], making it difficult to be sure it is an IL-4 homologue, although clearly related to Th2-type cytokines [24]. In fugu, T cell enriched PBL was found to express more IL-4/13A and IL-4/13B after stimulation with recombinant B7 molecules [142]. In zebrafish a recombinant IL4/13B was shown to increase the number of IgT-positive and CD209-positive cells in blood [159, 160], and in zebrafish spleen the expression of IL-4/13B and transcription factor related to Th2 immune response as GATA-3, and STAT6 was simultaneously enhanced after PHA stimulation [161]. The IL-4/13A gene was identified in trout and salmon [162], where the tissue distribution of salmonid IL-4/13A and GATA-3 expression were compared to the expression of IL-4, IL-13, and GATA-3 in mice. High levels of these transcripts were found in both salmonid and murine thymus, while constitutive IL-4/13A richness of skin and respiratory tissue was found in salmonids but not in mice. Experiments with isolated cells from gill and pronephros (head kidney) indicated that trout IL-4/13A is mainly expressed by surface IgM-negative cells, readily inducible by PHA but not by poly I:C, and regulated differently from the Th1 cytokine IFNγ gene. In mammals, IL-5 is also considered a Th-2 type cytokine and along with IL-3 and GM-CSF it signals through receptors with a common γ-chain (γC). None of these cytokines have been discovered in fish to date [24].

1.6.3. Interleukin 7

The cytokine IL-7 plays several important roles during lymphocyte development, survival, and homeostatic proliferation [163]. It is produced by many different stromal cell types, including epithelial cells of the thymus and the intestine [164-166]. There is only one report on IL-7 in fish, for the fugu molecule that was discovered using a gene synteny approach by searching with the mammalian IL-7 gene neighbours C8orf70 and PKIA. Fugu IL-7 shows constitutive expression in head kidney, spleen, liver, intestine, gill and muscle, with expression shown to increase in head kidney cultures stimulated with LPS, poly I:C or PHA [24, 167].

1.6.4. Interleukin 15

The central action of IL-15 cytokine is on T-cells, dendritic cells and NK cells. IL- 15 is an important regulator of the innate immune response to infection and autoimmune disease conditions. This gene shares activities with IL-2 and utilizes IL-2R β and γ units [45].

Two genes with homology to IL-15 have been discovered in fish. One shows similar gene organisation and synteny to mammalian and chicken IL-15, and has been termed IL-15. The second gene, which has a 4-exon structure and is in a different genome location, has been termed IL-15-like [168-170]. They show differential expression patterns in terms of the tissues where constitutive expression is apparent, and in terms of inducibility in PBL, with IL-15L being refractory to induction [168]. Two alternative splice variants of IL-15L (IL-15La and IL-15Lb) have also been described [170]. Trout IL-15, which has subsequently been cloned and sequenced, was strongly induced by rIFNγ in two trout cell lines (RTS-11 and RTG-2). rIL-15 could up-regulate IFNγ expression in splenic leucocytes, suggesting a positive feedback loop exists in fish between these two cytokines. Interestingly, unstimulated head kidney leucocytes were not responsive to rIL-15, at least in terms of the IFNγ expression level [171].

1.6.5. Interleukin 21

Interleukin 21 (IL-21) is a newly recognized member of IL-2 cytokine family that utilizes the common γ-chain receptor subunit for signal transduction [172-174]. In humans and other mammals, IL-21 is produced by both Th1 and Th2 cells [172, 175, 176]. IL-21 has pleiotropic effects on both innate and adaptive immune responses and can act on CD4+ and CD8+ T cells, B cells, NK cells, dendritic cells (DC), myeloid cells, and other tissue cells. IL-21 enhances the proliferation of anti–CD3-stimulated T cells and acts in concert with other γc cytokines to enhance the growth of CD4+ T cells [177]. IL-21–producing CD4+ T cells exhibit a stable phenotype of IL-21 production in the presence of IL-6 but retain the potential to produce IL-4 under Th2-polarizing conditions and IL-17A under Th17-polarizing conditions [178]. IL-21 stimulates CD8+ T cell proliferation and synergizes with IL-15 in promoting CD8+ T cell expansion in vitro and their antitumor effects in vivo [177, 179]. B cells that encounter IL-21 in the context of Ag-specific (BCR) stimulation and T cell co/stimulation undergo class-switch recombination and differentiate into Ab-producing plasma cells. In contrast, B cells encountering IL-21 during nonspecific TLR stimulation or without proper T cell help undergo apoptosis [180].

Since its discovery in fugu as a gene neighbour of IL-2 [141], IL-21 has been reported in tetraodon [181, 182] and rainbow trout [182]. Fugu IL-21 shows low constitutive expression. However, stimulation of isolated kidney leucocytes with PHA induced IL-21 expression. IL-21 was also up-regulated at mucosal sites as gill and gut when fish were injected with LPS or poly I:C [141]. Similarly, in tetraodon IL/21 expression is low but detectable in the gut, gonad and gills of healthy fish, and is induced in the kidney, spleen and skin following LPS injection [181]. In trout IL-21 expression is highest in gills and intestine, and is induced *in vivo* by bacterial (*Y. ruckeri*) and viral (VHSV) infection [182]. Relative to IL-2, induction of IL-21 expression in head kidney cells appears more rapidly but has shorter duration after stimulation. The trout rIL-21 has also been produced and shown to increase the expression of IL-10, IL-22 and IFNγ, and to a lesser extent IL-21, and to maintain the expression levels of key lymphocyte markers in primary cultures [182]. Thus, IL-21 may act as a survival factor for fish T and B cells [24].

1.7. The interleukin 10 family

Interleukin-IL-10 is an anti-inflammatory cytokine and a member of the class II cytokine family that also includes IL-19, IL-20, IL-22, IL-24, IL-26 and the interferons [183]. Although the predicted helical structure of these homodimeric molecules is conserved, certain receptor-binding residues are variable and define the interaction with specific heterodimers of different type-2 cytokine receptors. This leads to diverse biological effects through the activation of signal transducer and activator of transcription (STAT) factors [184].

1.7.1. Interleukin 10

Interleukin-10 (IL-10) was discovered initially as an inhibitory factor for the production of Th1 cytokines. Subsequently, pleiotropic inhibitory and stimulatory effects on various types of blood cells were described for IL-10, including its role as a survival and differentiation factor for B cells. IL-10, which is produced by activated monocytes, T cells and other cell types like keratinocytes, appears to be a crucial factor for at least some forms of peripheral tolerance and a major suppressor of the immune response and inflammation. The inhibitory function of IL-10 is mediated by the induction of regulatory T cells [185].

IL-10 was discovered in fish by searching the fugu genome. The translation showed 42–45% similarity to mammalian molecules with very low constitutive expression in tissues [186]. IL-10 has since been cloned in several other fish species including carp [187] zebrafish [188], rainbow trout [189], sea bass [190, 191] and cod [79]. Such studies have shown that IL-10 expression can be increased by LPS stimulation, by bacterial infection, by bath administration of immunostimulants [192] and by IPNV infection which may be associated with mechanisms of immune evasion [78].

1.7.2. Interleukin 20 (IL-20Like)

In mammals, IL-20 was discovered as a new member of the IL-10 family of cytokines. IL-20 shares the highest amino-acid sequence identity with IL-10, IL-24 and IL-19. It is secreted by immune cells and activated epithelial cells like keratinocytes. A high expression of the corresponding IL-20 receptor chains has been detected on epithelial cells. In terms of function, IL-20 might therefore mediate crosstalk between epithelial cells and tissue-infiltrating immune cells under inflammatory conditions [193].

In fish, the gene of IL-20 has been described in putterfish [183], zebrafish [194] and trout [195]. In the latter work, the IL-20 gene, called IL-20-like (IL-20L) has been described as having a high level of expression in immune related tissues and in the brain, suggesting an important role of the fish IL-20L molecule in both the immune and nervous systems. Although the exact cell types expressing IL-20L have yet to be defined, macrophages express IL-20L. Moreover, IL-20L expression in the macrophage cell line RTS-11 is modulated by pro-inflammatory cytokines, signalling pathway activators, microbial mimics and the immuno-suppressor dexamethasone. These data suggest that trout IL-20L plays an important role in the cytokine network. The increased expression of IL-20L was only detected at late stages (4–24 h) of LPS stimulation in RTS-11 cells and in spleen 24–72 h after infection with *Yersinia ruckeri*, which suggests that the increased expression of IL-20L by LPS and infection is via the rapid increase of pro-inflammatory cytokines (e.g., IL-1β) and other factors known to occur [195].

1.7.3. Interleukin 22/26

In mammals, interleukin-22 is secreted by Th17 cells [196], as well as by a subset of NK cells, designated as NK22 [197]; and even by some Th1 cells [198]. Studies have suggested there is a distinct Th22 cell lineage [199, 200]. Many of the same cytokines that induce differentiation and proliferation of IL-17-producing cells also lead to the secretion of IL-22 by Th17 cells, NK22 cells, and putative Th22 cells, including IL-6, IL-23, IL- 1β, TGF-β, and TNFα [201]. IL-17 and IL-22 are therefore frequently produced together in response to infections [202]. Interleukin-22 interacts with a heterodimeric receptor, IL-10R2/IL-22R1 [203], which is expressed on a variety of non-lymphoid cells, especially epithelial cells. Ligation of this receptor leads to both protective and detrimental effects. In synergy with IL-17, IL-22 induces pro-inflammatory cytokines in human bronchial epithelial cells against *Klebsiella pneumoniae* infection [204] and in colonic myofibroblasts [205]. Independently or in synergy with IL-17, IL-22 acts in defence against intestinal infection of mice with *Citrobacter rodentium* [206]. Moreover, IL-22 has been implicated in intestinal homeostasis keeping commensal bacteria contained in anatomical niches, which is key to our symbiotic relationship and normal intestinal physiology. However, the mechanisms that restrict colonization to specific niches are unclear. David Artis and colleagues have described a crucial role for IL-22-producing innate lymphoid cells (ILCs) in preventing lymphoid-resident commensal bacteria from escaping their niche and causing inflammation [207].

IL-26 can be produced by primary T cells, NK cells and T cell clones following stimulation with specific antigen or mitogenic lectins. IL-26 was initially shown by several groups to be co-expressed with IL-22 [208]. IL-26 is co-expressed with IFNγ and IL-22 by human Th1 clones, but not by Th2 clones. It was subsequently found that IL-26 is co-expressed with IL-17 and IL-22 by Th17 cells, an important subset of CD4+ T-helper cells that are distinct from Th1 and Th2 cells [209-211]. More recently, a novel subset of CD56+ NKp44+ NK cells was identified that co-expresses IL-22 and IL-26, especially following treatment with IL-23 [212]. Furthermore, a different subset of immature NK cells was described that do not express CD56 or NKp44 but do express CD117 and CD161 and constitutively express IL-22 and IL-26 [213].

The mechanisms that regulate transcription of the human IL-26 gene are so far largely undefined. It is possible and perhaps likely that expression of the IL-26 gene is induced in an IL-23-dependent manner because IL-23 is known to induce differentiation of Th17 cells, and IL-23 amplifies expression of IL-17 and IL-22 by Th17 cells [214].

In fish, the IFNγ locus was discovered using a gene synteny approach, and was first reported for fugu [215]. It contained a homologue of IL-22/26, that later studies of the zebrafish genome revealed to be two genes, one with clear homology to IL-22 and one with somewhat less clear homology to IL-26 [216]. The IL-22 gene was expressed constitutively in intestine and gills in all the treated and non-treated tissues. The gene was also expressed in kidney and spleen in LPS and PolyI:C-treated tissues, respectively, while IL-26 was expressed only in intestine treated with PolyI:C without expression [216]. IL-22 expression has been correlated with disease resistance in haddock vaccinated against *V. anguillarum*, with a strong constitutive expression in gills in vaccinated fish but not in control fish 24 hours post bath challenge, resulting in complete protection in fish vaccinated [217]. Moreover, IL-22, a cytokine released by Th-17 cells in mammals, is also interesting, and such responses are thought to be crucial for protection against extracellular microbes and at mucosal sites [218]. This coupled with the recent discovery of novel gill-associated immune tissue in fish [219] may provide a clue to a potential mechanism of resistance elicited by the *V. anguillarum* vaccination [24].

1.8. The interleukin 17 family

Interleukin-17 and a related family of genes are known to have pro-inflammatory actions and are associated with diseases [220]. After the discovery of the human IL- 17 gene [221], five cellular paralogs of IL-17 were identified, namely IL-17B, C, D, E and F [222-227]. These paralogs, identified by ESTs, genomics and proteomic databases, share identities of 20–50% with IL-17A gene. Human IL-17 A and F are present in tandem in opposite transcriptional orientation on the same chromosome 6p12, while IL-17B (Chr 5q24), IL-17C (Chr 16q24), IL-17D (Chr 13q11) and IL-17E (Chr 14q11) are dispersed. The structural similarities lead to the classification of IL-17 A, B, C, D, E, and F genes to a larger IL-17 sub-family [45]. Several IL-17 family members have been discovered in teleost fish, but homology to mammalian genes has not always been easy to assign. Two IL-17A or F homologue genes (IL-17A/F) have been found on the same chromosome. However, it has been difficult to determine which gene codes IL-17A and F. This gene in zebrafish was named IL-17A/F1 and 2. Furthermore, another IL-17A or F homologue gene (IL-17A/F3) has been found in zebrafish localized on a chromosome different from that of IL-17A/F1 and 2 [228]. In addition to those in zebrafish, IL-17A or F homologue genes have been found in rainbow trout [229], Atlantic salmon [230], pufferfish (IL-17A/F1, 2 and 3) [231], and medaka (IL-17A/F1, 2 and 3) [232].

The tissue distribution of the fugu IL-17 gene family also differs. In particular, IL-17 family genes are highly expressed in the head kidney and gills. Moreover, expression of IL-17 family genes is significantly up-regulated in the lipopolysaccharide-stimulated head kidney, suggesting that Fugu IL-17 family members are involved in inflammatory responses [231]. In Atlantic salmon IL-17D expression is widely distributed in tissues, with the highest levels of expression in testis, ovary and skin. Infection with *A. salmonicida* by injection increases IL-17D expression levels in the head kidney (but not the spleen) in a time-dependent manner. Skin and kidney showed an increased IL-17D expression level in fish given a cohabitation challenge with *A. salmonicida* [230]. The two trout IL-17C genes show some degree of differential expression within tissues, with IL-17C1 being more dominant in the gills and skin, whilst IL- 17C2 is more dominant in the spleen, head kidney and brain. Expression of both genes increases significantly with bacterial infection, although the increased expression of IL-17C2 is greater in terms of fold change. Similarly, both genes could be up-regulated in the trout RTS-11 cell line by LPS, poly I:C, calcium ionophore and rIL-1β, with IL-17C2 showing higher fold increases in all cases [229].

1.9. Interleukin 12

IL-12 is a heterodimeric cytokine composed of p35 and p40 subunits. It can mediate a number of different activities, including stimulation of IFNγ secretion from resting lymphocytes, NK cell stimulation and cytolytic T cell maturation. Perhaps most crucially, IL-12 also affects the progression of uncommitted T cells to either the Th1 lineage, which in general is characterized by secretion of lymphokines associated with cell-mediated rather than humoral immunity [233].

The p35 and p40 subunits were discovered in fish by analysis of the fugu genome [234]. The p35 locus is quite well conserved, with Schip1 being the immediate neighbour in all cases. This association has allowed p35 to be cloned by gene walking from Schip1 from fish species for which no genome sequence is available [24, 235]. The p40 subunit in fugu is constitutively expressed in all the tissues examined, except muscle, and no increases in expression were seen 3 h after injection with poly I:C or LPS. This constitutive and broad expression distribution of the p40 subunit suggests that it may be expressed in most cell types. The expression of the p35 subunit is more limited in its tissue expression and is induced after injection with poly I:C in the head kidney and the spleen, but not after injection with LPS. These results show that there are differences from the mammalian data in fugu IL-12 subunit expression. Further investigation will be required to show whether this is unique to fugu, if IL-12 is involved more in antiviral defence in fish and if the two subunits are regulated differently from their regulation in the mammalian system [234].

1.10. Transforming growth factor β (TGF-β)

TGF-β is a pleiotropic cytokine that regulates cell development, proliferation, differentiation, migration, and survival in various leukocyte lineages including lymphocytes, dendritic cells, NK cells, macrophages and granulocytes [236, 237]. In the mammalian immune system, TGF-β1 is a well-known suppressive cytokine and its dominant role is to maintain immune tolerance and suppress autoimmunity [238, 239]. The potent immunosuppressive effects of TGF-β1 are mediated predominantly through its multiple effects on T cells: TGF-β1 suppress Th1 and Th2 cell proliferation, while it promotes T regulatory cell generation by inducing Foxp3 expression. On the other hand, TGF-β also promotes immune responses by inducing the generation of Th17 cells [236, 240, 241]. Therefore, the regulatory roles of TGF-β as a positive or negative control device in immunity are widely acknowledged in mammals [238, 240, 241].

In teleost, despite the lack of extensive investigation on the functional role of TGF-β, some recent studies have revealed that TGF-β1 also exterts powerful immune depressing effects on activated leukocytes, as it does in mammals. For instance, TGF-β1 significantly blocks TNFα-induced activation of macrophage in goldfish and common carp, but induces the proliferation of the goldfish fibroblast cell line CCL71 [242, 243]. In grass carp, TGF-β1 downregulates LPS/PHA-stimulated the proliferation of peripheral blood lymphocyte by contrast with the stimulatory effect of TGF-b1 alone in the same cells [244]. In red sea bream, similar phenomenon was observed during leukocyte migration under TGF-β1 treatment, with or without LPS challenges [245]. These findings not only define TGF-β1 as an immune regulator in teleost, but also indicate that TGF-β1 may have retained similar functions in immunity during the evolution of vertebrates [246].

1.11. Interferons

Interferons genes are involved in mediating cellular resistance against viral pathogens and modulating innate and adaptive immune systems. Broadly, IFNs are classified into two main groups called type I and type II [45]. Type I IFN includes the classical IFNα/β, which is induced by viruses in most cells, whereas type II IFN is only composed of a single gene called IFNγ and is produced by NK cells (NK cells) and T lymphocytes in response to interleukin-12 (IL-12), IL-18, mitogens or antigens [247]. Structurally both IFN types belong to the class II a-helical cytokine family, but have different 3-dimensional structures and bind to different receptors [248].

Two IFNs (IFNα1 and IFNα2) have been cloned from Atlantic salmon and characterized with respect to sequence, gene structure, promoter, antiviral activity and induction of ISGs [249-252]. Salmon IFNα1 induces both Mx and ISG15 proteins in TO cells and thus has properties similar to mammalian IFNα/β and IFNλ [251, 252]. Furthermore, salmon IFNα1 induces potent antiviral activity against the IPNV *in vitro* [251], but this protection has not been observed *in vivo*, despite a high level of expression of IFNα detected in spleen and head kidney of Atlantic salmon challenged intraperitoneally with IPNV [78].

At least three type I IFNs have been discovered in rainbow trout. The IFN1 (rtIFN1) and rtIFN2 show high sequence similar to Atlantic salmon IFNα1 and IFNα2, which contains two cysteines. On the other hand, rtIFN3 contains four cysteines, which further confirms the relationship between mammalian IFNα and fish IFNs. Recombinant rtIFN1 and rtIFN2 have both been shown to up-regulate expression of Mx and inhibit VHSV replication in RTG-2 cells. In contrast, recombinant rtIFN3 has been found to be a poor inducer of Mx and antiviral activity. Interestingly, the three rtIFNs show differential expression in cells and tissues [253]. This suggests that the three trout IFNs have different functions in the immune system of fish, which is an interesting subject for further research [254].

IFNγ has been identified in several fish species, including rainbow trout and Atlantic salmon [215, 216, 248, 255-257]. In contrast to the type I IFNs, fish and mammalian IFNγ are similar in exon/intron structure and display gene synteny. However, some fish species also possess a second IFNγ subtype named IFN gamma rel, which is quite different from the classical IFNγ [258]. Rainbow trout and carp IFNγ have several functional properties in common with mammalian IFNγ, including the ability to enhance respiratory burst activity, nitric oxide production, and phagocytosis of bacteria in macrophages [257-259]. Far less is known about the antiviral properties of fish IFNγ. However, it has been reported that it induces antiviral activity against both IPNV and the Salmon Alpha Virus (SAV) in salmon cell lines [260]

1.12. Tools for fish cytokine analysis

The major strategy of functional genomics is to identify the types of responses to specific pathogens based on cytokines expression as a predictor of profile immune response, which began by using suppressive subtractive hybridization as major tools at the beginning of the immunogenomics and upgrade to platforms of wide screening that allow identify thousands of EST's that are differentially regulated in their expression and that allow identifying potential candidates as biomarkers in the progression of the immune response at differential environmental conditions, not only against pathogens, but also in captivity stress conditions that affect the fisheries production.

1.13. Suppressive subtractive hybridization (SSH)

One of the most important biological processes in higher eukaryotes against external stimuli is the response mediated by differential gene expression. To understand the molecular regulation of these processes, the relevant subsets of differentially expressed genes of interest must be identified, cloned, and studied in detail using specific molecular techniques. In this matter, subtractive cDNA hybridization has been a powerful approach to identify and isolate cDNAs of differentially expressed genes [261-263]. Numerous cDNA subtraction methods have been reported. In general, they involve hybridization of cDNA from one population (tester) to excess of mRNA (cDNA) from another population (driver) and then separation of the unhybridized fraction (target) from the common hybridized sequences. One of these tools is a PCR-based technique called representational difference analysis, which does not require physical separation of single-strand (ss) and double-strand (ds) cDNAs. Representational difference analysis has been applied to enrich genomic fragments that differ in size or representation [264] and to clone differentially expressed cDNAs [265]. However, representational difference analysis has the problem of the wide differences in abundance of individual mRNA species so that multiple rounds of subtraction are needed [265]. Other strategies, such as mRNA differential display [266] and RNA fingerprinting by arbitrary primed PCR [267], are potentially faster methods for identifying differentially expressed genes, but both of these methods have high levels of false positives [268] that bias high-copy-number mRNA [269], which can inappropriate in experiments where only a few genes are expected to vary [268]. One of the techniques most often used to establish differential expression pattern between two conditions is suppression subtractive hybridization (SSH), which selectively amplifies target cDNA fragments (differentially expressed) and simultaneously suppresses non-target DNA amplification. The method is based on the fact that long inverted terminal repeats attached to DNA fragments can selectively suppress amplification of undesirable sequences in PCR procedures [270]. This method overcomes the problem of differences in mRNA abundance by incorporating a hybridization step that normalizes (equalizes) sequence abundance during the course of subtraction by standard hybridization kinetics [271]. Two types of SSH are possible: forward SSH, when the reaction involves the hybridization of cDNA from one population indicated as the evaluated phenotype (tester) to excess of mRNA (cDNA) from a control phenotype (driver); and reverse

SSH, when the conditions described above are inverted. Together, the two processes are called reciprocal SSH.

Different works have been done with SSH to evaluate fish immune response at the gene expression level against challenges with bacteria-derived pathogen-associated molecular patterns (PAMP) like LPS [272, 273] and whole bacteria like *Aeromonas salmonicida* [274, 275], *Listonella anguillarum* [276], *Edwardsiella tarda* [277], and *Vibrio parahaemolyticus* [278] (Table 1).

A critical step in any immune response is the recognition of invading organisms. This is mediated by many proteins, including pattern recognition receptors (PRR), which recognize and bind to molecules present on the surface of microorganisms. LPS is an essential cell wall component of gram-negative bacteria and is recognized by PRR, triggering a series of responses that lead to the activation of the host defence system. These PRRs include a number of toll-like receptors, as well as other cell-surface and cytosolic receptors that, upon stimulation, modulate immunity [279, 280]. In LPS-stimulated yellow grouper spleen a subtracted cDNA library was constructed using SSH. The contigs and singlets obtained were analyzed and a low number of immune-related genes were found [272]. In Asian seabass the up-regulation of differentially expressed genes like pro-inflammatory cytokines and related receptors, such as TNF receptor super family member 14 (TNFRSF14), IL-31 receptor A (IL31RA), chemokine receptor-like 1 (CMKLR1), chemokine (C-X-C motif) receptor 3 (CXCR3), chemokine (C-C motif) receptor 7 (CCR7) and chemokine (C-C motif) ligand 25 (CCL25), was identified at 24h post-challenge by bacterial LPS in spleen Complement components were also identified [273]. These genes are a solid basis for a better understanding of immunity in Asian seabass and for developing effective strategies for immune protection against infections in that species.

Infection of Atlantic salmon by *A. salmonicida* was observed to stimulate an acute-phase response (APR) as part of the innate immune defence system to infection, whose gene expression pattern was remarkably observed in liver at 7 days post-infection [275] indicating that the liver appears to be the main source of APPs in fish, as in mammals. Not surprisingly, the liver gene expression pattern observed in other fish species against *L. anguillarum* [276], *E. tarda* [277], and *V. parahaemolyticus* [278]. The APR is characterized by alterations in the levels of plasma proteins referred to as acute-phase proteins (APPs), as well as the secretion of some other innate defence molecules important for innate immunity, such as complementary systems [281-283]. In Atlantic cod stimulated with atypical *A. salmonicida* (formalin-killed) interleukin-1β (IL-1- β), interleukin-8 (IL-8), CC chemokine type 3, interferon regulatory factor 1 (IRF1), ferritin heavy subunit, cathelicidin, and hepcidin were identified in the forward spleen SSH library. Atlantic cod IRF1 was constitutively expressed at low levels, and expression was significantly elevated in spleen and head kidney at 24 h following A. salmonicida stimulation, with the highest levels of induction observed in the spleen [274]. The target IRFI genes, as well as their importance in innate immune responses in fish, have not yet been determined, although the expression of IRF1 in teleost macrophages can be induced by both IFNγ and IL-1β, with IFNγ being a much more potent inducer of IRF1 than IL-1β [99].

Microorgan-ism	Fish	Pathogen	Tissue/Cell type	Infection route	Reference
Bacteria	Atlantic salmon	*Aeromonas salmonicida* A449	Liver, head kidney and spleen	Intraperito-neal	Tsoi et al., 2004
	Yellow grouper	*LPS (E. coli)*	Spleen	Intraperito-neal	Wang et al., 2007
	Atlantic cod	*Aeromonas salmonicida* (formalin-killed)	Head kidney, Spleen	Intraperito-neal	Feng et al., 2009
	Asian seabass	*LPS (E. coli)*	Spleen	Intraperito-neal	Xia and Yue 2010
	Ayu	*Listonella anguillarum*	Liver	Intraperito-neal	Li et al., 2011
	Japanese flounder	*Edwardsiella tarda* (NE8003)	PBL	Intraperito-neal	Matsuyama et al., 2011
	Marine medaka	*Vibrio parahaemolyticus*	Liver	Intraperito-neal	Bo et al., 2012
Virus	Mandarian fish	ISKNV	Spleen	Intramuscular	He et al., 2006
	Sea bream	Nodavirus (strain 475-9/99)	Brain	Intramuscular	Dios et al., 2007
	Atlantic cod	poly I:C	Spleen	Intraperito-neal	Rise et al., 2008
	Sea bass	SBNNV	Head kidney	Intramuscular	Poisa-Beiro et al., 2009
	Atlantic cod	ACNNV	Brain	Intraperito-neal	Rise et al., 2010
	Orange-spotted grouper	SGIV	Spleen	Intraperito-neal	Xu et al., 2010

Table 1. Transcriptomics studies on fish after treatments with bacteria or virus *in vivo* analyzed with SSH. LPS: Lipopolysaccharide; ISKNV: Infectious spleen and kidney necrosis virus; poly I:C: polyriboinosinic polyribocytidylic acid; SBNNV: Sea bass nervous necrosis virus; ACNNV: Atlantic cod nervous necrosis virus; SGIV: Singapure grouper iridovirus; PBL: Peripheral blood leukocytes.

SSH has been in several investigations to evaluate fish gene expression patterns against challenges with PAMPs, such as polyriboinosinic polyribocytidylic acid (poly I:C) [284], Infectious Spleen and Kidney Necrosis Virus (ISKNV) [285], Nodavirus [286-288], and Singapore grouper iridovirus (SGIV) [289].

Spleen gene expression in mandarin fish at 4 days post-infection with ISKNV of Mx protein, interferon-inducible protein Gig-2, and viperin (interferon-inducible and antiviral protein) was up-regulated, suggesting IFN pathway stimulation after ISKNV infection [285]. Also, two inflammatory cytokine genes, CC chemokine and IL-8, were found in the forward SSH library, whereas the CD59/Neurotoxin/Ly-6-like protein gene was down-regulated. In mammals, CD59 is a complement regulatory protein, which can inhibit complement activation and membrane attack complex (MAC) formation on autologous cells [290], suggesting that down-regulation in the ISKNV-infected host cells may make these cells more sensitive to complement attack, mounting an anti-virus mechanism of the host [285].

In orange-spotted grouper after 5 days of infection with Singapore grouper iridovirus (SGIV) novel genes were annotated as immune-related, such as C-type lectin, epinecidin, and complement components C3 and C9. Interestingly, the most abundant clone was C-type lectin, and the microarray results at 1, 5 and 9 days post-infection indicated that its expression was up-regulated in liver, spleen and kidney [289]. Lectins are multivalent carbohydrate-binding proteins that function as important pattern-recognition receptors (PRR) and have been isolated and characterized in fish [291-294]. C-type lectin represents a very large family, most members of which are able to bind PAMP and microorganisms themselves through sugar moieties and play important roles in non-self recognition and clearance of invading microorganisms. The up-regulation of C-type lectin in different organs with immunological functions confirmed as SSH as microarrays suggest an important role in the development of control strategies against SGIV infection.

The SSH method was used to generate a subtracted cDNA library enriched in gene transcripts differentially expressed after 1 day post-infection in the brains of sea bream infected with nodavirus. Most of the expressed sequence tags (ESTs) differentially expressed in infected tissues fell into gene categories related to cell structure, transcription, cell signalling or different metabolic routes. Other interesting putative homologies corresponded to genes expressed in stress responses, such as heat shock proteins (Hsp-70) and to immune-related genes such as the Fms-interacting protein, TNFα-induced protein, interferon-induced with helicase C domain protein (mda-5), which in mammals play an important role in the synthesis and secretion of IFN type I [295]. Another nodavirus, sea bass nervous necrosis virus (SBNNV) was studied to identify genes potentially involved in antiviral immune defence in sea bass head kidney using the SSH technique [287]. The results of up-regulated EST from sea bass head kidney SSH showed significant similarities with immune genes, such as β-2 microglobulin, heat shock protein 90 (Hsp-90), IgM, MHC class I and class II, and β-galactoside-binding lectin, identified as a member of the galectin family and closely related to the galectin-1 group (Sbgalectin-1). When the recombinant protein (rSbgalectin-1) was produced and functional assays were conducted, a decrease in IL-1β, TNFα, and Mx expression was observed in the brain of sea bass simultaneously injected with nodavirus and rSbgalectin-1 compared to those infected with the nodavirus alone, suggesting a potential anti-inflammatory protective role of Sbgalectin-1 during viral infection. A similar nodavirus, the Atlantic cod nervous necrosis virus (ACNNV), was studied to evaluate the transcript expression responses in the Atlantic cod (*Gadus morhua*) brain to asymptomatic high nodavirus carrier state [288]. In the forward brain SSH library was identified with significant similarity to genes with immune-relevant functional annotations the interferon stimulated gene 15 (ISG15), IL-8 variant 5, DEXH (Asp-Glu-X-His) box polypeptide 58 (DHX58; LGP2), radical Sadenosyl methionine domain-containing 2 (RSAD2; viperin), β-2-microglobulin (B2M), chemokine CXC-like protein, signal transducer and activator of transcription 1 (STAT1), and CC chemokine type 2. Interestingly, ISG15, DHX58, RSAD2, and sacsin (SACS) transcripts are all strongly upregulated by both high nodavirus carriage and intraperitoneal poly I:C stimulation, suggesting a similar host response is significantly induced in the brain by both nodavirus and poly I:C. This expression pattern is corroborated when the response of Atlantic cod spleen is evaluated against poly I:C stimulation, showing the up-regulation of ISG15, RSAD2, LGP2 and other transcripts such as MHC class I, and IRF1, 7, and 10, indicating that Atlantic cod recognize dsRNA and mount a interferon pathway response [284].

1.14. Microarrays

Microarray analysis measures the expression of large numbers of genes in parallel. This methodology, which combines hypotheses-driven and hypotheses-free research strategies, is used to infer molecular mechanisms, classify samples, and diagnose and search for novel biomarkers. With the use of standard platforms, laboratory protocols and procedures for processing of primary data, the results of microarrays analyses are well suited for database management and meta-analysis across multiple experiments, whilst data mining is based on powerful statistical procedures with support from functional and structural annotations of genes [296].

The Atlantic salmon is of particular importance to the global aquaculture industry. Salmonid cDNA microarrays were constructed shortly after large-scale sequencing of salmon and trout cDNA libraries by several research institutes. One of the projects related to salmon sequencing is GRASP (Genomics Research on Atlantic Salmon Project), an initiative funded by Genome Canada that is intended to improve understanding of physiological and evolutionary processes influencing the survival and phenotype of salmonids and other fish in natural and aquaculture environments. The first salmonid GRASP microarray platform (GRASP-1), containing 7356 salmonid elements representing 3557 different cDNAs (3.7K), was obtained from 80,388 ESTs, principally from cDNA libraries [298] of different salmon species such as Atlantic salmon, rainbow trout, Chinook salmon, sockeye salmon, and lake whitefish cDNA libraries. The second version of the GRASP microarray platform (GRASP-2) was developed and contained cDNAs representing 16,006 genes (16K). The genes identified in the array have been stringently selected from Atlantic salmon and rainbow trout EST databases representing a wide variety of different classes of genes [297]. Finally, a new expanded salmonid cDNA microarray (GRASP-3) of 32,000 features (32K) was created where 69% of the total EST collection used was from Atlantic salmon [298]. The Aleksei Krasnov's group designed the rainbow trout microarray (SFA1.0) by identifying a relatively small number of genes (1300 genes; 1.3K) using clones from normalized and subtracted cDNA libraries, as well as genes selected by the functional categories of Gene Ontology for inclusion in a microarray aimed at characterizing transcriptome responses to environmental stressors [299] to maximize the presence of transcripts related directly to immune response in rainbow trout, because of which this platform is also called Immunochip (SFA1.0 immunochip). The updated SFA platform (1.8K; SFA2.0 immunochip) was specially designed for studies of responses to pathogens and stressors and has substantially improved coverage of immune genes [300]. Another cDNA platform in commercial fish species has been designed in Japanese flounder [301] and European flounder [302], turbot [303], and sole [304].However despite impressive achievements, cDNA platforms suffer from limitations and disadvantages. At present most research groups working with salmonids and other aquaculture species do not have full access to clones required for fabrication of cDNA microarrays. Maintenance and PCR amplification of large clone sets is expensive and time consuming, while the risk of errors is high [296]. Probably the most important drawback of cDNA microarrays is their limited ability to discriminate paralogs since long probes cross-hybridize with highly similar transcripts from members of multi-gene families [305]. In salmonids this problem is aggravated by the large number of expressed gene duplicates. These complications can be resolved with oligonucleotide microarrays (ONM) that also provide greater accuracy and reproducibility of analyses. Until recently, the use of ONM platforms was hampered by the cost, but they are now rapidly replacing cDNA platforms. Construction of ONM platforms begins with establishment of mRNA sequence sets for comprehensive coverage of transcriptomes with low redundancy. The next stage is identifying genes by searching protein databases and annotating them according to functions, pathways and structural features. For successful development and use of ONM, it is necessary to define the gene composition and optimum number of spot replicates and to choose criteria for quality assessment [296].

Because of the commercial importance of salmonid species, there is special interest gene expression pattern against different pathogens. Initially salmonid (rainbow trout) ONM contained 1672 elements, representing more than 1400 genes [306]. Currently, one of the most often used ONM platforms to evaluate the response against different conditions and pathogens is the custom salmon ONM (SIQ-3), based on the Agilent Technology system (21K in 4x44K format). Because limited availability of peripheral blood leukocyte (PBL) markers is a well-recognised problem of fish immunology, this platform compares the transcriptomes of PBL and other tissues to search for genes with preferential expression in leukocytes [296], making it a very significant tool to evaluate the response to pathogens in Atlantic salmon. Another ONM platform based on 500K ESTs Atlantic salmon and 250K ESTs rainbow trout [298] is the cGRASP 44K salmonid oligo array (Agilent eArray), although no studies employing this platform have been published yet. Another ONM has been designed in fish model organisms like zebrafish and in commercial fish species such as channel catfish and turbot [307].

Functional genomic studies based on evaluating immune responses, also called immunogenomics, have been conducted *in vivo* to evaluate the response to different pathogens at the systemic level in different organs, especially the liver and head kidney. The functional genomic approach has been used with *Salmo salar* and *Oncorhynchus mykiss*, where PAMPs, whole bacteria and viruses are the most studied pathogens. Here we present different works in fish challenged by bacteria or viruses where differential gene expression profiles were evaluated using microarray platforms with special emphasis on *in vivo* fish immune response (Table 2).

1.15. Studies with bacterial pathogens, PAMPs and cytokine network interactions

One of the most commonly studied bacterial pathogens is *Aeromonas salmonicida*, a gram-negative bacteria and the causative agent of furunculosis. In fact prior to the development of species-specific cDNA microarrays a preliminary study used a human microarray (GENE-FILTERS GF211) to explore the liver response in Atlantic salmon infected using a cohabitation model [275]. Only 4 mRNAs were consistently up-regulated ($p < 0.01$) from the 241 positively identified spots with a clearly detectable hybridization signal, none of them related to cytokine expression. This was probably due to the lack of sequence homology, a problem commonly associated with cross-species cDNA hybridization. Thus the creation of species-specific platforms was a key step in fish immunology. Using a custom Atlantic salmon cDNA microarray (NRC-IMB) consisting of over 4,000 different cDNA amplicons, the first results for challenge with Aeromonas salmonicida were reported in 2005 [275]. The study described a cohabitation challenge and identified 16 up-regulated mRNAs in all three tissues studied (spleen, liver and head kidney), whereas 2 and 19 mRNAs were identified as down-regulated in the head kidney and liver, respectively. The authors found that genes related to the acute phase response were up-regulated in spleen and head kidney of infected salmon, indicating that the infected fish underwent a typical acute phase response to infection.

The effects of an *Aeromonas salmonicida* infection were recently reported in turbot, *Scolphtalmus maximus*, [307]. Using a custom designed oligonucleotide-microarray (8x15K), the authors identified a set of 48 differentially regulated mRNAs in the spleen of challenged fish at 3 dpi, mostly related to the acute-phase and the stress/defence immune response. A study using channel and blue catfish explored the effects of a gram-negative bacterial infection on the acute phase response (APR) [308]. The authors showed up-regulation of mRNA transcripts involved in iron homeostasis, transport proteins, complement components and inflammatory and humoral immune response, indicating that conserved APR occurs as part of the innate immune response in both catfish species. Interestingly, a more acute response was observed composed of several immune pathways in the blue but not the channel catfish. More studies are required to elucidate expression patterns resulting from gram-negative bacterial infection of phylogenetically similar and different fish are required to describe common and divergent responses. This could lead to the development of marker systems, consensus on the APR in fish and treatments tailored to certain species, all of which have significant applied interest.

The activity of LPS from gram-negative bacteria, a common membrane-associated PAMP used in immunological studies, has been explored in several fish species. These studies include effects on the spleen in channel catfish [285], rainbow trout head kidney [309], and liver in the Senegalese sole [310]. Using a 19K oligonucleotide microarray (ONM) it was observed that some pro-inflammatory mRNAs in the catfish spleen were up-regulated very quickly, principally between 2 and 4 hours post-injection with LPS, whereas immunoglobulin- (2h post-injection) and antigenic presentation-related mRNA transcripts were repressed 24h post-injection [311]. A similar inhibition was reported in head kidney of rainbow trout, where the suppression of major cellular processes, including immune function and an initial stress reaction, was followed by a proliferative hematopoietic-type/biogenesis response 3 dpi [309]. However, in the Senegalese sole a clear up-regulation of transcripts related to the immune response was reported 24 hpi in the liver [310]. These results collectively highlight the diversity of responses observed at the tissue level and reflect the nature of the immune system that is diffusely located throughout many organ compartments.

Microorgan-ism	Fish	Pathogen	Tissue/Cell type	Resource	Platform	Reference
Bacteria	Atlantic salmon	Aeromonas salmonicida (A449)	Liver	cDNA	GENE- FIL-TERS GF211	Tsoi et al., 2003
		Aeromonas salmonicida (A449)	Head kidney/liver/spleen	cDNA	NRC-IMB	Ewart et al., 2005
		Aeromonas salmonicida	Liver/spleen	cDNA	SFA-2	Skugor et al., 2009
		Aeromonas salmonicida (Brivax II)	Liver	cDNA	TRAITS/SPG	Martin et al., 2010
		Piscirickettsia salmonis	Head kidney	cDNA	GRASP-1	Rise et al., 2004
	Chinook salmon	Vibrio anguillarum	Head kidney	cDNA	GRASP-1	Ching et al., 2010
	Rainbow Trout	Vibrio anguillarum (FDKC)	Liver	ONM	OSUrbt	Gerwick et al., 2007
		LPS (E.coli 026:B6)	Head kidney	cDNA	SFA-1	MacKenzie et al., 2008
		LPS (E. coli 026:B6)	HK macrophages	cDNA	SFA-1	Mackenzie et al., 2008
		LPS, PGN (E. coli 0111:B4)	HK macrophages	cDNA	SFA-2	Boltaña et al., 2011
		PGN (E. coli 0111:B4, K12)	HK macrophages	cDNA	SFA-2	Boltaña et al., 2011
		LPS, PGN (E. coli 0111:B4)/poly I:C	Erythrocytes	cDNA	SFA-2	Morera et al., 2011
	Brook trout	LPS (E. coli 026:B6)	HK macrophages	cDNA	SFA-1	Mackenzie et al., 2006
	Channel catfish	LPS (E.coli 0127:B8)	Spleen	ONM	UMSMED-1	Li et al., 2006
		Edwardsiella ictaluri (MS-S97-773)	Gills/head kidney/liver/skin/spleen	ONM	UMSMED-2	Peatman et al., 2007

	Blue cat-fish	*Edwardsiella ictaluri* (MS-S97-773)	Liver	ONM	UMSMED-2	Peatman et al., 2008
	Japanese flounder	*Edwardsiella tarda* (NE9505)	Head kidney	cDNA	Japanese flounder custom-3	Yasuike et al., 2010
		Streptococcus iniae (02; FKC)	Head kidney cells	cDNA	Japanese flounder custom-3	Dumrongphol et al., 2008
		Mycobacterium bovis (TUMSAT-Msp001) FKC	Kidney cells	cDNA	Japanese flounder custom-3	Kato el al., 2010
	Zebra-fish	*Mycobacterium marinum* (M, E11)	Whole fish	ONM	MWG - Sigma Genosys - Affimetrix	Meijer et al., 2005
		Mycobacterium marinum (E11; Mma20)	Whole fish	ONM	ZF Agilent	Van der Sar et al., 2009
		Streptococcus suis (HA9801)	Whole fish	ONM	Affimetrix Zebrafish GeneChip	Wu et al., 2010
	Turbot	*Aeromonas salmonicida*	Spleen	ONM	Turbot custom Agilent	Millán et al., 2010
	Solea	LPS (*E. coli* 011:B4)	Liver	cDNA	GENIPOL-1	Osuna-Jimenez et al., 2009
Virus	Atlantic salmon	ISAV (Glesvaer 2/90)	Gills/heart/liver/spleen	cDNA	SFA-2	Jorgensen et al., 2008
		ISAV (Glesvaer2/90)	Heart/PBL	ONM	SIQ-3	Krasnov et al., 2011
	Chinook salmon	ISAV NA-HRP4 (970)	Head kidney	cDNA	GRASP-3	Leblanc et al., 2010
	Rainbow Trout	IHNV (32/87)/attenuated IHNV	Head kidney	cDNA	SFA-1	MacKenzie et al., 2008
		IHNV (strain 220-90)	Head kidney	cDNA	GRASP-2	Purcell et al., 2011
	Japanese flounder	VHSV (KRRV9822)	Head kidney cells	cDNA	Japanese flounder custom-1	Byon et al., 2005
		VHSV (KRRV9822)	Head kidney cells	cDNA	Japanese flounder custom-2	Byon et al., 2006

	HIRRV (8601H)	Kidney cells	cDNA	Japanese flounder custom-1	Yasuike et al., 2007
	HIRRV (8601H)	Kidney cells	cDNA	Japanese flounder custom-3	Yasuike et al., 2010
Zebra-fish	VHSV (07.71)	Fin/head kidney/ liver/spleen	ONM	ZF Agilent	Encinas et a., 2010
Turbot	Nodavirus (AH95)/poly (I:C)	Kidney	cDNA	Turbot custom	Park et al., 2009

Table 2. Transcriptomics studies on fish after treatments with bacteria or virus in vivo analyzed with microarrays. FDKC: Formaldehyde –killed cells; LPS: Lipopolysaccharide; FKC: Formalin-killed cells; ISAV: Infectious salmon anemia virus; PD: Pancreas disease; CMS: Cardiomyopathy syndrome; HSMI: Heart and skeletal muscle inflammation; IHNV: Infectious hematopoietic necrosis virus; VHSV: Viral hemorrhagic septicemia virus; HIRRV: Hirame rhabdovirus; poly I:C: polyriboinosinic polyribocytidylic acid; PBL: Peripheral blood leukocytes; ONM: oligonucleotide microarray.

For gram-positive infections in fish at the level of transcriptome analyses, infection of zebra-fish with *Streptococcus suis* is the only model reported [272]. *Streptococcus suis* is a pathogen associated with zoonosis reported in several countries [312, 313]. The Affymetrix Zebrafish GeneChip was used to identify 125 up-regulated transcripts where the most significant pathways were antigen processing and presentation, leukocyte trans-endothelial migration and the proteosome. The authors suggested that the target list obtained could serve as infection markers for gram-positive infection in fish.

Undoubtedly, the identification of prognostic biomarkers for disease resistance is a major aim for aquaculture. Functional genomics has the potential to identify such potential tools. Disease resistance is normally measured by challenge with the pathogen of interest and assessing the cumulative mortalities. Surviving fish or non-challenged siblings from the same family are then considered 'resistant'. Because this process is costly there is a need for non-lethal methodologies of measuring resistance, ideally based on molecular determinants of resistance. An initial example of this approach used the GRASP 3.7K cDNA array to identify *in vitro* macrophage and *in vivo* head kidney biomarkers in response to *Piscirickettsia salmonis* infection, yielding a number of 11 regulated genes common to both challenges. The researchers proposed 19 highly regulated transcripts as potential biomarkers to evaluate the efficacy of vaccines against *Piscirickettsia salmonis* [314]. C-type lectin 2-1, a gene whose product is involved in endocytosis and the C/EBP-driven inflammatory response [315] was identified and has been identified in almost all reports in which bacterial preparations have been used to challenge live fish [275, 309, 314, 316, 317]. Another study aimed at identifying biomarkers at the transcriptional level described differences between triploid and diploid Chinook salmon under live Vibrio anguillarum challenge using the GRASP 3.7K cDNA microarray [318]. Twelve annotated mRNAs were identified as showing significant differences between diploid and triploid fish. The authors however were unable to provide a description of the underlying mechanisms to explain the observed reduced immune function of triploid salmon.

Individual variation is a major hurdle for the development of prognostic markers as both genetic and epigenetic factors must be taken into account. The utility of, for example, C-type lectin in salmon, and other potential biomarkers in other species for bacterial disease resistance, requires further development. The future publication of several fish genomes coupled to array platforms with a much increased transcript representation could provide an exciting route to further develop this strategy by combining both functional and structural genomics for species of commercial interest with a sequenced genome. Several studies have attempted to correlate gene expression profiles with the activity of bacterins (killed bacteria preparations) used to vaccinate fish in culture. Most studies have concentrated on the rainbow trout and Japanese flounder [262, 277, 319, 320]. In trout, intraperitoneal administration of killed *V. anguillarum* resulted in identifying 36 differentially expressed transcripts [320]. Most of the identified targets are involved in inflammatory response and respond to a broad range of stimuli. This suggests that these targets have little use as markers for vaccination, contrary to previous descriptions in other studies. Both the second and fourth versions of the Japanese flounder cDNA microarray have been used to address vaccination [262, 277, 319]. The results of experimental infection with Gram-negative *E. tarda* indicated that a formalin-killed preparation reduced mortality in vaccinated fish from 90% to 20% [277]. However, a correlation between the transcriptome and the efficacy of vaccination could not be identified.

The effects of a commercial vaccine for Atlantic salmon (a six-component oil-adjuvant vaccine from PHARMAQ) were evaluated to correlate vaccine protection to high and low resistance to furunculosis. The authors did not find any association between either group and suggested that "outcomes of vaccination depend largely on the ability of host to prevent the negative impacts of immune response and to repair damages" [305]. Although this study did not identify correlations between vaccination and gene expression profiles, the potential of a functional genomics approach to evaluate the efficacy and underlying mechanisms of vaccination is highlighted. In terms of the immune response and the resulting complexity in expression patterns resulting from multiple cell types and different tissue responses, the investigator has the potential to obtain a clearer 'image' of the biological response from global expression data. A key objective is therefore to increase the available genomic resources much facilitated by next generation sequencing technologies to form a more robust representation of the immune system among different fish species. Furthermore, the increasing use of ONM platforms will also improve comparison across species as data sets become more easily comparable.

It remains difficult to compare microarray experiments across distinct platforms. In this respect, Meijer et al. 2005, evaluated host transcriptome profiling to *Mycobacterium marinum* infection of adult zebrafish employing three oligonucleotides platforms (MWG, Sigma Genosys, and Affimetrix). At a significance level of $P < 1.00E-5$, there were differences among the platforms in the total number of more than 2-fold up-regulated genes, whereas the 2-fold down-regulated genes were in a similar range. Evaluation of the distribution of infection-induced genes over different categories reveled some divergence in the set from MWG, probably due to the abundance of genes of the same UniGene cluster. As well, from the total overlap of 4,138 UniGene clusters among the three microarrays, only 66 and 93 genes were up- and down-regulated, respectively [321]. With this antecedent, the same group generated a new platform (Agilent 44K) that includes their 22K probes, a 16K set probes similar to the Sigma-Compugen oligonucleotide library, and 6K set of probes for selected genes of interest indentified by previous data mining of zebrafish transcript and genome databases [322], and they evaluated the transcriptome response to acute and chronic infection by *Mycobacterium marinum*. This important effort in combing different platforms makes it clear that not all relevant genes, including immune-related ones, are represented in all platforms. Consequntly, new efforts are necessary to broaden our understanding of the immune response in fish challenged with a pathogen of interest.

1.16. Wide screening in fish challenged with viral pathogens

Two studies have reported host responses to IHNV with the SFA and GRASP platforms [309, 323]. The potential mechanisms responsible for host-specific virulence were assessed in rainbow trout infected with high (M) and low virulence (U) strains of IHNV. A marked down-regulation in biological processes, including the immune response, lymphocyte activation, response to stress, transcription and translation, together with a greater viral load (M), suggest that the higher virulence is due to the ability to suppress the immune response via the transcriptional and translational machinery of cell [323]. Furthermore, in rainbow trout was compared the expression profiles of IHNV and attenuated IHNV were compared in rainbow trout over a short time frame of one and three days post-challenge. At 3 dpi, a significant change in the transcriptional program of head kidney revealed an immunological shift orientated toward the activation of adaptive immunity. This shift was IHNV-dependent as determined by differences between the attenuated and virulent IHNV specific expression profiles. The rapid systemic spreading of IHNV inhibited TNFα, MHC class I, and several macrophage and cell cycle/differentiation markers and favored a MHC class II, immunoglobulin and MMP/TBX4-enhanced immune response [309].

Parallel studies were conducted with a cDNA microarray enriched with 213 immune-related genes to study the immune response and the efficacy of DNA vaccines containing the viral G proteins of VHSV and HIRRV administered intramuscularly in the Japanese flounder [301, 324]. As expected, all DNA vaccines containing the viral G glycoprotein conferred specific protection to the challenged fish one month after vaccination. It is suggested that the protection occurs via the IFN type I system due to the number of IFN-related genes up-regulated in both studies, ISG-15, interferon-stimulated gene 56kDa (ISG56) and the Mx protein. In both studies, VHSV and HIRRV in the Japanese flounder, the majority of differentially up-regulated genes were identified between 3 and 7 days post-vaccination (dpv), including the less effective DNA vaccine containing N protein of HIRRV. Interestingly, Mx, an antiviral protein commonly used as a marker for antiviral activity in animal species, was consistently up-regulated across vaccinations [301]. In a similar observation, IRF-3, Mx, Vig-1 and Vig-8 were up-regulated in trout at the site of DNA vaccination against IHNV at 7 dpv [323].

In turbot challenged with nodavirus, both Mx and IFN-inducible proteins were identified 24 hpi [303]. These and the previously described results suggest that both the host-expressed viral glycoprotein and the virulent rhadovirus induce a systemic anti-viral state indicative of non-specific IFN type1 innate immune response and that this canonical response is conserved among all fish. However, the mechanisms to develop a specific cytotoxic T or B lymphocyte-mediated humoral response in fish vaccinated with plasmid DNA-IHNV G that confers protective immunity have not been identified [323].

In direct relation to the above, a significant increase in transcript markers for adaptive immunity was reported in Atlantic salmon during ISA virus (ISAV) infection [325]. Importantly, a progressive increase was observed in IgZ mRNA parallel to a decrease in IgM expression that peaked > 30 days post-infection. This coordinated increase in a group of genes related to B lymphocyte differentiation and maturation and activation of T lymphocyte-mediated immunity, including CD4, TGF-β, CD8α and IFNγ, provides strong evidence for the coordinated regulation of the two arms of the immune system in response to viral infection. An important technological contribution derived from the above study was the development of an ONM for Atlantic salmon (SIQ-3). The first assessment of the performance of these arrays was carried out in Atlantic salmon for the study of virus-responsive genes from samples infected with ISA, salmonid alpha virus/PD-virus, cardiomyopathy syndrome (CMS) agent, heart and skeletal muscle inflammation (HSMI) and PBL from fish infected with ISAV. Some 95 up-regulated transcripts were identified. Most of the regulated transcripts are related directly to the immune response or associated with antiviral response [296]. As previously mentioned, the creation of species-specific platforms has been a key challenge for the study of the immune response against pathogens. Despite impressive achievements, cDNA microarrays suffer from limitations and disadvantages, the most important drawback being the limited ability to discriminate between paralogs as long cDNA probes cross-hybridize with highly similar transcripts from members of multi-gene families [305]. Furthermore this information needs to be supplemented to establish if the increased level of detected transcripts is consistent with specific protein synthesis. Recently, a study employed a combined proteomic and transcriptomic approach to evaluate the immune response against VHS [326]. In the fins of infected fish a series of mRNA transcripts principally related to complement components, immunoglobulin-related proteins, and macrophages were up-regulated (> 2-fold), whereas in parallel using two dimensional differential gel electrophoresis (2D-DIGE), enzymes of the glycolytic pathway and some proteins related to cytoskeletal remodelling and apoptosis (such as annexin A1a) increased with infection. However, very few proteins related to anti-viral response were identified.

2. Concluding remarks

A complex network exists to regulate the innate and adaptive immune responses of fish from the various cytokines that have been reported. The study of the functional activity of these cytokines is in progress and it will be interesting to know whether mammalian Th1, Th2, Th17 and Treg responses are present in fish, regulating specific cell-mediated immunity. The recombinant production of these cytokines and antibodies against them will be the next challenge in understanding the balance of such immune responses and aid in the effective design of therapeutic strategies to manipulate the fish immune system. towards humoral or cellular immunity in response to specific antigen stimulation, vaccine strategies, functional diets to increase the quality of fishery production and predict the health of cultured fish.

The study of functional genomics in fish has provided substantial data on species of commercial interest. The major aim has been to functionally identify the intensity of responses to specific pathogens and their associated molecular components and to identify transcripts in a whole organism or specific tissue that contribute to such responses. However, the complex biology of the immune response, in which different spatial-temporal expression occurs in multiple cell types at distinct body locations, makes complete mapping of a response difficult and expensive. Moreover, considering that arrays are only as good as the transcripts represented upon them. Thus, the representation of transcripts relevant to the immune response is intimately linked to gene discovery efforts through large-scale sequencing projects, where strategies like SSH contribute not only to understanding transcriptomic response against specific pathogens but also to gene discovery. In this area, access to high-throughput NGS technology has increased in recent years and promises to make an important contribution to understanding immune response in fish. The major task now is the meta-analysis of transcriptomic data to delineate responses common among fish species to specific pathogen groups and highly specific responses. This approach will reveal host specific expression profiles and facilitate the identification of prognostic markers for diseases.

Acknowledgments

The authors are grateful for the support of VRID-USACH, INNOVA-CORFO 07CN13PBT-90, INNOVA-CORFO 09MCSS6691, INNOVA-CORFO 09MCSS6698 and CONICYT Fellowships to K. Maisey and D. Toro-Ascuy.

Author details

Sebastián Reyes-Cerpa[1*], Kevin Maisey[2], Felipe Reyes-López[3], Daniela Toro-Ascuy[1], Ana María Sandino[4] and Mónica Imarai[2]

*Address all correspondence to: sebastian.reyesc@usach.cl

1 Dept. Biology, Laboratory of Virology, University of Santiago of Chile, Santiago, Chile

2 Dept. Biology, Laboratory of Immunology, University of Santiago of Chile, Santiago, Chile

3 Dept. Cell Biology, Physiology & Immunology, Unit of Animal Physiology, Autonomous University of Barcelona, Barcelona, Spain

4 Dept. Biology, Laboratory of Virology, University of Santiago of Chile, Santiago, Chile; Activaq S.A., Santiago, Chile

References

[1] Flajnik MF, Du Pasquier L. Evolution of the Immune System. 6 th ed. Paul WE, editor: Lippincott Williams & Wilkins; 2008.

[2] Akirav EM, Liao S, Ruddle NH. Lymphoid Tissues and Organs. 6 th ed. Paul WE, editor: Lippincott Williams & Wilkins; 2008.

[3] Zapata AG, Chibá A, Varas A. Cells and tissues of the immune system of fish. In: Iwama G, Nakanishi T, editors. The fish immune system Organism, pathogen and environment: Academic Press; 1996. p. 1-62.

[4] Meseguer J, Lopez-Ruiz A, Garcia-Ayala A. Reticulo-endothelial stroma of the head-kidney from the seawater teleost gilthead seabream (Sparus aurata L.): an ultrastructural and cytochemical study. Anat Rec. 1995 Mar;241(3):303-9.

[5] Danneving BH, Lauve A, Press MC, Landsverk T. Receptor-mediated endocytosis and phagocytosis by rainbow trout head kidney sinusoidal cells. Fish Shellfish Immunol. 1994;4:3-18.

[6] Brattgjerd S, Evensen O. A sequential light microscopic and ultrastructural study on the uptake and handling of Vibrio salmonicida in phagocytes of the head kidney in experimentally infected Atlantic salmon (Salmo salar L.). Vet Pathol. 1996 Jan;33(1): 55-65.

[7] Kaattari SL, Irwin MJ. Salmonid spleen and anterior kidney harbor populations of lymphocytes with different B cell repertoires. Dev Comp Immunol. 1985 Summer; 9(3):433-44.

[8] Herraez MP, Zapata AG. Structure and function of the melano-macrophage centres of the goldfish Carassius auratus. Vet Immunol Immunopathol. 1986 Jun;12(1-4): 117-26.

[9] Tsujii T, Seno S. Melano-macrophage centers in the aglomerular kidney of the sea horse (teleosts): morphologic studies on its formation and possible function. Anat Rec. 1990 Apr;226(4):460-70.

[10] Bowden TJ, Cook P, Rombout JH. Development and function of the thymus in teleosts. Fish Shellfish Immunol. 2005 Nov;19(5):413-27.

[11] Zapata AG, Amemiya CT. Phylogeny of lower vertebrates and their immunological structures. Curr Top Microbiol Immunol. 1999;248:67-107.

[12] Espenes A, Press CM, Dannevig BH, Landsverk T. Immune-complex trapping in the splenic ellipsoids of rainbow trout (Oncorhynchus mykiss). Cell Tissue Res. 1995 Oct; 282(1):41-8.

[13] Vallejo AN, Miller NW, Harvey NE, Cuchens MA, Warr GW, Clem LW. Cellular pathway(s) of antigen processing and presentation in fish APC: endosomal involvement and cell-free antigen presentation. Dev Immunol. 1992;3(1):51-65.

[14] Vallejo AN, Miller NW, Clem LW. Cellular pathway(s) of antigen processing in fish APC: effect of varying in vitro temperatures on antigen catabolism. Dev Comp Immunol. 1992 Sep-Oct;16(5):367-81.

[15] Ellis AE. Antigen trapping in the spleen and kidney of the plaice, Pleuronectes platessa. J Fish Dis. 1980;3:413-26.

[16] Agius C, Roberts RJ. Melano-macrophage centres and their role in fish pathology. J Fish Dis. 2003 Sep;26(9):499-509.

[17] Medzhitov R. The Innate Immune System. 6 th ed. Paul WE, editor: Lippincott Williams & Wilkins; 2008.

[18] Janeway CA, Jr. How the immune system protects the host from infection. Microbes Infect. 2001 Nov;3(13):1167-71.

[19] McHeyzer-Williams LJ, McHeyzer-Williams MG. Antigen-specific memory B cell development. Annu Rev Immunol. 2005;23:487-513.

[20] Sallusto F, Lanzavecchia A. Heterogeneity of CD4+ memory T cells: functional modules for tailored immunity. Eur J Immunol. 2009 Aug;39(8):2076-82.

[21] Dixon B, Stet RJ. The relationship between major histocompatibility receptors and innate immunity in teleost fish. Dev Comp Immunol. 2001 Oct-Dec;25(8-9):683-99.

[22] Abbas AK, Lichtman AH, Pillai S. Cells and Tissues of the Adaptive Immune System. 6 th ed: Saunders; 2007.

[23] Castro R, Bernard D, Lefranc MP, Six A, Benmansour A, Boudinot P. T cell diversity and TcR repertoires in teleost fish. Fish Shellfish Immunol. 2011 Nov;31(5):644-54.

[24] Secombes CJ, Wang T, Bird S. The interleukins of fish. Dev Comp Immunol. 2011 Dec;35(12):1336-45.

[25] Wang T, Gorgoglione B, Maehr T, Holland JW, Vecino JL, Wadsworth S, et al. Fish Suppressors of Cytokine Signaling (SOCS): Gene Discovery, Modulation of Expression and Function. J Signal Transduct. 2011;2011:905813.

[26] Alejo A, Tafalla C. Chemokines in teleost fish species. Dev Comp Immunol. 2011 Dec;35(12):1215-22.

[27] Nakanishi T, Toda H, Shibasaki Y, Somamoto T. Cytotoxic T cells in teleost fish. Dev Comp Immunol. 2011 Dec;35(12):1317-23.

[28] Boltana S, Donate C, Goetz FW, MacKenzie S, Balasch JC. Characterization and expression of NADPH oxidase in LPS-, poly(I:C)- and zymosan-stimulated trout (Oncorhynchus mykiss W.) macrophages. Fish Shellfish Immunol. 2009 Apr;26(4):651-61.

[29] Hanington PC, Hitchen SJ, Beamish LA, Belosevic M. Macrophage colony stimulating factor (CSF-1) is a central growth factor of goldfish macrophages. Fish Shellfish Immunol. 2009 Jan;26(1):1-9.

[30] Salinas I, Zhang YA, Sunyer JO. Mucosal immunoglobulins and B cells of teleost fish. Dev Comp Immunol. 2011 Dec;35(12):1346-65.

[31] Zhang YA, Salinas I, Oriol Sunyer J. Recent findings on the structure and function of teleost IgT. Fish Shellfish Immunol. 2011 Nov;31(5):627-34.

[32] Randelli E, Buonocore F, Scapigliati G. Cell markers and determinants in fish immunology. Fish Shellfish Immunol. 2008 Oct;25(4):326-40.

[33] Rombout JH, Huttenhuis HB, Picchietti S, Scapigliati G. Phylogeny and ontogeny of fish leucocytes. Fish Shellfish Immunol. 2005 Nov;19(5):441-55.

[34] Araki K, Akatsu K, Suetake H, Kikuchi K, Suzuki Y. Characterization of CD8+ leukocytes in fugu (Takifugu rubripes) with antiserum against fugu CD8alpha. Dev Comp Immunol. 2008;32(7):850-8.

[35] Toda H, Saito Y, Koike T, Takizawa F, Araki K, Yabu T, et al. Conservation of characteristics and functions of CD4 positive lymphocytes in a teleost fish. Dev Comp Immunol. 2011 Jun;35(6):650-60.

[36] Takizawa F, Dijkstra JM, Kotterba P, Korytar T, Kock H, Kollner B, et al. The expression of CD8alpha discriminates distinct T cell subsets in teleost fish. Dev Comp Immunol. 2011 Jul;35(7):752-63.

[37] Kaattari S, Evans D, Klemer J. Varied redox forms of teleost IgM: an alternative to isotypic diversity? Immunol Rev. 1998 Dec;166:133-42.

[38] Wilson M, Bengten E, Miller NW, Clem LW, Du Pasquier L, Warr GW. A novel chimeric Ig heavy chain from a teleost fish shares similarities to IgD. Proc Natl Acad Sci U S A. 1997 Apr 29;94(9):4593-7.

[39] Stenvik J, Jorgensen TO. Immunoglobulin D (IgD) of Atlantic cod has a unique structure. Immunogenetics. 2000 May;51(6):452-61.

[40] Danilova N, Bussmann J, Jekosch K, Steiner LA. The immunoglobulin heavy-chain locus in zebrafish: identification and expression of a previously unknown isotype, immunoglobulin Z. Nat Immunol. 2005 Mar;6(3):295-302.

[41] Hansen JD, Landis ED, Phillips RB. Discovery of a unique Ig heavy-chain isotype (IgT) in rainbow trout: Implications for a distinctive B cell developmental pathway in teleost fish. Proc Natl Acad Sci U S A. 2005 May 10;102(19):6919-24.

[42] Li J, Barreda DR, Zhang YA, Boshra H, Gelman AE, Lapatra S, et al. B lymphocytes from early vertebrates have potent phagocytic and microbicidal abilities. Nat Immunol. 2006 Oct;7(10):1116-24.

[43] Overland HS, Pettersen EF, Ronneseth A, Wergeland HI. Phagocytosis by B-cells and neutrophils in Atlantic salmon (Salmo salar L.) and Atlantic cod (Gadus morhua L.). Fish Shellfish Immunol. 2010 Jan;28(1):193-204.

[44] Wang T, Huang W, Costa MM, Secombes CJ. The gamma-chain cytokine/receptor system in fish: more ligands and receptors. Fish Shellfish Immunol. 2011 Nov;31(5): 673-87.

[45] Savan R, Sakai M. Genomics of fish cytokines. Comp Biochem Physiol Part D Genomics Proteomics. 2006 Mar;1(1):89-101.

[46] Salazar-Mather TP, Hokeness KL. Cytokine and chemokine networks: pathways to antiviral defense. Curr Top Microbiol Immunol. 2006;303:29-46.

[47] Svanborg C, Godaly G, Hedlund M. Cytokine responses during mucosal infections: role in disease pathogenesis and host defence. Curr Opin Microbiol. 1999 Feb;2(1): 99-105.

[48] DeVries ME, Ran L, Kelvin DJ. On the edge: the physiological and pathophysiological role of chemokines during inflammatory and immunological responses. Semin Immunol. 1999 Apr;11(2):95-104.

[49] Whyte SK. The innate immune response of finfish--a review of current knowledge. Fish Shellfish Immunol. 2007 Dec;23(6):1127-51.

[50] Rahman MM, McFadden G. Modulation of tumor necrosis factor by microbial pathogens. PLoS Pathog. 2006 Feb;2(2):e4.

[51] Benedict CA, Banks TA, Ware CF. Death and survival: viral regulation of TNF signaling pathways. Curr Opin Immunol. 2003 Feb;15(1):59-65.

[52] Hirono I, Nam BH, Kurobe T, Aoki T. Molecular cloning, characterization, and expression of TNF cDNA and gene from Japanese flounder Paralychthys olivaceus. J Immunol. 2000 Oct 15;165(8):4423-7.

[53] Laing KJ, Wang T, Zou J, Holland J, Hong S, Bols N, et al. Cloning and expression analysis of rainbow trout Oncorhynchus mykiss tumour necrosis factor-alpha. Eur J Biochem. 2001 Mar;268(5):1315-22.

[54] Zou J, Wang T, Hirono I, Aoki T, Inagawa H, Honda T, et al. Differential expression of two tumor necrosis factor genes in rainbow trout, Oncorhynchus mykiss. Dev Comp Immunol. 2002 Mar;26(2):161-72.

[55] Garcia-Castillo J, Pelegrin P, Mulero V, Meseguer J. Molecular cloning and expression analysis of tumor necrosis factor alpha from a marine fish reveal its constitutive expression and ubiquitous nature. Immunogenetics. 2002 Jun;54(3):200-7.

[56] Saeij JP, Stet RJ, de Vries BJ, van Muiswinkel WB, Wiegertjes GF. Molecular and functional characterization of carp TNF: a link between TNF polymorphism and trypanotolerance? Dev Comp Immunol. 2003 Jan;27(1):29-41.

[57] Zou J, Secombes CJ, Long S, Miller N, Clem LW, Chinchar VG. Molecular identification and expression analysis of tumor necrosis factor in channel catfish (Ictalurus punctatus). Dev Comp Immunol. 2003 Dec;27(10):845-58.

[58] Praveen K, Evans DL, Jaso-Friedmann L. Constitutive expression of tumor necrosis factor-alpha in cytotoxic cells of teleosts and its role in regulation of cell-mediated cytotoxicity. Mol Immunol. 2006 Feb;43(3):279-91.

[59] Ordas MC, Costa MM, Roca FJ, Lopez-Castejon G, Mulero V, Meseguer J, et al. Turbot TNFalpha gene: molecular characterization and biological activity of the recombinant protein. Mol Immunol. 2007 Jan;44(4):389-400.

[60] Grayfer L, Walsh JG, Belosevic M. Characterization and functional analysis of goldfish (Carassius auratus L.) tumor necrosis factor-alpha. Dev Comp Immunol. 2008;32(5):532-43.

[61] Garcia-Castillo J, Chaves-Pozo E, Olivares P, Pelegrin P, Meseguer J, Mulero V. The tumor necrosis factor alpha of the bony fish seabream exhibits the in vivo proinflammatory and proliferative activities of its mammalian counterparts, yet it functions in a species-specific manner. Cell Mol Life Sci. 2004 Jun;61(11):1331-40.

[62] Roca FJ, Mulero I, Lopez-Munoz A, Sepulcre MP, Renshaw SA, Meseguer J, et al. Evolution of the inflammatory response in vertebrates: fish TNF-alpha is a powerful activator of endothelial cells but hardly activates phagocytes. J Immunol. 2008 Oct 1;181(7):5071-81.

[63] Hardie LJ, Chappell LH, Secombes CJ. Human tumor necrosis factor alpha influences rainbow trout Oncorhynchus mykiss leucocyte responses. Vet Immunol Immunopathol. 1994 Jan;40(1):73-84.

[64] Iliev DB, Liarte CQ, MacKenzie S, Goetz FW. Activation of rainbow trout (Oncorhynchus mykiss) mononuclear phagocytes by different pathogen associated molecular pattern (PAMP) bearing agents. Mol Immunol. 2005 Jun;42(10):1215-23.

[65] Wang WL, Hong JR, Lin GH, Liu W, Gong HY, Lu MW, et al. Stage-specific expression of TNFalpha regulates bad/bid-mediated apoptosis and RIP1/ROS-mediated secondary necrosis in Birnavirus-infected fish cells. PLoS One. 2011;6(2):e16740.

[66] Dinarello C, Arend W, Sims J, Smith D, Blumberg H, O'Neill L, et al. IL-1 family nomenclature. Nat Immunol. 2010 Nov;11(11):973.

[67] Dinarello CA. Interleukin-1. Cytokine Growth Factor Rev. 1997 Dec;8(4):253-65.

[68] Zou J, Grabowski PS, Cunningham C, Secombes CJ. Molecular cloning of interleukin 1beta from rainbow trout Oncorhynchus mykiss reveals no evidence of an ice cut site. Cytokine. 1999 Aug;11(8):552-60.

[69] Fujiki K, Shin DH, Nakao M, Yano T. Molecular cloning and expression analysis of carp (Cyprinus carpio) interleukin-1 beta, high affinity immunoglobulin E Fc receptor gamma subunit and serum amyloid A. Fish Shellfish Immunol. 2000 Apr;10(3): 229-42.

[70] Scapigliati G, Buonocore F, Bird S, Zou J, Pelegrin P, Falasca C, et al. Phylogeny of cytokines: molecular cloning and expression analysis of sea bass Dicentrarchus labrax interleukin-1beta. Fish Shellfish Immunol. 2001 Nov;11(8):711-26.

[71] Pelegrin P, Garcia-Castillo J, Mulero V, Meseguer J. Interleukin-1beta isolated from a marine fish reveals up-regulated expression in macrophages following activation with lipopolysaccharide and lymphokines. Cytokine. 2001 Oct 21;16(2):67-72.

[72] Corripio-Miyar Y, Bird S, Tsamopoulos K, Secombes CJ. Cloning and expression analysis of two pro-inflammatory cytokines, IL-1 beta and IL-8, in haddock (Melanogrammus aeglefinus). Mol Immunol. 2007 Feb;44(6):1361-73.

[73] Lee DS, Hong SH, Lee HJ, Jun LJ, Chung JK, Kim KH, et al. Molecular cDNA cloning and analysis of the organization and expression of the IL-1beta gene in the Nile tilapia, Oreochromis niloticus. Comp Biochem Physiol A Mol Integr Physiol. 2006 Mar; 143(3):307-14.

[74] Pleguezuelos O, Zou J, Cunningham C, Secombes CJ. Cloning, sequencing, and analysis of expression of a second IL-1beta gene in rainbow trout (Oncorhynchus mykiss). Immunogenetics. 2000 Oct;51(12):1002-11.

[75] Cerretti DP, Kozlosky CJ, Mosley B, Nelson N, Van Ness K, Greenstreet TA, et al. Molecular cloning of the interleukin-1 beta converting enzyme. Science. 1992 Apr 3;256(5053):97-100.

[76] Secombes CJ, Wang T, Hong S, Peddie S, Crampe M, Laing KJ, et al. Cytokines and innate immunity of fish. Dev Comp Immunol. 2001 Oct-Dec;25(8-9):713-23.

[77] Lu DQ, Bei JX, Feng LN, Zhang Y, Liu XC, Wang L, et al. Interleukin-1beta gene in orange-spotted grouper, Epinephelus coioides: molecular cloning, expression, biological activities and signal transduction. Mol Immunol. 2008 Feb;45(4):857-67.

[78] Reyes-Cerpa S, Reyes-Lopez FE, Toro-Ascuy D, Ibanez J, Maisey K, Sandino AM, et al. IPNV modulation of pro and anti-inflammatory cytokine expression in Atlantic salmon might help the establishment of infection and persistence. Fish Shellfish Immunol. 2012 Feb;32(2):291-300.

[79] Seppola M, Larsen AN, Steiro K, Robertsen B, Jensen I. Characterisation and expression analysis of the interleukin genes, IL-1beta, IL-8 and IL-10, in Atlantic cod (Gadus morhua L.). Mol Immunol. 2008 Feb;45(4):887-97.

[80] Zou J, Holland J, Pleguezuelos O, Cunningham C, Secombes CJ. Factors influencing the expression of interleukin-1 beta in cultured rainbow trout (Oncorhynchus mykiss) leucocytes. Dev Comp Immunol. 2000 Sep-Oct;24(6-7):575-82.

[81] Buonocore F, Forlenza M, Randelli E, Benedetti S, Bossu P, Meloni S, et al. Biological activity of sea bass (Dicentrarchus labrax L.) recombinant interleukin-1beta. Mar Biotechnol (NY). 2005 Nov-Dec;7(6):609-17.

[82] Hong S, Zou J, Crampe M, Peddie S, Scapigliati G, Bols N, et al. The production and bioactivity of rainbow trout (Oncorhynchus mykiss) recombinant IL-1 beta. Vet Immunol Immunopathol. 2001 Aug 30;81(1-2):1-14.

[83] Peddie S, Zou J, Cunningham C, Secombes CJ. Rainbow trout (Oncorhynchus mykiss) recombinant IL-1beta and derived peptides induce migration of head-kidney leucocytes in vitro. Fish Shellfish Immunol. 2001 Nov;11(8):697-709.

[84] Peddie S, Zou J, Secombes CJ. Immunostimulation in the rainbow trout (Oncorhynchus mykiss) following intraperitoneal administration of Ergosan. Vet Immunol Immunopathol. 2002 May;86(1-2):101-13.

[85] Dinarello CA, Fantuzzi G. Interleukin-18 and host defense against infection. J Infect Dis. 2003 Jun 15;187 Suppl 2:S370-84.

[86] Gracie JA, Robertson SE, McInnes IB. Interleukin-18. J Leukoc Biol. 2003 Feb;73(2): 213-24.

[87] Sugawara I. Interleukin-18 (IL-18) and infectious diseases, with special emphasis on diseases induced by intracellular pathogens. Microbes Infect. 2000 Aug;2(10):1257-63.

[88] Micallef MJ, Ohtsuki T, Kohno K, Tanabe F, Ushio S, Namba M, et al. Interferon-gamma-inducing factor enhances T helper 1 cytokine production by stimulated human T cells: synergism with interleukin-12 for interferon-gamma production. Eur J Immunol. 1996 Jul;26(7):1647-51.

[89] Okamura H, Nagata K, Komatsu T, Tanimoto T, Nukata Y, Tanabe F, et al. A novel costimulatory factor for gamma interferon induction found in the livers of mice causes endotoxic shock. Infect Immun. 1995 Oct;63(10):3966-72.

[90] Dao T, Mehal WZ, Crispe IN. IL-18 augments perforin-dependent cytotoxicity of liver NK-T cells. J Immunol. 1998 Sep 1;161(5):2217-22.

[91] Leung BP, Culshaw S, Gracie JA, Hunter D, Canetti CA, Campbell C, et al. A role for IL-18 in neutrophil activation. J Immunol. 2001 Sep 1;167(5):2879-86.

[92] Xu D, Trajkovic V, Hunter D, Leung BP, Schulz K, Gracie JA, et al. IL-18 induces the differentiation of Th1 or Th2 cells depending upon cytokine milieu and genetic background. Eur J Immunol. 2000 Nov;30(11):3147-56.

[93] Ghayur T, Banerjee S, Hugunin M, Butler D, Herzog L, Carter A, et al. Caspase-1 processes IFN-gamma-inducing factor and regulates LPS-induced IFN-gamma production. Nature. 1997 Apr 10;386(6625):619-23.

[94] Gu Y, Kuida K, Tsutsui H, Ku G, Hsiao K, Fleming MA, et al. Activation of interferon-gamma inducing factor mediated by interleukin-1beta converting enzyme. Science. 1997 Jan 10;275(5297):206-9.

[95] Akita K, Ohtsuki T, Nukada Y, Tanimoto T, Namba M, Okura T, et al. Involvement of caspase-1 and caspase-3 in the production and processing of mature human interleukin 18 in monocytic THP.1 cells. J Biol Chem. 1997 Oct 17;272(42):26595-603.

[96] Sugawara S, Uehara A, Nochi T, Yamaguchi T, Ueda H, Sugiyama A, et al. Neutrophil proteinase 3-mediated induction of bioactive IL-18 secretion by human oral epithelial cells. J Immunol. 2001 Dec 1;167(11):6568-75.

[97] Huising MO, Stet RJ, Savelkoul HF, Verburg-van Kemenade BM. The molecular evolution of the interleukin-1 family of cytokines; IL-18 in teleost fish. Dev Comp Immunol. 2004 May 3;28(5):395-413.

[98] Zou J, Bird S, Truckle J, Bols N, Horne M, Secombes C. Identification and expression analysis of an IL-18 homologue and its alternatively spliced form in rainbow trout (Oncorhynchus mykiss). Eur J Biochem. 2004 May;271(10):1913-23.

[99] Martin SA, Zou J, Houlihan DF, Secombes CJ. Directional responses following recombinant cytokine stimulation of rainbow trout (Oncorhynchus mykiss) RTS-11 macrophage cells as revealed by transcriptome profiling. BMC Genomics. 2007;8:150.

[100] Wang T, Secombes CJ. Identification and expression analysis of two fish-specific IL-6 cytokine family members, the ciliary neurotrophic factor (CNTF)-like and M17 genes, in rainbow trout Oncorhynchus mykiss. Mol Immunol. 2009 Jul;46(11-12):2290-8.

[101] Taga T, Kishimoto T. Gp130 and the interleukin-6 family of cytokines. Annu Rev Immunol. 1997;15:797-819.

[102] Hirano T. Interleukin 6 and its receptor: ten years later. Int Rev Immunol. 1998;16(3-4):249-84.

[103] Inoue K, Takano H, Shimada A, Morita T, Yanagisawa R, Sakurai M, et al. Cytoprotection by interleukin-6 against liver injury induced by lipopolysaccharide. Int J Mol Med. 2005 Feb;15(2):221-4.

[104] Kaplanski G, Marin V, Montero-Julian F, Mantovani A, Farnarier C. IL-6: a regulator of the transition from neutrophil to monocyte recruitment during inflammation. Trends Immunol. 2003 Jan;24(1):25-9.

[105] Bird S, Zou J, Savan R, Kono T, Sakai M, Woo J, et al. Characterisation and expression analysis of an interleukin 6 homologue in the Japanese pufferfish, Fugu rubripes. Dev Comp Immunol. 2005;29(9):775-89.

[106] Iliev DB, Castellana B, Mackenzie S, Planas JV, Goetz FW. Cloning and expression analysis of an IL-6 homolog in rainbow trout (Oncorhynchus mykiss). Mol Immunol. 2007 Mar;44(7):1803-7.

[107] Nam BH, Byon JY, Kim YO, Park EM, Cho YC, Cheong J. Molecular cloning and characterisation of the flounder (Paralichthys olivaceus) interleukin-6 gene. Fish Shellfish Immunol. 2007 Jul;23(1):231-6.

[108] Castellana B, Iliev DB, Sepulcre MP, MacKenzie S, Goetz FW, Mulero V, et al. Molecular characterization of interleukin-6 in the gilthead seabream (Sparus aurata). Mol Immunol. 2008 Jul;45(12):3363-70.

[109] Costa MM, Maehr T, Diaz-Rosales P, Secombes CJ, Wang T. Bioactivity studies of rainbow trout (Oncorhynchus mykiss) interleukin-6: effects on macrophage growth and antimicrobial peptide gene expression. Mol Immunol. 2011 Sep;48(15-16): 1903-16.

[110] Czupryn MJ, McCoy JM, Scoble HA. Structure-function relationships in human interleukin-11. Identification of regions involved in activity by chemical modification and site-directed mutagenesis. J Biol Chem. 1995 Jan 13;270(2):978-85.

[111] Bartz H, Buning-Pfaue F, Turkel O, Schauer U. Respiratory syncytial virus induces prostaglandin E2, IL-10 and IL-11 generation in antigen presenting cells. Clin Exp Immunol. 2002 Sep;129(3):438-45.

[112] Kernacki KA, Goebel DJ, Poosch MS, Hazlett LD. Early cytokine and chemokine gene expression during Pseudomonas aeruginosa corneal infection in mice. Infect Immun. 1998 Jan;66(1):376-9.

[113] Nandurkar HH, Robb L, Begley CG. The role of IL-II in hematopoiesis as revealed by a targeted mutation of its receptor. Stem Cells. 1998;16 Suppl 2:53-65.

[114] Trepicchio WL, Bozza M, Pedneault G, Dorner AJ. Recombinant human IL-11 attenuates the inflammatory response through down-regulation of proinflammatory cytokine release and nitric oxide production. J Immunol. 1996 Oct 15;157(8):3627-34.

[115] Orazi A, Du X, Yang Z, Kashai M, Williams DA. Interleukin-11 prevents apoptosis and accelerates recovery of small intestinal mucosa in mice treated with combined chemotherapy and radiation. Lab Invest. 1996 Jul;75(1):33-42.

[116] Liu Q, Du XX, Schindel DT, Yang ZX, Rescorla FJ, Williams DA, et al. Trophic effects of interleukin-11 in rats with experimental short bowel syndrome. J Pediatr Surg. 1996 Aug;31(8):1047-50; discussion 50-1.

[117] Huising MO, Kruiswijk CP, van Schijndel JE, Savelkoul HF, Flik G, Verburg-van Kemenade BM. Multiple and highly divergent IL-11 genes in teleost fish. Immunogenetics. 2005 Jul;57(6):432-43.

[118] Wang T, Holland JW, Bols N, Secombes CJ. Cloning and expression of the first non-mammalian interleukin-11 gene in rainbow trout Oncorhynchus mykiss. FEBS J. 2005 Mar;272(5):1136-47.

[119] Santos MD, Yasuike M, Kondo H, Hirono I, Aoki T. Teleostean IL11b exhibits complementing function to IL11a and expansive involvement in antibacterial and antiviral responses. Mol Immunol. 2008 Jul;45(12):3494-501.

[120] Ribeiro CM, Hermsen T, Taverne-Thiele AJ, Savelkoul HF, Wiegertjes GF. Evolution of recognition of ligands from Gram-positive bacteria: similarities and differences in

the TLR2-mediated response between mammalian vertebrates and teleost fish. J Immunol. 2010 Mar 1;184(5):2355-68.

[121] Murphy PM, Baggiolini M, Charo IF, Hebert CA, Horuk R, Matsushima K, et al. International union of pharmacology. XXII. Nomenclature for chemokine receptors. Pharmacol Rev. 2000 Mar;52(1):145-76.

[122] Mukaida N, Harada A, Matsushima K. Interleukin-8 (IL-8) and monocyte chemotactic and activating factor (MCAF/MCP-1), chemokines essentially involved in inflammatory and immune reactions. Cytokine Growth Factor Rev. 1998 Mar;9(1):9-23.

[123] Hebert CA, Vitangcol RV, Baker JB. Scanning mutagenesis of interleukin-8 identifies a cluster of residues required for receptor binding. J Biol Chem. 1991 Oct 5;266(28): 18989-94.

[124] Clark-Lewis I, Schumacher C, Baggiolini M, Moser B. Structure-activity relationships of interleukin-8 determined using chemically synthesized analogs. Critical role of NH2-terminal residues and evidence for uncoupling of neutrophil chemotaxis, exocytosis, and receptor binding activities. J Biol Chem. 1991 Dec 5;266(34):23128-34.

[125] Laing KJ, Zou JJ, Wang T, Bols N, Hirono I, Aoki T, et al. Identification and analysis of an interleukin 8-like molecule in rainbow trout Oncorhynchus mykiss. Dev Comp Immunol. 2002 Jun;26(5):433-44.

[126] Lee EY, Park HH, Kim YT, Choi TJ. Cloning and sequence analysis of the interleukin-8 gene from flounder (Paralichthys olivaceous). Gene. 2001 Aug 22;274(1-2): 237-43.

[127] Sangrador-Vegas A, Lennington JB, Smith TJ. Molecular cloning of an IL-8-like CXC chemokine and tissue factor in rainbow trout (Oncorhynchus mykiss) by use of suppression subtractive hybridization. Cytokine. 2002 Jan 21;17(2):66-70.

[128] Chen L, He C, Baoprasertkul P, Xu P, Li P, Serapion J, et al. Analysis of a catfish gene resembling interleukin-8: cDNA cloning, gene structure, and expression after infection with Edwardsiella ictaluri. Dev Comp Immunol. 2005;29(2):135-42.

[129] Najakshin AM, Mechetina LV, Alabyev BY, Taranin AV. Identification of an IL-8 homolog in lamprey (Lampetra fluviatilis): early evolutionary divergence of chemokines. Eur J Immunol. 1999 Feb;29(2):375-82.

[130] Laing KJ, Secombes CJ. Chemokines. Dev Comp Immunol. 2004 May 3;28(5):443-60.

[131] Osborne LC, Abraham N. Regulation of memory T cells by gammac cytokines. Cytokine. 2010 May;50(2):105-13.

[132] Sogo T, Kawahara M, Ueda H, Otsu M, Onodera M, Nakauchi H, et al. T cell growth control using hapten-specific antibody/interleukin-2 receptor chimera. Cytokine. 2009 Apr;46(1):127-36.

[133] Morgan DA, Ruscetti FW, Gallo R. Selective in vitro growth of T lymphocytes from normal human bone marrows. Science. 1976 Sep 10;193(4257):1007-8.

[134] Smith KA. T-cell growth factor. Immunol Rev. 1980;51:337-57.

[135] Smith KA. Interleukin-2: inception, impact, and implications. Science. 1988 May 27;240(4856):1169-76.

[136] Granucci F, Vizzardelli C, Pavelka N, Feau S, Persico M, Virzi E, et al. Inducible IL-2 production by dendritic cells revealed by global gene expression analysis. Nat Immunol. 2001 Sep;2(9):882-8.

[137] Walker E, Leemhuis T, Roeder W. Murine B lymphoma cell lines release functionally active interleukin 2 after stimulation with Staphylococcus aureus. J Immunol. 1988 Feb 1;140(3):859-65.

[138] Gaffen SL, Wang S, Koshland ME. Expression of the immunoglobulin J chain in a murine B lymphoma is driven by autocrine production of interleukin 2. Cytokine. 1996 Jul;8(7):513-24.

[139] Yu TK, Caudell EG, Smid C, Grimm EA. IL-2 activation of NK cells: involvement of MKK1/2/ERK but not p38 kinase pathway. J Immunol. 2000 Jun 15;164(12):6244-51.

[140] Gold MR, DeFranco AL. Biochemistry of B lymphocyte activation. Adv Immunol. 1994;55:221-95.

[141] Bird S, Zou J, Kono T, Sakai M, Dijkstra JM, Secombes C. Characterisation and expression analysis of interleukin 2 (IL-2) and IL-21 homologues in the Japanese pufferfish, Fugu rubripes, following their discovery by synteny. Immunogenetics. 2005 Mar;56(12):909-23.

[142] Sugamata R, Suetake H, Kikuchi K, Suzuki Y. Teleost B7 expressed on monocytes regulates T cell responses. J Immunol. 2009 Jun 1;182(11):6799-806.

[143] Diaz-Rosales P, Bird S, Wang TH, Fujiki K, Davidson WS, Zou J, et al. Rainbow trout interleukin-2: cloning, expression and bioactivity analysis. Fish Shellfish Immunol. 2009 Sep;27(3):414-22.

[144] Zhang YA, Hikima J, Li J, LaPatra SE, Luo YP, Sunyer JO. Conservation of structural and functional features in a primordial CD80/86 molecule from rainbow trout (Oncorhynchus mykiss), a primitive teleost fish. J Immunol. 2009 Jul 1;183(1):83-96.

[145] Laing KJ, Bols N, Secombes CJ. A CXC chemokine sequence isolated from the rainbow trout Oncorhynchus mykiss resembles the closely related interferon-gamma-inducible chemokines CXCL9, CXCL10 and CXCL11. Eur Cytokine Netw. 2002 Oct-Dec;13(4):462-73.

[146] Banchereau J, Bidaud C, Fluckiger AC, Galibert L, Garrone P, Malisan F, et al. Effects of interleukin 4 on human B-cell growth and differentiation. Res Immunol. 1993 Oct; 144(8):601-5.

[147] Mosmann TR, Cherwinski H, Bond MW, Giedlin MA, Coffman RL. Two types of murine helper T cell clone. I. Definition according to profiles of lymphokine activities and secreted proteins. J Immunol. 1986 Apr 1;136(7):2348-57.

[148] Finkelman FD, Urban JF, Jr. The other side of the coin: the protective role of the TH2 cytokines. J Allergy Clin Immunol. 2001 May;107(5):772-80.

[149] Yokota T, Otsuka T, Mosmann T, Banchereau J, DeFrance T, Blanchard D, et al. Isolation and characterization of a human interleukin cDNA clone, homologous to mouse B-cell stimulatory factor 1, that expresses B-cell- and T-cell-stimulating activities. Proc Natl Acad Sci U S A. 1986 Aug;83(16):5894-8.

[150] Noma Y, Sideras P, Naito T, Bergstedt-Lindquist S, Azuma C, Severinson E, et al. Cloning of cDNA encoding the murine IgG1 induction factor by a novel strategy using SP6 promoter. Nature. 1986 Feb 20-26;319(6055):640-6.

[151] Lee F, Yokota T, Otsuka T, Meyerson P, Villaret D, Coffman R, et al. Isolation and characterization of a mouse interleukin cDNA clone that expresses B-cell stimulatory factor 1 activities and T-cell- and mast-cell-stimulating activities. Proc Natl Acad Sci U S A. 1986 Apr;83(7):2061-5.

[152] Heussler VT, Eichhorn M, Dobbelaere DA. Cloning of a full-length cDNA encoding bovine interleukin 4 by the polymerase chain reaction. Gene. 1992 May 15;114(2): 273-8.

[153] Li-Weber M, Krammer PH. Regulation of IL4 gene expression by T cells and therapeutic perspectives. Nat Rev Immunol. 2003 Jul;3(7):534-43.

[154] Ansel KM, Djuretic I, Tanasa B, Rao A. Regulation of Th2 differentiation and Il4 locus accessibility. Annu Rev Immunol. 2006;24:607-56.

[155] Ohtani M, Hayashi N, Hashimoto K, Nakanishi T, Dijkstra JM. Comprehensive clarification of two paralogous interleukin 4/13 loci in teleost fish. Immunogenetics. 2008 Jul;60(7):383-97.

[156] Jaillon O, Aury JM, Brunet F, Petit JL, Stange-Thomann N, Mauceli E, et al. Genome duplication in the teleost fish Tetraodon nigroviridis reveals the early vertebrate proto-karyotype. Nature. 2004 Oct 21;431(7011):946-57.

[157] Li JH, Shao JZ, Xiang LX, Wen Y. Cloning, characterization and expression analysis of pufferfish interleukin-4 cDNA: the first evidence of Th2-type cytokine in fish. Mol Immunol. 2007 Mar;44(8):2078-86.

[158] Secrist H, Egan M, Peters MG. Tissue-specific regulation of IL-4 mRNA expression in human tonsil. J Immunol. 1994 Feb 1;152(3):1120-6.

[159] Lin AF, Xiang LX, Wang QL, Dong WR, Gong YF, Shao JZ. The DC-SIGN of zebrafish: insights into the existence of a CD209 homologue in a lower vertebrate and its involvement in adaptive immunity. J Immunol. 2009 Dec 1;183(11):7398-410.

[160] Hu YL, Xiang LX, Shao JZ. Identification and characterization of a novel immunoglobulin Z isotype in zebrafish: implications for a distinct B cell receptor in lower vertebrates. Mol Immunol. 2010 Jan;47(4):738-46.

[161] Mitra S, Alnabulsi A, Secombes CJ, Bird S. Identification and characterization of the transcription factors involved in T-cell development, t-bet, stat6 and foxp3, within the zebrafish, Danio rerio. FEBS J. 2010 Jan;277(1):128-47.

[162] Takizawa F, Koppang EO, Ohtani M, Nakanishi T, Hashimoto K, Fischer U, et al. Constitutive high expression of interleukin-4/13A and GATA-3 in gill and skin of salmonid fishes suggests that these tissues form Th2-skewed immune environments. Mol Immunol. 2011 Jul;48(12-13):1360-8.

[163] Fry TJ, Mackall CL. The many faces of IL-7: from lymphopoiesis to peripheral T cell maintenance. J Immunol. 2005 Jun 1;174(11):6571-6.

[164] Alves NL, Richard-Le Goff O, Huntington ND, Sousa AP, Ribeiro VS, Bordack A, et al. Characterization of the thymic IL-7 niche in vivo. Proc Natl Acad Sci U S A. 2009 Feb 3;106(5):1512-7.

[165] Mazzucchelli RI, Warming S, Lawrence SM, Ishii M, Abshari M, Washington AV, et al. Visualization and identification of IL-7 producing cells in reporter mice. PLoS One. 2009;4(11):e7637.

[166] Repass JF, Laurent MN, Carter C, Reizis B, Bedford MT, Cardenas K, et al. IL7-hCD25 and IL7-Cre BAC transgenic mouse lines: new tools for analysis of IL-7 expressing cells. Genesis. 2009 Apr;47(4):281-7.

[167] Kono T, Bird S, Sonoda K, Savan R, Secombes CJ, Sakai M. Characterization and expression analysis of an interleukin-7 homologue in the Japanese pufferfish, Takifugu rubripes. FEBS J. 2008 Mar;275(6):1213-26.

[168] Bei JX, Suetake H, Araki K, Kikuchi K, Yoshiura Y, Lin HR, et al. Two interleukin (IL)-15 homologues in fish from two distinct origins. Mol Immunol. 2006 Mar;43(7): 860-9.

[169] Fang W, Xiang LX, Shao JZ, Wen Y, Chen SY. Identification and characterization of an interleukin-15 homologue from Tetraodon nigroviridis. Comp Biochem Physiol B Biochem Mol Biol. 2006 Mar;143(3):335-43.

[170] Gunimaladevi I, Savan R, Sato K, Yamaguchi R, Sakai M. Characterization of an interleukin-15 like (IL-15L) gene from zebrafish (Danio rerio). Fish Shellfish Immunol. 2007 Apr;22(4):351-62.

[171] Wang T, Holland JW, Carrington A, Zou J, Secombes CJ. Molecular and functional characterization of IL-15 in rainbow trout Oncorhynchus mykiss: a potent inducer of IFN-gamma expression in spleen leukocytes. J Immunol. 2007 Aug 1;179(3):1475-88.

[172] Parrish-Novak J, Dillon SR, Nelson A, Hammond A, Sprecher C, Gross JA, et al. Interleukin 21 and its receptor are involved in NK cell expansion and regulation of lymphocyte function. Nature. 2000 Nov 2;408(6808):57-63.

[173] Asao H, Okuyama C, Kumaki S, Ishii N, Tsuchiya S, Foster D, et al. Cutting edge: the common gamma-chain is an indispensable subunit of the IL-21 receptor complex. J Immunol. 2001 Jul 1;167(1):1-5.

[174] Vosshenrich CA, Di Santo JP. Cytokines: IL-21 joins the gamma(c)-dependent network? Curr Biol. 2001 Mar 6;11(5):R175-7.

[175] Chtanova T, Tangye SG, Newton R, Frank N, Hodge MR, Rolph MS, et al. T follicular helper cells express a distinctive transcriptional profile, reflecting their role as non-Th1/Th2 effector cells that provide help for B cells. J Immunol. 2004 Jul 1;173(1):68-78.

[176] Wurster AL, Rodgers VL, Satoskar AR, Whitters MJ, Young DA, Collins M, et al. Interleukin 21 is a T helper (Th) cell 2 cytokine that specifically inhibits the differentiation of naive Th cells into interferon gamma-producing Th1 cells. J Exp Med. 2002 Oct 7;196(7):969-77.

[177] Leonard WJ, Zeng R, Spolski R. Interleukin 21: a cytokine/cytokine receptor system that has come of age. J Leukoc Biol. 2008 Aug;84(2):348-56.

[178] Suto A, Kashiwakuma D, Kagami S, Hirose K, Watanabe N, Yokote K, et al. Development and characterization of IL-21-producing CD4+ T cells. J Exp Med. 2008 Jun 9;205(6):1369-79.

[179] Frederiksen KS, Lundsgaard D, Freeman JA, Hughes SD, Holm TL, Skrumsager BK, et al. IL-21 induces in vivo immune activation of NK cells and CD8(+) T cells in patients with metastatic melanoma and renal cell carcinoma. Cancer Immunol Immunother. 2008 Oct;57(10):1439-49.

[180] Konforte D, Simard N, Paige CJ. IL-21: an executor of B cell fate. J Immunol. 2009 Feb 15;182(4):1781-7.

[181] Wang HJ, Xiang LX, Shao JZ, Jia S. Molecular cloning, characterization and expression analysis of an IL-21 homologue from Tetraodon nigroviridis. Cytokine. 2006 Aug;35(3-4):126-34.

[182] Wang T, Diaz-Rosales P, Costa MM, Campbell S, Snow M, Collet B, et al. Functional characterization of a nonmammalian IL-21: rainbow trout Oncorhynchus mykiss IL-21 upregulates the expression of the Th cell signature cytokines IFN-gamma, IL-10, and IL-22. J Immunol. 2011 Jan 15;186(2):708-21.

[183] Lutfalla G, Roest Crollius H, Stange-Thomann N, Jaillon O, Mogensen K, Monneron D. Comparative genomic analysis reveals independent expansion of a lineage-specific gene family in vertebrates: the class II cytokine receptors and their ligands in mammals and fish. BMC Genomics. 2003 Jul 17;4(1):29.

[184] Fickenscher H, Hor S, Kupers H, Knappe A, Wittmann S, Sticht H. The interleukin-10 family of cytokines. Trends Immunol. 2002 Feb;23(2):89-96.

[185] Moore KW, de Waal Malefyt R, Coffman RL, O'Garra A. Interleukin-10 and the interleukin-10 receptor. Annu Rev Immunol. 2001;19:683-765.

[186] Zou J, Clark MS, Secombes CJ. Characterisation, expression and promoter analysis of an interleukin 10 homologue in the puffer fish, Fugu rubripes. Immunogenetics. 2003 Aug;55(5):325-35.

[187] Savan R, Igawa D, Sakai M. Cloning, characterization and expression analysis of interleukin-10 from the common carp, Cyprinus carpio L. Eur J Biochem. 2003 Dec; 270(23):4647-54.

[188] Zhang DC, Shao YQ, Huang YQ, Jiang SG. Cloning, characterization and expression analysis of interleukin-10 from the zebrafish (Danio rerion). J Biochem Mol Biol. 2005 Sep 30;38(5):571-6.

[189] Inoue Y, Kamota S, Ito K, Yoshiura Y, Ototake M, Moritomo T, et al. Molecular cloning and expression analysis of rainbow trout (Oncorhynchus mykiss) interleukin-10 cDNAs. Fish Shellfish Immunol. 2005 Apr;18(4):335-44.

[190] Buonocore F, Randelli E, Bird S, Secombes C, Facchiano A, Constantini S, et al. Interleukin-10 expression by real-time PCR and homology modelling analysis in the European sea bass (Dicentrarchus Labrax L.). Aquaculture. 2007;270(1-4):512 - 22.

[191] Pinto RD, Nascimento DS, Reis MI, do Vale A, Dos Santos NM. Molecular characterization, 3D modelling and expression analysis of sea bass (Dicentrarchus labrax L.) interleukin-10. Mol Immunol. 2007 Mar;44(8):2056-65.

[192] Zhang Z, Swain T, Bogwald J, Dalmo RA, Kumari J. Bath immunostimulation of rainbow trout (Oncorhynchus mykiss) fry induces enhancement of inflammatory cytokine transcripts, while repeated bath induce no changes. Fish Shellfish Immunol. 2009 May;26(5):677-84.

[193] Wegenka UM. IL-20: biological functions mediated through two types of receptor complexes. Cytokine Growth Factor Rev. 2010 Oct;21(5):353-63.

[194] Stein C, Caccamo M, Laird G, Leptin M. Conservation and divergence of gene families encoding components of innate immune response systems in zebrafish. Genome Biol. 2007;8(11):R251.

[195] Wang T, Diaz-Rosales P, Martin SA, Secombes CJ. Cloning of a novel interleukin (IL)-20-like gene in rainbow trout Oncorhynchus mykiss gives an insight into the evolution of the IL-10 family. Dev Comp Immunol. 2010 Feb;34(2):158-67.

[196] Liang SC, Tan XY, Luxenberg DP, Karim R, Dunussi-Joannopoulos K, Collins M, et al. Interleukin (IL)-22 and IL-17 are coexpressed by Th17 cells and cooperatively enhance expression of antimicrobial peptides. J Exp Med. 2006 Oct 2;203(10):2271-9.

[197] Cupedo T, Crellin NK, Papazian N, Rombouts EJ, Weijer K, Grogan JL, et al. Human fetal lymphoid tissue-inducer cells are interleukin 17-producing precursors to RORC + CD127+ natural killer-like cells. Nat Immunol. 2009 Jan;10(1):66-74.

[198] Duhen T, Geiger R, Jarrossay D, Lanzavecchia A, Sallusto F. Production of interleukin 22 but not interleukin 17 by a subset of human skin-homing memory T cells. Nat Immunol. 2009 Aug;10(8):857-63.

[199] Eyerich S, Eyerich K, Pennino D, Carbone T, Nasorri F, Pallotta S, et al. Th22 cells represent a distinct human T cell subset involved in epidermal immunity and remodeling. J Clin Invest. 2009 Dec;119(12):3573-85.

[200] Fujita H, Nograles KE, Kikuchi T, Gonzalez J, Carucci JA, Krueger JG. Human Langerhans cells induce distinct IL-22-producing CD4+ T cells lacking IL-17 production. Proc Natl Acad Sci U S A. 2009 Dec 22;106(51):21795-800.

[201] Wolk K, Haugen HS, Xu W, Witte E, Waggie K, Anderson M, et al. IL-22 and IL-20 are key mediators of the epidermal alterations in psoriasis while IL-17 and IFN-gamma are not. J Mol Med (Berl). 2009 May;87(5):523-36.

[202] Feinen B, Russell MW. Contrasting Roles of IL-22 and IL-17 in Murine Genital Tract Infection by Neisseria gonorrhoeae. Front Immunol. 2012;3:11.

[203] Wolk K, Kunz S, Witte E, Friedrich M, Asadullah K, Sabat R. IL-22 increases the innate immunity of tissues. Immunity. 2004 Aug;21(2):241-54.

[204] Aujla SJ, Chan YR, Zheng M, Fei M, Askew DJ, Pociask DA, et al. IL-22 mediates mucosal host defense against Gram-negative bacterial pneumonia. Nat Med. 2008 Mar; 14(3):275-81.

[205] Andoh A, Zhang Z, Inatomi O, Fujino S, Deguchi Y, Araki Y, et al. Interleukin-22, a member of the IL-10 subfamily, induces inflammatory responses in colonic subepithelial myofibroblasts. Gastroenterology. 2005 Sep;129(3):969-84.

[206] Zheng Y, Valdez PA, Danilenko DM, Hu Y, Sa SM, Gong Q, et al. Interleukin-22 mediates early host defense against attaching and effacing bacterial pathogens. Nat Med. 2008 Mar;14(3):282-9.

[207] Bird L. Mucosal immunology: IL-22 keeps commensals in their place. Nat Rev Immunol. 2012 Jul 6.

[208] Wolk K, Kunz S, Asadullah K, Sabat R. Cutting edge: immune cells as sources and targets of the IL-10 family members? J Immunol. 2002 Jun 1;168(11):5397-402.

[209] Wilson NJ, Boniface K, Chan JR, McKenzie BS, Blumenschein WM, Mattson JD, et al. Development, cytokine profile and function of human interleukin 17-producing helper T cells. Nat Immunol. 2007 Sep;8(9):950-7.

[210] Manel N, Unutmaz D, Littman DR. The differentiation of human T(H)-17 cells requires transforming growth factor-beta and induction of the nuclear receptor RORgammat. Nat Immunol. 2008 Jun;9(6):641-9.

[211] Pene J, Chevalier S, Preisser L, Venereau E, Guilleux MH, Ghannam S, et al. Chronically inflamed human tissues are infiltrated by highly differentiated Th17 lymphocytes. J Immunol. 2008 Jun 1;180(11):7423-30.

[212] Cella M, Fuchs A, Vermi W, Facchetti F, Otero K, Lennerz JK, et al. A human natural killer cell subset provides an innate source of IL-22 for mucosal immunity. Nature. 2009 Feb 5;457(7230):722-5.

[213] Hughes T, Becknell B, McClory S, Briercheck E, Freud AG, Zhang X, et al. Stage 3 immature human natural killer cells found in secondary lymphoid tissue constitutively and selectively express the TH 17 cytokine interleukin-22. Blood. 2009 Apr 23;113(17):4008-10.

[214] Donnelly RP, Sheikh F, Dickensheets H, Savan R, Young HA, Walter MR. Interleukin-26: an IL-10-related cytokine produced by Th17 cells. Cytokine Growth Factor Rev. 2010 Oct;21(5):393-401.

[215] Zou J, Yoshiura Y, Dijkstra JM, Sakai M, Ototake M, Secombes C. Identification of an interferon gamma homologue in Fugu, Takifugu rubripes. Fish Shellfish Immunol. 2004 Oct;17(4):403-9.

[216] Igawa D, Sakai M, Savan R. An unexpected discovery of two interferon gamma-like genes along with interleukin (IL)-22 and -26 from teleost: IL-22 and -26 genes have been described for the first time outside mammals. Mol Immunol. 2006 Mar;43(7): 999-1009.

[217] Corripio-Miyar Y, Zou J, Richmond H, Secombes CJ. Identification of interleukin-22 in gadoids and examination of its expression level in vaccinated fish. Mol Immunol. 2009 Jun;46(10):2098-106.

[218] Kolls JK. Th17 cells in mucosal immunity and tissue inflammation. Semin Immunopathol. 2010 Mar;32(1):1-2.

[219] Haugarvoll E, Bjerkas I, Nowak BF, Hordvik I, Koppang EO. Identification and characterization of a novel intraepithelial lymphoid tissue in the gills of Atlantic salmon. J Anat. 2008 Aug;213(2):202-9.

[220] Moseley TA, Haudenschild DR, Rose L, Reddi AH. Interleukin-17 family and IL-17 receptors. Cytokine Growth Factor Rev. 2003 Apr;14(2):155-74.

[221] Yao Z, Timour M, Painter S, Fanslow W, Spriggs M. Complete nucleotide sequence of the mouse CTLA8 gene. Gene. 1996 Feb 12;168(2):223-5.

[222] Li H, Chen J, Huang A, Stinson J, Heldens S, Foster J, et al. Cloning and characterization of IL-17B and IL-17C, two new members of the IL-17 cytokine family. Proc Natl Acad Sci U S A. 2000 Jan 18;97(2):773-8.

[223] Shi Y, Ullrich SJ, Zhang J, Connolly K, Grzegorzewski KJ, Barber MC, et al. A novel cytokine receptor-ligand pair. Identification, molecular characterization, and in vivo immunomodulatory activity. J Biol Chem. 2000 Jun 23;275(25):19167-76.

[224] Hymowitz SG, Filvaroff EH, Yin JP, Lee J, Cai L, Risser P, et al. IL-17s adopt a cystine knot fold: structure and activity of a novel cytokine, IL-17F, and implications for receptor binding. EMBO J. 2001 Oct 1;20(19):5332-41.

[225] Lee J, Ho WH, Maruoka M, Corpuz RT, Baldwin DT, Foster JS, et al. IL-17E, a novel proinflammatory ligand for the IL-17 receptor homolog IL-17Rh1. J Biol Chem. 2001 Jan 12;276(2):1660-4.

[226] Starnes T, Robertson MJ, Sledge G, Kelich S, Nakshatri H, Broxmeyer HE, et al. Cutting edge: IL-17F, a novel cytokine selectively expressed in activated T cells and monocytes, regulates angiogenesis and endothelial cell cytokine production. J Immunol. 2001 Oct 15;167(8):4137-40.

[227] Starnes T, Broxmeyer HE, Robertson MJ, Hromas R. Cutting edge: IL-17D, a novel member of the IL-17 family, stimulates cytokine production and inhibits hemopoiesis. J Immunol. 2002 Jul 15;169(2):642-6.

[228] Gunimaladevi I, Savan R, Sakai M. Identification, cloning and characterization of interleukin-17 and its family from zebrafish. Fish Shellfish Immunol. 2006 Oct;21(4): 393-403.

[229] Wang T, Martin SA, Secombes CJ. Two interleukin-17C-like genes exist in rainbow trout Oncorhynchus mykiss that are differentially expressed and modulated. Dev Comp Immunol. 2010 May;34(5):491-500.

[230] Kumari J, Larsen AN, Bogwald J, Dalmo RA. Interleukin-17D in Atlantic salmon (Salmo salar): molecular characterization, 3D modelling and promoter analysis. Fish Shellfish Immunol. 2009 Nov;27(5):647-59.

[231] Korenaga H, Kono T, Sakai M. Isolation of seven IL-17 family genes from the Japanese pufferfish Takifugu rubripes. Fish Shellfish Immunol. 2010 May-Jun;28(5-6): 809-18.

[232] Kono T, Korenaga H, Sakai M. Genomics of fish IL-17 ligand and receptors: a review. Fish Shellfish Immunol. 2011 Nov;31(5):635-43.

[233] Huang D, Cancilla MR, Morahan G. Complete primary structure, chromosomal localisation, and definition of polymorphisms of the gene encoding the human interleukin-12 p40 subunit. Genes Immun. 2000 Dec;1(8):515-20.

[234] Yoshiura Y, Kiryu I, Fujiwara A, Suetake H, Suzuki Y, Nakanishi T, et al. Identification and characterization of Fugu orthologues of mammalian interleukin-12 subunits. Immunogenetics. 2003 Aug;55(5):296-306.

[235] Nascimento DS, do Vale A, Tomas AM, Zou J, Secombes CJ, dos Santos NM. Cloning, promoter analysis and expression in response to bacterial exposure of sea bass (Dicentrarchus labrax L.) interleukin-12 p40 and p35 subunits. Mol Immunol. 2007 Mar;44(9):2277-91.

[236] Li MO, Flavell RA. TGF-beta: a master of all T cell trades. Cell. 2008 Aug 8;134(3): 392-404.

[237] Li MO, Wan YY, Sanjabi S, Robertson AK, Flavell RA. Transforming growth factor-beta regulation of immune responses. Annu Rev Immunol. 2006;24:99-146.

[238] Saxena V, Lienesch DW, Zhou M, Bommireddy R, Azhar M, Doetschman T, et al. Dual roles of immunoregulatory cytokine TGF-beta in the pathogenesis of autoimmunity-mediated organ damage. J Immunol. 2008 Feb 1;180(3):1903-12.

[239] Zhang L, Yi H, Xia XP, Zhao Y. Transforming growth factor-beta: an important role in CD4+CD25+ regulatory T cells and immune tolerance. Autoimmunity. 2006 Jun; 39(4):269-76.

[240] Li B, Samanta A, Song X, Furuuchi K, Iacono KT, Kennedy S, et al. FOXP3 ensembles in T-cell regulation. Immunol Rev. 2006 Aug;212:99-113.

[241] Wan YY, Flavell RA. 'Yin-Yang' functions of transforming growth factor-beta and T regulatory cells in immune regulation. Immunol Rev. 2007 Dec;220:199-213.

[242] Haddad G, Hanington PC, Wilson EC, Grayfer L, Belosevic M. Molecular and functional characterization of goldfish (Carassius auratus L.) transforming growth factor beta. Dev Comp Immunol. 2008;32(6):654-63.

[243] Kadowaki T, Yasui Y, Takahashi Y, Kohchi C, Soma G, Inagawa H. Comparative immunological analysis of innate immunity activation after oral administration of wheat fermented extract to teleost fish. Anticancer Res. 2009 Nov;29(11):4871-7.

[244] Yang M, Wang Y, Wang X, Chen C, Zhou H. Characterization of grass carp (Ctenopharyngodon idellus) Foxp1a/1b/2: evidence for their involvement in the activation of peripheral blood lymphocyte subpopulations. Fish Shellfish Immunol. 2010 Feb; 28(2):289-95.

[245] Cai Z, Gao C, Li L, Xing K. Bipolar properties of red seabream (Pagrus major) transforming growth factor-beta in induction of the leucocytes migration. Fish Shellfish Immunol. 2010 Apr;28(4):695-700.

[246] Yang M, Wang X, Chen D, Wang Y, Zhang A, Zhou H. TGF-beta1 exerts opposing effects on grass carp leukocytes: implication in teleost immunity, receptor signaling and potential self-regulatory mechanisms. PLoS One. 2012;7(4):e35011.

[247] Samuel CE. Antiviral actions of interferons. Clin Microbiol Rev. 2001 Oct;14(4): 778-809, table of contents.

[248] Robertsen B. The interferon system of teleost fish. Fish Shellfish Immunol. 2006 Feb; 20(2):172-91.

[249] Bergan V, Steinsvik S, Xu H, Kileng O, Robertsen B. Promoters of type I interferon genes from Atlantic salmon contain two main regulatory regions. FEBS J. 2006 Sep; 273(17):3893-906.

[250] Kileng O, Brundtland MI, Robertsen B. Infectious salmon anemia virus is a powerful inducer of key genes of the type I interferon system of Atlantic salmon, but is not inhibited by interferon. Fish Shellfish Immunol. 2007 Aug;23(2):378-89.

[251] Robertsen B, Bergan V, Rokenes T, Larsen R, Albuquerque A. Atlantic salmon interferon genes: cloning, sequence analysis, expression, and biological activity. J Interferon Cytokine Res. 2003 Oct;23(10):601-12.

[252] Rokenes TP, Larsen R, Robertsen B. Atlantic salmon ISG15: Expression and conjugation to cellular proteins in response to interferon, double-stranded RNA and virus infections. Mol Immunol. 2007 Feb;44(5):950-9.

[253] Zou J, Tafalla C, Truckle J, Secombes CJ. Identification of a second group of type I IFNs in fish sheds light on IFN evolution in vertebrates. J Immunol. 2007 Sep 15;179(6):3859-71.

[254] Robertsen B. Expression of interferon and interferon-induced genes in salmonids in response to virus infection, interferon-inducing compounds and vaccination. Fish Shellfish Immunol. 2008 Oct;25(4):351-7.

[255] Milev-Milovanovic I, Long S, Wilson M, Bengten E, Miller NW, Chinchar VG. Identification and expression analysis of interferon gamma genes in channel catfish. Immunogenetics. 2006 Feb;58(1):70-80.

[256] Stolte EH, Savelkoul HF, Wiegertjes G, Flik G, Lidy Verburg-van Kemenade BM. Differential expression of two interferon-gamma genes in common carp (Cyprinus carpio L.). Dev Comp Immunol. 2008;32(12):1467-81.

[257] Zou J, Carrington A, Collet B, Dijkstra JM, Yoshiura Y, Bols N, et al. Identification and bioactivities of IFN-gamma in rainbow trout Oncorhynchus mykiss: the first Th1-type cytokine characterized functionally in fish. J Immunol. 2005 Aug 15;175(4): 2484-94.

[258] Grayfer L, Garcia EG, Belosevic M. Comparison of macrophage antimicrobial responses induced by type II interferons of the goldfish (Carassius auratus L.). J Biol Chem. 2010 Jul 30;285(31):23537-47.

[259] Arts JA, Tijhaar EJ, Chadzinska M, Savelkoul HF, Verburg-van Kemenade BM. Functional analysis of carp interferon-gamma: evolutionary conservation of classical phagocyte activation. Fish Shellfish Immunol. 2010 Nov;29(5):793-802.

[260] Sun B, Skjaeveland I, Svingerud T, Zou J, Jorgensen J, Robertsen B. Antiviral activity of salmonid gamma interferon against infectious pancreatic necrosis virus and salmonid alphavirus and its dependency on type I interferon. J Virol. 2011 Sep;85(17): 9188-98.

[261] Duguid JR, Dinauer MC. Library subtraction of in vitro cDNA libraries to identify differentially expressed genes in scrapie infection. Nucleic Acids Res. 1990 May 11;18(9):2789-92.

[262] Hara E, Kato T, Nakada S, Sekiya S, Oda K. Subtractive cDNA cloning using oligo(dT)30-latex and PCR: isolation of cDNA clones specific to undifferentiated human embryonal carcinoma cells. Nucleic Acids Res. 1991 Dec;19(25):7097-104.

[263] Hedrick SM, Cohen DI, Nielsen EA, Davis MM. Isolation of cDNA clones encoding T cell-specific membrane-associated proteins. Nature. 1984 Mar 8-14;308(5955):149-53.

[264] Lisitsyn N, Wigler M. Cloning the differences between two complex genomes. Science. 1993 Feb 12;259(5097):946-51.

[265] Hubank M, Schatz DG. Identifying differences in mRNA expression by representational difference analysis of cDNA. Nucleic Acids Res. 1994 Dec 25;22(25):5640-8.

[266] Liang P, Pardee AB. Differential display of eukaryotic messenger RNA by means of the polymerase chain reaction. Science. 1992 Aug 14;257(5072):967-71.

[267] Welsh J, Chada K, Dalal SS, Cheng R, Ralph D, McClelland M. Arbitrarily primed PCR fingerprinting of RNA. Nucleic Acids Res. 1992 Oct 11;20(19):4965-70.

[268] Sompayrac L, Jane S, Burn TC, Tenen DG, Danna KJ. Overcoming limitations of the mRNA differential display technique. Nucleic Acids Res. 1995 Nov 25;23(22):4738-9.

[269] Bertioli DJ, Schlichter UH, Adams MJ, Burrows PR, Steinbiss HH, Antoniw JF. An analysis of differential display shows a strong bias towards high copy number mRNAs. Nucleic Acids Res. 1995 Nov 11;23(21):4520-3.

[270] Siebert PD, Chenchik A, Kellogg DE, Lukyanov KA, Lukyanov SA. An improved PCR method for walking in uncloned genomic DNA. Nucleic Acids Res. 1995 Mar 25;23(6):1087-8.

[271] Diatchenko L, Lau YF, Campbell AP, Chenchik A, Moqadam F, Huang B, et al. Suppression subtractive hybridization: a method for generating differentially regulated or tissue-specific cDNA probes and libraries. Proc Natl Acad Sci U S A. 1996 Jun 11;93(12):6025-30.

[272] Wang L, Wu X. Identification of differentially expressed genes in lipopolysaccharide-stimulated yellow grouper Epinephelus awoara spleen. Fish Shellfish Immunol. 2007 Aug;23(2):354-63.

[273] Xia JH, Yue GH. Identification and analysis of immune-related transcriptome in Asian seabass Lates calcarifer. BMC Genomics. 2010;11:356.

[274] Feng CY, Johnson SC, Hori TS, Rise M, Hall JR, Gamperl AK, et al. Identification and analysis of differentially expressed genes in immune tissues of Atlantic cod stimulated with formalin-killed, atypical Aeromonas salmonicida. Physiol Genomics. 2009 May 13;37(3):149-63.

[275] Tsoi SC, Ewart KV, Penny S, Melville K, Liebscher RS, Brown LL, et al. Identification of immune-relevant genes from atlantic salmon using suppression subtractive hybridization. Mar Biotechnol (NY). 2004 May-Jun;6(3):199-214.

[276] Li CH, Chen J, Shi YH, Lu XJ. Use of suppressive subtractive hybridization to identify differentially expressed genes in ayu (Plecoglossus altivelis) associated with Listonella anguillarum infection. Fish Shellfish Immunol. 2011 Sep;31(3):500-6.

[277] Matsuyama T, Fujiwara A, Takano T, Nakayasu C. Suppression subtractive hybridization coupled with microarray analysis to examine differential expression of genes in Japanese flounder Paralichthys olivaceus leucocytes during Edwardsiella tarda and viral hemorrhagic septicemia virus infection. Fish Shellfish Immunol. 2011 Oct; 31(4):524-32.

[278] Bo J, Giesy JP, Ye R, Wang KJ, Lee JS, Au DW. Identification of differentially expressed genes and quantitative expression of complement genes in the liver of marine medaka Oryzias melastigma challenged with Vibrio parahaemolyticus. Comp Biochem Physiol Part D Genomics Proteomics. 2012 Jun;7(2):191-200.

[279] Purcell MK, Smith KD, Hood L, Winton JR, Roach JC. Conservation of Toll-Like Receptor Signaling Pathways in Teleost Fish. Comp Biochem Physiol Part D Genomics Proteomics. 2006 Mar;1(1):77-88.

[280] Trinchieri G, Sher A. Cooperation of Toll-like receptor signals in innate immune defence. Nat Rev Immunol. 2007 Mar;7(3):179-90.

[281] Bayne CJ, Gerwick L. The acute phase response and innate immunity of fish. Dev Comp Immunol. 2001 Oct-Dec;25(8-9):725-43.

[282] Bayne CJ, Gerwick L, Fujiki K, Nakao M, Yano T. Immune-relevant (including acute phase) genes identified in the livers of rainbow trout, Oncorhynchus mykiss, by means of suppression subtractive hybridization. Dev Comp Immunol. 2001 Apr; 25(3):205-17.

[283] Holland MC, Lambris JD. The complement system in teleosts. Fish Shellfish Immunol. 2002 May;12(5):399-420.

[284] Rise ML, Hall J, Rise M, Hori T, Gamperl AK, Kimball J, et al. Functional genomic analysis of the response of Atlantic cod (Gadus morhua) spleen to the viral mimic polyriboinosinic polyribocytidylic acid (pIC). Dev Comp Immunol. 2008;32(8):916-31.

[285] He W, Yinlt ZX, Li Y, Huo WL, Guan HJ, Weng SP, et al. Differential gene expression profile in spleen of mandarin fish Siniperca chuatsi infected with ISKNV, derived from suppression subtractive hybridization. Dis Aquat Organ. 2006 Dec 14;73(2): 113-22.

[286] Dios S, Poisa-Beiro L, Figueras A, Novoa B. Suppression subtraction hybridization (SSH) and macroarray techniques reveal differential gene expression profiles in brain of sea bream infected with nodavirus. Mol Immunol. 2007 Mar;44(9):2195-204.

[287] Poisa-Beiro L, Dios S, Ahmed H, Vasta GR, Martinez-Lopez A, Estepa A, et al. Nodavirus infection of sea bass (Dicentrarchus labrax) induces up-regulation of galectin-1 expression with potential anti-inflammatory activity. J Immunol. 2009 Nov 15;183(10):6600-11.

[288] Rise ML, Hall JR, Rise M, Hori TS, Browne MJ, Gamperl AK, et al. Impact of asymptomatic nodavirus carrier state and intraperitoneal viral mimic injection on brain

transcript expression in Atlantic cod (Gadus morhua). Physiol Genomics. 2010 Jul 7;42(2):266-80.

[289] Xu D, Wei J, Cui H, Gong J, Yan Y, Lai R, et al. Differential profiles of gene expression in grouper Epinephelus coioides, infected with Singapore grouper iridovirus, revealed by suppression subtractive hybridization and DNA microarray. J Fish Biol. 2010 Aug;77(2):341-60.

[290] Fisicaro N, Aminian A, Hinchliffe SJ, Morgan BP, Pearse MJ, D'Apice AJ, et al. The pig analogue of CD59 protects transgenic mouse hearts from injury by human complement. Transplantation. 2000 Sep 27;70(6):963-8.

[291] Janeway CA, Jr., Medzhitov R. Innate immune recognition. Annu Rev Immunol. 2002;20:197-216.

[292] Tasumi S, Ohira T, Kawazoe I, Suetake H, Suzuki Y, Aida K. Primary structure and characteristics of a lectin from skin mucus of the Japanese eel Anguilla japonica. J Biol Chem. 2002 Jul 26;277(30):27305-11.

[293] Tsutsui S, Iwamoto K, Nakamura O, Watanabe T. Yeast-binding C-type lectin with opsonic activity from conger eel (Conger myriaster) skin mucus. Mol Immunol. 2007 Feb;44(5):691-702.

[294] Zhang H, Robison B, Thorgaard GH, Ristow SS. Cloning, mapping and genomic organization of a fish C-type lectin gene from homozygous clones of rainbow trout (Oncorhynchus mykiss). Biochim Biophys Acta. 2000 Nov 15;1494(1-2):14-22.

[295] Honda K, Yanai H, Takaoka A, Taniguchi T. Regulation of the type I IFN induction: a current view. Int Immunol. 2005 Nov;17(11):1367-78.

[296] Krasnov A, Timmerhaus G, Afanasyev S, Jorgensen SM. Development and assessment of oligonucleotide microarrays for Atlantic salmon (Salmo salar L.). Comp Biochem Physiol Part D Genomics Proteomics. 2011 Mar;6(1):31-8.

[297] von Schalburg KR, Rise ML, Cooper GA, Brown GD, Gibbs AR, Nelson CC, et al. Fish and chips: various methodologies demonstrate utility of a 16,006-gene salmonid microarray. BMC Genomics. 2005;6:126.

[298] Koop BF, von Schalburg KR, Leong J, Walker N, Lieph R, Cooper GA, et al. A salmonid EST genomic study: genes, duplications, phylogeny and microarrays. BMC Genomics. 2008;9:545.

[299] Krasnov A, Koskinen H, Pehkonen P, Rexroad CE, 3rd, Afanasyev S, Molsa H. Gene expression in the brain and kidney of rainbow trout in response to handling stress. BMC Genomics. 2005;6:3.

[300] Schiotz BL, Jorgensen SM, Rexroad C, Gjoen T, Krasnov A. Transcriptomic analysis of responses to infectious salmon anemia virus infection in macrophage-like cells. Virus Res. 2008 Sep;136(1-2):65-74.

[301] Kurobe T, Yasuike M, Kimura T, Hirono I, Aoki T. Expression profiling of immune-related genes from Japanese flounder Paralichthys olivaceus kidney cells using cDNA microarrays. Dev Comp Immunol. 2005;29(6):515-23.

[302] Williams TD, Diab AM, George SG, Godfrey RE, Sabine V, Conesa A, et al. Development of the GENIPOL European flounder (Platichthys flesus) microarray and determination of temporal transcriptional responses to cadmium at low dose. Environ Sci Technol. 2006 Oct 15;40(20):6479-88.

[303] Park KC, Osborne JA, Montes A, Dios S, Nerland AH, Novoa B, et al. Immunological responses of turbot (Psetta maxima) to nodavirus infection or polyriboinosinic poly-ribocytidylic acid (pIC) stimulation, using expressed sequence tags (ESTs) analysis and cDNA microarrays. Fish Shellfish Immunol. 2009 Jan;26(1):91-108.

[304] Cerda J, Mercade J, Lozano JJ, Manchado M, Tingaud-Sequeira A, Astola A, et al. Genomic resources for a commercial flatfish, the Senegalese sole (Solea senegalensis): EST sequencing, oligo microarray design, and development of the Soleamold bioinformatic platform. BMC Genomics. 2008;9:508.

[305] Skugor S, Jorgensen SM, Gjerde B, Krasnov A. Hepatic gene expression profiling reveals protective responses in Atlantic salmon vaccinated against furunculosis. BMC Genomics. 2009;10:503.

[306] Tilton SC, Gerwick LG, Hendricks JD, Rosato CS, Corley-Smith G, Givan SA, et al. Use of a rainbow trout oligonucleotide microarray to determine transcriptional patterns in aflatoxin B1-induced hepatocellular carcinoma compared to adjacent liver. Toxicol Sci. 2005 Dec;88(2):319-30.

[307] Millan A, Gomez-Tato A, Fernandez C, Pardo BG, Alvarez-Dios JA, Calaza M, et al. Design and performance of a turbot (Scophthalmus maximus) oligo-microarray based on ESTs from immune tissues. Mar Biotechnol (NY). 2010 Aug;12(4):452-65.

[308] Peatman E, Baoprasertkul P, Terhune J, Xu P, Nandi S, Kucuktas H, et al. Expression analysis of the acute phase response in channel catfish (Ictalurus punctatus) after infection with a Gram-negative bacterium. Dev Comp Immunol. 2007;31(11):1183-96.

[309] MacKenzie S, Balasch JC, Novoa B, Ribas L, Roher N, Krasnov A, et al. Comparative analysis of the acute response of the trout, O. mykiss, head kidney to in vivo challenge with virulent and attenuated infectious hematopoietic necrosis virus and LPS-induced inflammation. BMC Genomics. 2008;9:141.

[310] Prieto-Alamo MJ, Abril N, Osuna-Jimenez I, Pueyo C. Solea senegalensis genes responding to lipopolysaccharide and copper sulphate challenges: large-scale identification by suppression subtractive hybridization and absolute quantification of transcriptional profiles by real-time RT-PCR. Aquat Toxicol. 2009 Mar 9;91(4):312-9.

[311] Li RW, Waldbieser GC. Production and utilization of a high-density oligonucleotide microarray in channel catfish, Ictalurus punctatus. BMC Genomics. 2006;7:134.

[312] Fittipaldi N, Gottschalk M, Vanier G, Daigle F, Harel J. Use of selective capture of transcribed sequences to identify genes preferentially expressed by Streptococcus suis upon interaction with porcine brain microvascular endothelial cells. Appl Environ Microbiol. 2007 Jul;73(13):4359-64.

[313] Lun ZR, Wang QP, Chen XG, Li AX, Zhu XQ. Streptococcus suis: an emerging zoonotic pathogen. Lancet Infect Dis. 2007 Mar;7(3):201-9.

[314] Rise ML, von Schalburg KR, Brown GD, Mawer MA, Devlin RH, Kuipers N, et al. Development and application of a salmonid EST database and cDNA microarray: data mining and interspecific hybridization characteristics. Genome Res. 2004 Mar; 14(3):478-90.

[315] Matsumoto M, Tanaka T, Kaisho T, Sanjo H, Copeland NG, Gilbert DJ, et al. A novel LPS-inducible C-type lectin is a transcriptional target of NF-IL6 in macrophages. J Immunol. 1999 Nov 1;163(9):5039-48.

[316] MacKenzie S, Iliev D, Liarte C, Koskinen H, Planas JV, Goetz FW, et al. Transcriptional analysis of LPS-stimulated activation of trout (Oncorhynchus mykiss) monocyte/macrophage cells in primary culture treated with cortisol. Mol Immunol. 2006 Mar;43(9):1340-8.

[317] Martin SA, Blaney SC, Houlihan DF, Secombes CJ. Transcriptome response following administration of a live bacterial vaccine in Atlantic salmon (Salmo salar). Mol Immunol. 2006 Apr;43(11):1900-11.

[318] Ching B, Jamieson S, Heath JW, Heath DD, Hubberstey A. Transcriptional differences between triploid and diploid Chinook salmon (Oncorhynchus tshawytscha) during live Vibrio anguillarum challenge. Heredity (Edinb). 2010 Feb;104(2):224-34.

[319] Dumrongphol Y, Hirota T, Kondo H, Aoki T, Hirono I. Identification of novel genes in Japanese flounder (Paralichthys olivaceus) head kidney up-regulated after vaccination with Streptococcus iniae formalin-killed cells. Fish Shellfish Immunol. 2009 Jan;26(1):197-200.

[320] Gerwick L, Corley-Smith G, Bayne CJ. Gene transcript changes in individual rainbow trout livers following an inflammatory stimulus. Fish Shellfish Immunol. 2007 Mar; 22(3):157-71.

[321] Meijer AH, Verbeek FJ, Salas-Vidal E, Corredor-Adamez M, Bussman J, van der Sar AM, et al. Transcriptome profiling of adult zebrafish at the late stage of chronic tuberculosis due to Mycobacterium marinum infection. Mol Immunol. 2005 Jun;42(10): 1185-203.

[322] van der Sar AM, Spaink HP, Zakrzewska A, Bitter W, Meijer AH. Specificity of the zebrafish host transcriptome response to acute and chronic mycobacterial infection and the role of innate and adaptive immune components. Mol Immunol. 2009 Jul; 46(11-12):2317-32.

[323] Purcell MK, Nichols KM, Winton JR, Kurath G, Thorgaard GH, Wheeler P, et al. Comprehensive gene expression profiling following DNA vaccination of rainbow trout against infectious hematopoietic necrosis virus. Mol Immunol. 2006 May;43(13): 2089-106.

[324] Byon JY, Ohira T, Hirono I, Aoki T. Use of a cDNA microarray to study immunity against viral hemorrhagic septicemia (VHS) in Japanese flounder (Paralichthys oliva-ceus) following DNA vaccination. Fish Shellfish Immunol. 2005 Feb;18(2):135-47.

[325] Jorgensen SM, Afanasyev S, Krasnov A. Gene expression analyses in Atlantic salmon challenged with infectious salmon anemia virus reveal differences between individu-als with early, intermediate and late mortality. BMC Genomics. 2008;9:179.

[326] Encinas P, Rodriguez-Milla MA, Novoa B, Estepa A, Figueras A, Coll J. Zebrafish fin immune responses during high mortality infections with viral haemorrhagic septice-mia rhabdovirus. A proteomic and transcriptomic approach. BMC Genomics. 2010;11:518.

Chapter 3

Cytokine Regulation of Teleost Inflammatory Responses

Leon Grayfer and Miodrag Belosevic

Additional information is available at the end of the chapter

1. Introduction

The inflammatory response is a highly regulated process initiated by tissue damage, infiltrating pathogens or both. The primary role of inflammation is the resolution of tissue damage, including the elimination of damaged or dead cells and any infiltrating pathogens, and restoration of homeostasis. The initial recognition of tissue damage and/or pathogens is mediated by tissue resident macrophages primarily through various sentinel pattern recognition receptors such as toll-like receptors. In response to and in accordance with distinct stimuli, macrophages become activated to produce a wide range of bioactive molecules, some of which attract other cells to the site of inflammation, and others that dictate the course of an inflammatory response and eventual tissue repair. Recent evidence suggests that there are at least two activation states of monocytes/ macrophages [4, 130, 211, 214]. The classically activated monocytes/macrophages possess significant antimicrobial armamentarium, secrete a plethora of factors that propagate and enhance the microbicidal activities and in general mediate pathogen clearance. The non-classically or alternatively activated monocytes/macrophages secrete factors that ablate the destructive components of the inflammatory response and promote tissue healing, repair and angiogenesis and will not be addressed further here.

The processes involved in the onset, progression and resolution of inflammation are complex and remain to be fully elucidated in vertebrates. However, it is widely believed that myeloid lineage cells are intimately involved in inflammatory reactions and their function is controlled by cytokines. This review focuses on the recent advancements in the understanding of the biology of hallmark fish pro-inflammatory cytokines, tumor necrosis factor alpha (TNFα), interferon gamma (IFNγ) and interleukin-1 beta (IL-1β) and their receptors.

2. Antimicrobial responses of fish phagocytes

It is well established that fish phagocytes possess oxidative burst responses, comparable to those of mammals. Absence of readily available fish cytokines limited the early research of fish phagocytes to employing pathogen products and/or crude activated cell supernatants, presumed to contain hallmark "activating" agents. This fundemental work of fish phagocyte-mediated inflammatory processes has been comprehensively reviewed [138]. Since then the specific genes encoding the components of the fish NADPH oxidase complex (Fig. 1) have been cloned in various fish species [13, 84, 119] and their expression correlated with reactive oxygen radical production [13, 63, 141]. The priming of the fish phagocyte ROI responses by recombinant fish cytokines such as TNFα [67, 133, 217], IFNγ [62, 63, 215] and IL-1β [54, 97, 144] has also been reported. Similar to the mammalian monocyte/macrophage paradigm [22], fish monocytes have greater ability to generate reactive oxygen intermediates (ROI) following short stimulation [67, 136, 152], whereas mature fish macrophages require relatively prolonged immune stimulations to achieve comparable magnitudes of this response [58, 59, 136, 171].

Phagocytes (primarily mature macrophages) also produce microbicidal/tumoricidal reactive nitrogen intermediates in a stimulus-specific manner. This response, catalyzed by the inducible nitric oxide synthase enzyme (iNOS, Fig. 2), involves the conversion of arginine to citruline and results in the production of nitric oxide (NO) and other products including nitrite, nitrate, and nitrosamines [86, 134, 184]. The NADPH oxidase produced superoxide anion may also react with NO to form the peroxynitrite intermediate [ONOO⁻] that also has potent microbiacidal activity [35, 156, 191, 213]. The biology of the iNOS enzyme has been reviewed in references [2, 111, 126].

The ability of fish phagocytes to produce microbicidal NOs has been well established (reviewed in reference [138]). The iNOS gene transcript has been identified in several fish species [101, 102, 163, 200] and fish macrophages have been demonstrated to up-regulate iNOS expression and produce copious amounts of NO in response to a plethora of immune stimuli [62-64, 66, 67, 85, 88, 140, 158, 180, 210]. The inflammatory cytokine regulation of fish iNOS and NO production is described below.

3. Cytokine regulation of inflammatory responses

The onset, progression and resolution of the inflammatory responses are tightly regultaed through soluble mediators (cytokines, monokines, chemokines) that orchestrate inflammation. Pro-inflammatory cytokines such as TNFα, IFNγ, and IL-1β enhance antimicrobial functions of immune cells and facilitate the pathogen clearance, while anti-inflammatory cytokines such as TGFβ and IL-10 down-regulate inflammatory processes and skew cell functions towards tissue repair mechanisms. It is important to note that in addition to the hallmark cytokines discussed in this review, other cytokines such as chemokines and growth factors also participate in the regulation of an inflammatory response.

The last decade has yeilded significant advances in the understanding of inflammatory responses of lower vertebrates, such as bony fish. The genes encoding hallmark cytokines have been identified and characterized in a number of fish species. Interestingly, many of these exhibit structural similarities and gene synteny organization comparable to their higher vertebrate counterparts. Conversely, multiple isoforms of certain cytokines are present in distinct fish species.

3.1. Tumor necrosis factor alpha (TNFα)

Tumor necrosis factor alpha is a central inflammatory mediator, initially identifed as a serum component capable of eliciting "hemorrhagic necrosis" of certain tumors [20]. Since discovery, TNFα has been found to be produced by many cell-types and confer an increadible range of immune processess [44, 203, 209]. In the context of inflammatory responses, TNFα promotes the chemotaxis of neutrophils and monocytes/macrophages [123, 207], enhances their phagocitic capacity [95, 105, 194], primes ROI and NO reposnses [42, 135], chemoattracts fibroblasts [168] and elicits platelet activating factor production [19, 73, 104].

The mammaian TNFα functions as a 26 kDa type II trans-membrane protein as well as a 17 kDa soluble moiety, released by the TNFα cleaving metalloproteinase enzyme (TACE)-mediated cleavage [98, 128, 148]. A homotrimerized TNFα (soluble or membrane bound) engages one of two cognate receptors, TNF-R1 or TNF-R2, which in turn trimerize around the ligand [7, 45]. Currently, there is no consensus as to the respective contribution of these receptors to the biological effects caused by TNFα. Some evidence suggests that TNF-R1 propagates the signal from the soluble TNFα, while the membrane-bound TNFα acts exclusively through TNF-R2 [70]. Other evidence suggest that TNF-R1 is primarily involved in induction of apoptosis while the TNF-R2 functions in proliferation and cell survival [129], while other contributions suggest cooperation between the two receptors [204]. The prevailing theory proposes that TNF-R1 confers signal propagation, while TNF-R2 binds and redistributes TNFα to TNF-R1 in a process coined "ligand passing" [28, 43, 197]. Despite this, more recent literature suggests that TNF-R2 is directly involved in many inflammatory processes including the activation of T lymphocytes [93, 94], stimulation of myofibroblasts [190], as well as tumor suppression [212]. The TNF-R1 and TNF-R2 utilize largely non-overlapping signaling mechanisms, relying on recruitment of numerous downstream signaling molecules, the relative abundance of which ultimately dictates signaling outcomes (for a current review of TNFα signal transduction see recent review [143]). It is likely that the roles of respective TNF receptors in the biological outcomes of TNFα stimulation are cell type and cell activation state dependent.

3.1.1. Identification of TNFα in fish

Presence of an endogenous bony fish TNF system was first suggested in the early 1990s where the human recombinant TNFα elicited ROI production in trout leukocytes while administration of a monoclonal anti-TNF-R1 antibody blocked this response [75, 87]. Hirono *et al.* [76] identified and characterized the first cDNA transcript encoding a Japanese flounder TNFα, which had only 20-30% amino acid identity with the mammalian TNFs, but had very similar intron/exon organization. The expression of this flounder TNFα gene increased in PBLs following LPS, ConA or PMA stimulation, suggesting a conserved role for this cytokine in fish inflammatory responses. Shortly thereafter, the rainbow trout TNFα was identified, shown to be constitutively expressed in the gill and kidney tissues, and upregulated in LPS-stimulated or IL-1β-treated kidney leukocytes and in the trout macrophage cell line, RTS11 [103]. The catfish TNFα was also constitutively expressed in healthy fish and several catfish immune cell lines including macrophage and T cell lines, but not in B cell or fibroblast cell lines [218]. Notably, all fish TNFα proteins possess the TNF family signature, [LV]-x-[LIVM]-x$_3$-G-[LIVMF]-Y-[LIMVMFY]$_2$-x$_2$-[QEKHL] [103], underlining the evolutionary conservation of this cytokine.

3.1.2. Isoforms of TNFα in fish

Trout were reported to possess an additional TNFα isoform [219] and the presence of multiple bony fish TNFs was confirmed in common carp, in which first two [162], and soon thereafter a third carp TNFα isoform [167] were identified. Interestingly, polymorphisms in the gene encoding the carp TNFα2 isoform were linked to carp resistance to the protozoan parasite, *T. borreli* [162].

3.1.3. TNFα receptors of fish

Tumor necrosis factor alpha has now been identified in a number of fish species including flounder [76], trout [103, 114], catfish [218], carp [162, 167], sea bream [18, 56], tilapia [151], turbot [142], ayu [193], fugu [165], zebrafish [165], sea bass [133], goldfish [67] and tuna [89]. Additionally, novel TNF-like molecules with unique intron/exon organization have been identified in zebrafish and flounder [165]. Despite this, knowledge regarding the cognate receptors for the teleost TNF proteins has been relatively limited.

A death domain containing TNF receptor was identified in zebrafish ovarian tissues and coined the ovarian TNF receptor (OTR) [12]. Predicted zebrafish TNF-R1 and TNF-R2 sequences are in the NCBI database, with zebrafish TNF-R1 showing high sequence homology to the OTR. The goldfish TNF-R1 and TNF-R2 cDNAs were identified based on the zebrafish sequences [61] and display conserved regions included cysteine residues and predicted docking sites for downstream signaling. Also, the goldfish TNF-R1 possesses a conserved death domain including the highly conserved motif (W/E)-X$_{31}$-L-X$_2$-W-X$_{12}$-L-X$_3$-L (with R in the W/E position) and six conserved or semi-conserved residues, crucial for the mammalian TNF-R1-mediated cytotoxicity [189]. Interestingly, while the mammalian TNF-R1 and TNF-R2 and the fish TNF-R2 contain four TNF homology domains (THD, defined by specific cysteine residues), the fish TNF-R1 proteins exhibit only 3 full THDs [61].

Recombinant extracellular domains of goldfish TNF-R1 and TNF-R2 bound to both goldfish recombinant TNFα1 or TNFα2 in *in vitro* binding assays [61]. As with the sea bream TNFα [56], the goldfish recombinant TNFα ligands and receptors all adopted dimeric conformations and interacted as dimers rather than trimers *in vitro*. Interetsingly, there have also been several reports indicateding dimerizion of the mammalian TNF-R1 [131, 132, 147]. Furthermore, the mammalian p75/NTR neurotrophin receptor, a member of the TNF superfamily of proteins with many structural similarities to teleost and mammalian TNF-R1, binds to its cognate NTR ligand as a dimer [26, 57].

By examining the TNF system in lower vertebrates such as teleost fish, we can gain insight into the evolutionary origins of our own immune systems and the selective pressures that shaped them. As in mammals, TNFα appears to be central to the regulation of inflammatory responses of bony fish. Further examination of biological effects of this molecule using different lineages of fish immune cells will yield a more concrete understanding of this system in teleosts.

3.1.4. Inflammatory roles of the fish TNFα

The first functional characterization of a fish TNFα was reported in 2003 when Zou et al. [217] demonstrated that recombinant trout TNFα1 and TNFα2 (isoforms) both induced the IL1β, TNFα1, TNFα2, IL-8 and COX-2 gene expression in primary kidney leukocytes and in the RTS11 trout macrophage cell line [217]. These recombinant TNFα isoforms also elicited dose-dependent chemotaxis of trout kidney leukocytes and phagocytosis of yeast particles. SDS-PAGE analysis of the recombinant TNFα1 and TNFα2 suggested that both of the recombinant molecules existed in monomeric, dimeric and trimeric states. Together, these findings suggested conservation in the pro-inflammatory roles for teleost TNFα.

Intraperitoneal administration of a mature recombinant (cleaved) form of gilthead sea bream TNFα to fish resulted in rapid recruitment of phagocytic granulocytes to the sites of injection, granulopoeisis and the priming for enhanced ROI of peritoneal and primary kidney leukocytes [56]. Intriguingly, size exclusion chromatography of this recombinant sea bream TNFα suggested that this protein existed primarily in a dimeric state [56], in contrast to the trimeric state of the mammalian TNFα.

Subsequent reports indicated that the pro-inflammatory effects of sea bream and zebrafish TNFαs were not direct but resulted from the stimulation of endothelial cells [155]. When sea bream peritoneal and head kidney leukocytes were primed with recombinant TNFα (100 ng/mL) or bacterial DNA (*Vibrio anguillarum*, 50 μg/mL) for 16 hours, TNFα elicited significant ROI responses, albeit modest when compared to those induced by *V. anguillarum* DNA. Intraperitoneal injections of the recombinant TNFα resulted in increased expression of several pro-inflammatory genes of sea bream peritoneal leukocytes, while TNFα treatments of sea bream endocardium endothelial cells (EECs) *in vitro* also increased their immune gene expression. Notably, while the *in vitro* TNFα stimulation of EECs and macrophages resulted in substantially elevated pro-inflammatory gene expression in both cell types, the stimulated macrophages exhibited significantly more robust transcriptional responses. Although this TNFα failed to chemoattract leukocytes, TNFα-stimulated EEC-conditioned medium and supernatants from TNFα-injected peritoneal exudate cells elicited leukocyte chemoattraction. Furthermore, the zebrafish TNFα conferred neutrophil recruitment but also increased fish susceptibility to bacterial and viral infections. Accordingly, the authors of these studies proposed that unlike the mammalian cytokine, the fish TNFα elicits inflammatory functions indirectly, through non-immune cells.

In similar studies, supernatants from recombinant carp TNFα-stimulated cardiac endothelial cells primed kidney phagocyte ROI responses while recombinant carp TNFα1 and 2 did not enhance phagocyte antimicrobial functions [155]. Notably, the recombinant forms of the zebrafish, trout, sea bream and carp TNFα proteins exhibited lytic activity towards *Trypanoplasma borreli*, akin to the trypanolytic capacity of the mammalian TNFs [52], where the membrane form of the fish TNFα was thought to be responsible for these effects. In summation of the above observations, Florenza *et al.* (2009) suggested that while the trypanolyitic roles of TNFα are evolutionarily conserved, the pro-inflammatory mechanisms elicited by this molecule were not [155].

Contrary to the sea bream and carp studies, other literature suggests that akin to the mammalian cytokine, teleost TNFα also directly elicits pro-inflammatory functions. The tilapia non-specific cytotoxic cells (NCCs) constitutively express membrane bound as well as soluble forms of TNFα (in addition to granzymes and Fas ligand) to confer cytolytic activity towards target cells, and when stimulated with recombinant tilapia TNFα, become protected from activation-induced apoptosis [151]. While the turbot recombinant TNFα did not enhance macrophage ROI, it elicited *in vitro* NO production and *in vivo* inflammatory cell recruitment and activation [142]. The Ayu fish, recombinant TNFα induced ROI production by kidney cells [193] and the blue fin tuna recombinant TNFα1 and 2 enhanced the phagocytic responses of tuna PBLs [89].

Zebrafish TNFα elicited cell signaling and conferred increased resistance to *Mycobacterium marinum* [30] while knockdown of the TNF-R1 led to enhanced mycobacterial disease progression, increased fish mortality, accelerated bacterial growth, granuloma breakdown and necrotic macrophage cell death [30]. Thus, it appears that the zebrafish TNFα is pivotal in the maintenance of encapsulated *M. marinum* granulomas and the restriction of the growth of this pathogen. Our recent work supports these findings, where the pre-treatment of goldfish macrophages with recombinant goldfish TNFα2 ablated the *M. marinum*-mediated down-regulation of NO production by these cells and reduced the survival of intracellular bacteria [65].

We recently identified two isoforms of the goldfish TNFα and functionally characterized a recombinant goldfish TNFα2 (rgTNFα2) in the context of primary kidney-derived goldfish macrophage cultures (PKMs) [67]. This rgTNFα2 induced dose-dependent chemotaxis of goldfish macrophages, enhanced their phagocytic abilities, NO production and primed the ROI responses of PKMs.

The extent of the conservation in the biology of teleost TNFα will become more evident with increased availability of tools, reagents and cell culture systems. While some literature proposes a lack of conservation of the inflammatory roles of teleost TNFα, others strongly implicate this molecule in the regulation of fish antimicrobial functions. It is also possible that the discrepancies in the above findings stem from the culture systems employed. For example, both the sea bream and carp studies that failed to observe direct effects of fish TNFα utilized freshly isolated adherent kidney phagocyte populations [52, 155]. Notably, freshly isolated mammalian myeloid populations are highly variable in their antimicrobial capabilities [116]. Similarly, goldfish PKMs exhibit temporal gain and loss of antimicrobial capabilities after prolonged cvultivation [67, 136]. Specifically, ROI is primarily mediated by PKM-monocytes while NO is predominantly produced by mature macrophages [67, 136].

3.2. Interferon-gamma (IFNγ)

Interferon gamma is a pleiotropic, pro-inflammatory and anti-viral cytokine, identified in the supernatants of PHA-activated lymphocyes for its unique anti-viral properties [205]. IFNγ is primarily produced by activated Th1 phenotype CD4+ cells [127], CD8+ cells [161] and natural killer (NK) cells [149] and is of central importance in host defense against intracellular and extracellular pathogens [8, 91, 99, 179, 182]. For example, IFNγ gene knock-out mice are incapable of controlling infections with *Leishmania major* [202], *Listeria monocytogenes* [81], and *Mycobacterium* [32], underlining the importance of this cytokine in the regulation of antimicrobial responses [8, 10, 21, 51, 115].

The mammalian IFNγ dimer ligates the interferon gamma receptor 1 (IFNGR1), which then associates with IFNGR2, forming a signaling complex and activating the Janus kinases (Jak) 1 and 2, associated with the receptor chains 1 and 2, respectively [83]. Jak1 and Jak2 in turn activate the IFNGR1-associated signal transducer of activation-1 (Stat1) transcription factor [33]. The IFNGR ligation may also activate and utilize Stat2 [187], albeit to lesser extent than Stat1. Subsequent transcriptional regulation of several other genes then ensues through homodimeric Stat1, heterodimeric Stat1: Stat2, through the transcription factor complexes ISGF3 and Stat1-p48, composed of Stat1: Stat2:IRF-9 and Stat1: Stat1:IRF-9, respectively [11, 118, 187, 188]. These confer transcriptional changes through recognition of IFNγ-activated sequences (GAS) in the promoter regions of target genes [188]. Within 30 minutes of IFNγ receptor ligation, there are increased transcript levels for several interferon regulatory factors (IRFs), which then modulate subsequent waves of gene expression in the IFNγ signaling cascade [208].

3.2.1. Identification of IFNγ in bony fish

Trout mitogen-simulated leukocyte supernatants possess macrophage activating capabilities (MAFs) akin to the mammalian IFNγ [58, 59], suggesting the existence of an IFNγ counterpart(s) in fish. It was also established that downstream signaling factors employed by the mammalian IFNγ (Stats), were present in fish [160] where the antibody-purified fish Stat-like factor was shown to bind the mammalian IFNγ promoter [160].

The initial fish IFNγ homolog discovery came from examination of fugu gene scaffolds [220]. Fugu homologs of mammalian genes syntenic to IFNγ were also present on the same fugu gene scaffold. This fugu IFNγ had 4 exon / 3 intron organization similar to its mammalian counterpart and shared 32.3-32.7 % and 34.9-43.3% identities with bird and mammalian IFNγ sequences, respectively. An identified trout IFNγ sequence exhibited low sequence identity with other vertebrate IFNγ proteins, but possessed the conserved signature motif ([IV]-Q-X-[KQ]-A-X_2-E-[LF]-X_2-[IV]) and C-terminal nuclear localization signal (NLS), characteristic of the mammalian IFNγ proteins [215].

3.2.2. Inflammatory roles of the fish IFNγ

The recombinant trout IFNγ (rtIFNγ) elicited increased immune gene expression in the RTS-11 monocyte-macrophage cell line [199], while this response was pharmacologically abrogated with ERK (transcription factor) or protein kinase C (PKC) inhibitors as well as by deleting the C-terminal NLS on the rtIFNγ [215]. Also, rtIFNγ stimulation of trout kidney leukocytes primed their ROI production, suggesting functional similarities between the trout and mammalian type II IFNs [215].

Adult zebrafish microinjected with recombinant IFNγ did not exhibit increased immune gene expression or enhanced protection against *Streptococcus iniae* and the spring viremia carp virus (SVCV) [110]. However, this lack of response could stem from IFNγ being bound up by cells expressing only the IFNGR1 without eliciting a detectable level of response from the relatively few cells expressing both IFNGR1 and IFNGR2. If this is the case, these results underline the localized nature of the zebrafish IFNγ function rather then the efficacy of regulation of antimicrobial functions by this cytokine.

We [62] and others [3] have functionally characterized the cyprinid IFNγ. A recombinant goldfish IFNγ (rgIFNγ) primed goldfish monocytes ROI response in a concentration dependent manner [62] and, as in mammals [72, 157], at lower concentrations rgIFNγ conferred additive ROI priming effects. The rgIFNγ also enhanced the expression of the ROI enzyme, NADPH oxidase, catalytic subunits, p67^{phox} and gp91$^{phox.}$ Similarly, the recombinant carp IFNγ (rcIFNγ) also primed carp kidney phagocytes for enhanced ROI production [3].

The goldfish rIFNγ elicited modest but significant enhancement of phagocytosis and NO production by goldfish monocytes and macrophages, respectively [62]. This was paralled with increased iNOS gene expression in rgIFNγ-stimulated macrophages. In contarst, carp kidney phagocytes only displayed significant iNOS gene expression and NO production when treated with a combination of carp recombinant IFNγ and 30 μg/mL LPS, but not following rcIFNγ treatments alone [3].

Since carp and goldfish are closely related species, the above discrepancies presumably stem from differences in experimental systems. Akin to bone-marrow-derived macrophages, which acquire antimicrobial capabilities with culture time, we have observed that the antimicrobial capabilities of goldfish kidney-derived phagocytes are dynamic [136, 137]. For this reason, we used culture-derived cells, while the carp studies used freshly isolated phagocytes. Additionally, it is well established that in mammals the production of ROIs is seen as early as 1 hour after immune stimulation, whereas the production RNIs is first detected approximately 24 hours post stimulation [169]. This is very similar to what we have observed with rgIFNγ-elicited iNOS expression and NO responses.

The treatment of mature goldfish macrophages with rgIFNγ resulted in increased expression of TNFα and IL-1β isoforms; IL-12 p35 and p40; IFNγ; IL-8 (CXCL-8]; CCL-1; and viperin (an anti-viral molecule) [62]. The treatment of carp phagocytes with a combination of carp IFNγ and LPS resulted in increased expression of TNFα; IL-1β; IL-12 subunits p35 and isoforms of IL-12 p40 subunit [3]. The carp IFNγ also induced the expression of the CXCL-10 like chemokine, CXCLb, while inhibiting the LPS-induced expression of CXCL-8 isoforms,

CXCL-8_L1 and CXCL-8_L2 [3]. The latter is reminiscent of the mammalian IFNγ and CXCL-8 relationship, where shorter treatments of mammalian granulocytes with IFNγ causes decreased CXCL-8 production [23, 90, 120], while prolonged treatments increase the CXCL-8 mRNA and protein levels [90]. Additionally, mammalian blood monocytes and macrophage cell lines stimulated with IFNγ up-regulate CXCL-8 mRNA transcripts and protein levels [14, 36] due to post-transcriptional stabilization of the CXCL-8 mRNAs rather than gene expression [14]. Together, these findings suggest that the pro-inflammatory roles of IFNγ, including synergism with LPS, are conserved in cyprinid fish, while it is probable that the discrepancies in gene expression profiles may be due to distinct immune cell model systems used (primary kidney phagocytes versus cultured mature macrophages).

3.2.3. Interferon-gamma-related (IFNγrel) cytokine

Using gene synteny analysis, Igawa et al. (2006) discovered two tandem IFNγ isoforms next to the fish IL-22 and IL-26 genes [82]. The corresponding IFNγ sequences, coined IFNγ1 and IFNγ2, were later renamed IFNγ-related (rel) and FNγ respectively [166]. These proteins share only 17 % amino acid identity, but exhibit exon/intron organization similar to that of mammalian and fugu IFNγ, and have the IFNγ signature motif ([IV]-Q-X-[KQ]-A-X_2-E-[LF]-X_2-[IV]). Interestingly, only IFNγ but not IFNγrel has a C-terminal nuclear localization signal (NLS).

The existence of two IFNγ isoforms, sharing all of the above characteristics was soon confirmed in siluriformes and other cypriniformes when IFNγ and IFNγrel were identified in the catfish [122] and common carp [183], respectively. Catfish and carp were both reported to possess two distinct, alternatively spliced IFNγ transcripts, as well as a single transcript of IFNγrel [122, 183]. The pattern of tissue gene expression of catfish IFNγ and IFNγrel differed while both cytokines were expressed in various cell lines and immune cells, with highest expression observed in macrophages, T cells and NK cells [122]. Interestingly, low transcript levels of IFNγrel but not IFNγ were also reported in a catfish B-cell line. Expression of the carp IFNγ increased in T cells after stimulation with PHA, while the mRNA levels of carp IFNγrel increased in IgM⁺, B lymphocyte-enriched immune cell fractions, stimulated with high doses of LPS [183]. Furthermore, increased transcript levels of the carp IFNγ, but not IFNγrel, were observed in the head kidneys of fish infected with *T. borreli*. The goldfish IFNγ and IFNγrel exhibited similar mRNA levels across tissues and immune cell types [63].

3.2.4. Inflammatory roles of fish IFNγrel

While the siluriforme IFNγrel has yet to be functionally characterized, there has been some insight into the functions of cyprinid IFNγrel. Zebrafish IFNγrel mRNA is detectable in freshly laid eggs, suggesting maternal supply of this transcript [174]. The IFNγrel mRNA levels persist throughout embryonic development, while the expression of IFNγ is not detected until later stages of development. Embryos injected with *in vitro* transcribed IFNγ or

IFNγrel mRNAs individually, exhibited similar immune gene expression changes, while combined injections further increased certain expression profiles, suggesting non-overlapping roles for the respective IFNγ proteins.

Morpholino knock-downs of either IFNγ or IFNγrel alone had negligible effects on zebrafish embryo survival following *Escherichia coli* challenge, while knock-down of both IFNγs resulted in substantially diminished survival following infection [174]. Interestingly, individual morpholino knockdowns of IFNγ or IFNγrel caused decreased survival rates of embryos infected with *Yersinia ruckeri,* while double knockdowns had a further deleterious effect on embryo survival. Presumably IFNγ and IFNγrel elicit some overlapping and some distinct antimicrobial mechanisms such that the presence of one cytokine may be sufficient for dealing with certain pathogens but not others. It is noteworthy that while the *E. coli* (strain DH5α) is not a natural fish pathogen, the *Y. ruckeri* is [159].

A comprehensive functional investigation of recombinant goldfish (rg) IFNγrel and rgIFNγ revealed that these molecules differed in respective capacities to modulate pro-inflammatory responses of goldfish PKMs [63]. While rgIFNγ conferred long-lasting ROI priming effects, rgIFNγrel induced short-lived ROI priming, causing subsequent unresponsiveness to ROI priming by other recombinant cytokines (rgIFNγ or rgTNFα2). While rgIFNγ elicited modest phagocytosis and nitric oxide responses in goldfish monocytes and macrophages, respectively [62, 63], rgIFNγrel was a highly potent inducer of both responses. Interestingly, rgIFNγ and rgIFNγrel induced different gene expression profiles in goldfish monocytes, where rgIFNγrel elicited significantly greater expression of key inflammatory genes. Notably, while both cytokines induced the phosphorylation of Stat1, its nuclear translocation was only observed following rgIFNγ treatment. Together, these findings suggest a functional segregation of the goldfish type II interferons in the regulation macrophage antimicrobial functions.

Further confirmation of this functional dichotomy between the fish type II IFNs is warranted using *in vivo* and other *in vitro* fish models. Notably, the zebrafish IFNγrel has recently also been demonstrated to elicit more robust pro-infalmmatory gene expression than IFNγ in larvae microinjected with respective IFN expression constructs [110]. Furthermore, these zebrafish IFNγrel-mediated effects were dependent on the myeloid transcription factor SP1, underlying the specificity of this cytokine for macrophag-lineage cells.

3.2.5. IFNγ receptors in fish

Despite the growing knowledge regarding the teleost type II IFNs, the receptor systems employed by these fish molecules remain poorly understood. The IFNGR1 and IFNGR2 chains were recently identified in the rainbow trout [55]. The expression of the IFNGR1 was generally greater than that of IFNGR2 and was subject to decrease following rIFNγ or rIL1β stimulation. Furthermore RTG-2 trout fibroblast cells transfected with an IFNGR1 construct, or CHO cells transfected with constructs expressing IFNGR1 and IFNGR2, both bound rtIFNγ, where, as in mammals, the expression of the IFNGR2 chain was essential for the IFNγ-induced activity. It should be noted that a reliable trout IFNγ reporter cell line has been established and was effectively demonstrated to specifically increase luciferase reporter expression following IFNγ stimulation but not in response to a range of other stimuli [24, 25]. Hence, a system is now in place to elucidate the distinct binding and signaling mechanisms involved in the salmonid IFNγ biological processes.

In light of the functional differences between IFNγ and IFNγrel, we postulated that these cytokines might function through distinct receptors. When we performed gene synteny analysis of IFNGR1, we observed that while some genes were localized to the chromosome bearing the known zebrafish IFNGR1 gene, other syntenic neighbours of the mammalian IFNGR1 were present on a distinct zebrafish chromosome. Further analysis of the chromosomal region flanked by these genes revealed a second gene, encoding a distinct IFNGR1 protein [60], (these genes were denoted IFNGR1-1 and IFNGR1-2). The corresponding goldfish receptor cDNA transcripts were identified. The fish IFNGR1 sequences displayed putative Jak1 and Stat1 binding sites, pivotal for the biological functions of the mammalian IFNγ [49, 68, 69]. While the zebrafish receptors displayed comparable tissue expression, the goldfish IFNGR1-1 exhibited substantially greater mRNA levels than the IFNGR1-2 in all tissues and immune cell types examined. In order to elucidate possible binding partners for the goldfish IFNGR1-1 and IFNGR1-2, recombinant forms of their extracellular domains were produced and *in vitro* binding assays were performed. While IFNGR1-1 bound exclusively to IFNγrel (IFNγ1), IFNGR1-2 bound strictly to IFNγ (IFNγ2, receptors were named after the fact). It has recently been reported that morpholino knockdowns of IFNGR1-1, IFNGR1-2 or a putative IFNGR2 abolished zebrafish IFNγ-induced gene expression [1]. In contrast, only the knockdown of IFNGR1-1, but not the knockdown of IFNGR1-2 or IFNGR2, abrogated gene expression elicited by IFNγrel. It was suggested that IFNγ might signal through a heterodimer of IFNGR1-1 and IFNGR1-2 and a homodimer of IFNGR2 while the IFNγrel would ligate with a homodimeric IFNGR1-1 and an as of yet unidentified receptor 2 chains. Alternatively, since IFNγrel is present in cyprinids early in development, the knockdown of its putative receptor, IFNGR1-1 in embryos might effect development of the components required for IFNγ function. A biological relationship of such nature would be phenotypically manifest as a loss of IFNγ function (as seen in the above zebrafish study).

It would appear that certain teleost species possess receptor signaling systems to facilitate a dichotomy of type II IFN functions. Presumably, the cyprinid IFNGR1 genes arose from duplications of an ancestral IFNGR1 and subsequently diverged in respective signaling mechanisms used, where the IFNγrel-induced Stat1-phosphorylation might be an artifact, remnant of the ancestral gene. Indeed, the importance of the C-terminal NLS of fish (and mammalian) IFNγ has been demonstrated [217] while the lack of this NLS on IFNγrel proteins is the key distinguishing feature of the latter cytokine. The leading model for mammalian IFNγ signaling [185] suggests that after IFNγ receptor ligation, Stat1 is delivered into the nucleus via the IFNγ NLS in a complex consisting of Stat1:IFNGR1:IFNγ. Therefore, due to a lack of an NLS, IFNγrel may have evolved to utilize distinct signaling mechanism.

3.3. Interleukin-1 beta (IL-1β)

3.3.1. Interleukin-1 cytokine family

The interleukin-1 cytokine family is becoming increasingly diverse, with new family members being discovered and/or assigned to this family. In addition to the well characterized IL-1α, IL-1β, IL-1 receptor antagonist (IL-1Ra) and IL-18 the new members of the family now include the IL-1F5-10, and IL-1F11 [34, 192]. Accordingly, it has been proposed that IL-1α, IL-1β, IL-1 receptor antagonist (IL-1Ra) should be renamed IL-1F1, IL-1F2, IL-1F3 and IL-1F4, respectively [177]. The exact biological roles of these new IL-1 cytokines (IL-1F5-11) have not been fully elucidated, and it was reported that concentrations of 100-1000 fold greater than those of IL-1β are required to them to induce biological effects [177, 178]. These new family members will not be discussed further, instead the remainder of this section will deal with the classical IL-1 cytokines and primarily IL-1β.

The IL-1α [106] and IL-1β [6] were initially identified as monocyte transcripts with only 23% amino acid identitiy but with structurally similarities. Both of these cytokines are produced as leaderless 31 kDa pro-peptides that are cleaved to generate mature 17 kDa molecules, which mediated their respective effects by binding to the IL-1RI [37, 38, 40]. The synthesis of IL-1 has been reported in several cell types, including keratinocytes, Langerhan's cells, synovial fibroblasts, mesangial cells, astrocytes, microglia, corneal cells, gingival cells, thymic epithelial cells, in addition to myeloid and some lymphoid cell types (reviewed in references [38, 40]). Although IL-1α and IL-1β signal through the same receptor, they induce different biological functions. It has been suggested that IL-1α is produced primarily by epithelial cells and keratinocytes [125] and is involved in mediating local inflammatory processes. In contrast, IL-1β is synthesized by cells such as monocytes, macrophages, Langerhan's cells and dendritic cells [38], and mediates systemic inflammatory responses [37]. This is corroborated by the fact that while IL-1α may act as a membrane-bound pro-IL-1α through myristolation of the protein [100, 181], IL-1β requires intracellular processing for activation.

As mentioned, IL-1α and IL-1β are both produced as pro-peptides and while IL-1α can mediate biological effects as pro-IL-1α or as a mature IL-1α following enzymatic cleavage (e.g. with calpain) [96], only the proteolytically cleaved mature IL-1β, but not the pro-IL-1β can elicit immune functions. In fact, the release of IL-1β appears to be a two-step process, whereby the first stimulus such myeloid cell encounter of a pathogen results in increased synthesis and cytosolic accumulation of pro-IL-1β, while a second (as of yet poorly defined) stimulus induces the proteolytic processing of pro-IL-1β by caspase-1/IL-1β-converting enzyme (ICE), and the subsequent release of the biologically active mature IL-1β [39]. This processing event is further enhanced by the presence of extracellular ATP, which is recognized by the ATP receptor $P2X_7R$, causing an efflux of K^+ and a concomitant activation of ICE [50]. A number of mechanisms have been proposed for the release of the mature IL-1β into the extracellular milieu. These include: exocytosis of IL-1β containing lysosomes; through microvesicular budding of the plasma membrane; release in exosomes by fusion of multivesicular bodies with the plasma membrane; export through specific membrane transporters; and cell-lysis-mediated release (reviewed in [46]). Possibly, these mechanisms are mutually exclusive and/or cell type dependent. Further research in multiple model organism and cell types is warranted before specific mechanism(s) can be defined.

Both IL-1α and IL-1β bind to IL-1RI, resulting in recruitment of the IL-1R associated protein (IL-1RAcP), which amplifies the signal transduction [176]. The IL-1RI is structurally related to the toll-like pattern recognition receptors where the signal propagation through the IL-1RI involves many of the same downstream signaling components (MyD88, IL-1R associated kinase (IRAK), and NFκB) as those employed in TLR signaling [139]. IL-1α and IL-1β also bind to the IL-1RII, but because this receptor lacks the intracellular signaling components of IL-1RI, it functionally serves as a "decoy" receptor by dampening the IL-1α/β signal transduction [31]. Additionally, the IL-1 receptor antagonist (IL-1Ra) exhibits competitive inhibition of the IL-1α/β signaling by interacting with the IL-1RI without eliciting the downstream signaling events [31].

Unlike most other cyokines and growth factors IL-1α and IL-1β (primarily IL-1β) target nearly every cell type and induce a range of biological processes (reviewed in references [37, 38, 41]). Some of the numerous pro-inflammatory roles of IL-1β include increases in collagen and pro-collagenase synthesis; increase of chondrocyte protease and proteoglycan release; increase in osteoclast-activating factor release and hence bone resorption; induction of the synthesis of lipid mediators such as PGE_2; and enhancement of the proliferation of fibroblasts, keratinocytes, mesangial, glial cells and smooth muscle cells. IL-1β also induces chemotaxis of T and B cells, the synthesis of thromboxane by neutrophils and monocytes, basophil histamine release and eosinophil degranulation. Additionally, IL-1β induces synthesis of type I IFNs, endothelial plasminogen activator inhibitors, and expression of leukocyte adherence receptors on endothelial cell surfaces. Based on the above, it is not surprising that IL-1β has been implicated in numerous disorders including cardiac disease [15], rheumatoid arthritis [92], and neurodegenerative diseases [71].

3.3.2. Identification of IL-1 in fish

Fish were first suspected of possessing an IL-1 homolog when the mammalian PBL-derived IL-1 was demonstrated to enhance the proliferation of catfish T-lymphocytes in response to ConA [74], while carp epithelial cells [175], carp macrophages and granulocytes [195] and catfish monocytes [47], were shown to produce factor(s) with properties similar to those of the mammalian IL-1. Gel filtration analysis of catfish monocyte revealed two distinct bands of approximately 70 kDa and 15 kDa, recognized by polyclonal antibodies raised against mammalian IL-1α and IL-1β. Paradoxically, the catfish 70 kDa molecule activated catfish (but not mouse) T cells, while the 15 kDa molecule activated mouse (but not catfish) T cells [47].

The first fish IL-1β cDNA sequence was identified in trout and exhibited 49-56% amino acid identity to the mammalian IL-1β [170]. Notably, the trout IL-1β did not possess a putative ICE cleavage site required for the maturation-cleavage of the mammalian IL-1β, while the gene expression of this trout cytokine could be induced in tissues and head kidney leukocytes following LPS stimulation [170, 216], suggesting its pro-inflammatory nature.

Although there have been no reports of a fish IL-1α, recently a novel IL-1 family member, nIL-1F has been identified in trout and *Tetraodon* spp. and was reported to have IL-1 family signature motif as well as a putative ICE cleavage site [198]. The expression of nIL-1F increased after activation of macrophages with LPS or recombinant IL-1β. Furthermore, a recombinant form of the C-terminal nIL-1F abrogated the rIL-1β-induced immune gene expression in RTS-11 trout macrophage-like cells, suggesting a possible competitive inhibition through the trout IL-1RI. This mechanism would be analogous to the mammalian IL-1Ra, which also competes with IL-1β for IL-1RI binding, without inducing downstream events.

3.3.3. Isoforms of IL-1β in fish

Because many fish species are tetraploid and have undergone genome duplication events, it is not surprising that additional isforms of IL-1β exist in certain fish species. An identified second trout IL-1β, denoted IL-1β2, also has the 6 exon/5 intron organization, 82% amino acid identity to the trout IL-1β1 and no putative ICE cleavage site [150]. In catfish, two IL-1β genes have been described and shown to undergo distinct expression patterns following challenge with *Edwardsiella ictaluri* [201]. Additionally identified cDNA of another carp IL-1β, IL-1β2, exhibited 74% identity with the carp IL-1β1 and 95-99% identity across individual IL-1β2 transcripts [48]. Since several distinct IL-1β2 sequences were identified in a homozygous individual, it was suggested that there might be multiple IL-1β2 genes. The expression of the carp IL-1β1 and IL-1β2 differed following immune stimuli where the expression of IL-1β1 gene was on average at least ten fold greater than that of IL-1β2. It was also reported that transcripts of IL-1β2 had high substitution numbers in the coding regions, including key areas predicted to be involved in receptor binding. In light of the above, it was suggested that the IL-1β2 may be a pseudogene [48]. Our observations in the goldfish, a close carp relative, support this theory. Notably, the predicted goldfish IL-1β2 protein is truncated compared to the IL-1β1, and while the expression of both IL-1β isoforms is subject to change following immune stimuli, specific treatments elicit more robust changes in the expression of IL-1β1 compared to IL-1β2 (unpublished observations).

3.3.4. Maturation cleavage of the fish IL-1β

As alluded to above, despite the functional similarities of the fish IL-1βs and their mammalian counterpart, all fish IL-1β proteins identified to date lack the typical ICE cleavage site necessary for the functional maturation of the mammalian IL-1β1. Despite this, evidence suggests that indeed the fish IL-1β proteins undergo cleavage. Early fish IL-1 studies using anti-mammalian IL-1α/β serum to surveys supernatants of activated fish cells demonstrated the recognition of distinct, multiple IL-1 protein species in these supernatants. While the catfish monocyte supernatants contained 70 kDa and 15 kDa molecules with IL-1-like activity [47], carp macrophages secreted 22 kDa and 15 kDa IL-1-like factors that were recognized by mammalian anti-IL-1 antibodies [195]. The immunoprecipitation experiments using the anti-mammalian IL-1 polyclonal antibodies recognized the 15kDa factor from the macrophage supernatants, suggesting possible maturation events of the carp IL-1 [195]. In an independent survey of PHA-stimulated carp leukocytes using a monoclonal anti-carp IL-1β antibody, the 15 kDa protein was confirmed to be a mature carp IL-1β, with the putative carp IL-1β cleavage site situated approximately 15 amino acids downstream of the mammalian ICE site [117]. Analysis of the supernatants of the trout macrophage cell line, RTS-11, using an anti-trout IL-1β polyclonal antibody detected putative native, as well as potentially cleaved trout IL-1β proteins of 29 kDa and 24 kDa, respectively [79]. When the RTS-11 cells were transfected with a plasmid encoding a C-terminally His[6] tagged trout IL-1β, and the RTS-11 supernatants assayed using an Ni-NTA-column specific for the His[6] tagged IL-1β, the authors detected primarily a mature, 24 kDa IL-1β protein, confirming that the trout IL-1β underwent maturation cleavage.

In an elegant set of studies it was demonstrated that combined immune stimuli of sea bream head kidney leukocytes resulted in an accumulation of a 30 kDa, pro-IL-1β protein, which, unlike the mammalian counterpart [113], did not exhibit a maturation cleavage or secretion following stimulation of the cells with extra-cellular ATP [146]. Furthermore, sea bream peritoneal acidophilic granulocytes and peripheral blood leukocytes accumulated the 30 kDa form of IL-1β following challenge with *Vibrio anguillarum* [27]. In contrast, the sea bream SAF-1 fibroblast cell line shed a mature, 22 kDa IL-1β protein through microvesicular plasma membrane budding within 30 minutes of treatment with extracellular ATP [113]. Interestingly this IL-1β maturation/shedding process could be ablated with a pharmacological inhibitor of the mammalian ICE, suggesting a presence of an orthologous sea bream enzyme responsible for this process.

The P2X₇R receptor is the primary receptor responsible for the recognition of extracellular ATP and the concomitant release of mature IL-1β [50]. Upon ATP treatment and activation of the HEK 293 mammalian cell line expressing the rat P2X₇R, human ICE and sea bream IL-1β, a non-cleaved (30 kDa) sea bream IL-1β was secreted by these cells [109]. Interestingly, neither sea bream nor zebrafish P2X₇R expression in ATP-stimulated HEK 293 cells resulted in sea bream or mammalian IL-1β secretion. In contrast, the expression of a chimeric P2X₇R bearing the sea bream ATP-binding and rat intracellular domains led to maturation/secretion of the mammalian IL-1β, while the outcomes of this combination on the sea bream IL-1β were not addressed [109]. The authors of this work suggest that the mechanisms involved in IL-1β secretion are conserved across vertebrates while the distinct stimuli that elicit the maturation events are not. We propose that alternatively, functional specificity may stem from the P2X₇R intracellular signaling, activation of species specific ICE (or alternative maturation mechanisms), and the substrate specificity of the fish Caspase1/ICE. Notably, the above group also identified the sea bream Caspase 1 and demonstrated that the recombinant form of this fish enzyme cleaved a commercially available substrate, for which the mammalian counterpart holds specificity [107]. Additional studies are needed to establish the association between the sea bream Caspase I/ICE, P2X₇R and IL-1β.

3.3.5. IL-1 receptors of fish

Majority of the IL-1 receptor signaling components have been described in fish. A fish IL-1RI was first described in Atlantic salmon, with 43-44% similarity to chicken and 31% similarity to the human IL-1RI, respectively [186]. The expression of this gene increased in fish tissues following LPS injection. The identification of a carp IL-1R1 was also described and the receptor characterized in the context of acute stress conditions [121].

The trout IL-1RII was identified through a selective subtractive hybridization of genes up-regulated following immune stimulation [164]. This receptor displayed low sequence identity with the mammalian IL-1RII, but exhibited surprisingly similar overall gene organization including a very short intracellular domain. The gene expression of the identified sea bream IL-1RII increased in stimulated macrophages to levels 15 times greater than IL-1β expression [108], suggesting a conservation in the roles of the sea bream IL-1RII as a "decoy" receptor. Furthermore, sea bream IL-1RII expressed on HEK 293 cells bound the recombinant IL-1β, confirming the specificity of this receptor-ligand pair.

Enhanced understanding of the evolutionary mechanisms responsible for shaping the vertebrate IL-1 biological pathways will be achieved through studies that examine the IL-1 receptor-ligand relationships and biological outcomes of these interactions in teleosts.

3.3.6. Inflammatory roles of the fish IL-1β

Using the selective subtractive hybridization technique, a carp IL-1β was identified [53] and the C-terminus of the cytokine produced as a recombinant protein [206]. This recombinant carp IL-1β dimerized and enhanced carp antibody responses to *Aeromonas hydrophila*, where sera from carp co-injected with the IL-1β and *A. hydrophila* had a greater agglutinating capacity than the respective controls.

A number of studies have since utilized recombinant technology to investigate the functions of the fish IL-1β. A recombinant form of the mature trout IL-1β (rtIL-1β) was produced by Hong and co-workers [80] and shown to enhance the expression of the MHCII β chain, IL-1β and COX-2 genes in trout head kidney leukocytes and the macrophage cell line, RTS-11. Functionally, rtIL-1β elicited the proliferation of trout head kidney cells as well as the proliferation of a murine cell line, D10.G4.1, known for its dependence on the mammalian IL-1β [80]. Also, rtIL-1β enhanced the phagocytosis of yeast particles by trout head kidney cells [80] while peritoneal adminnistrations of rtIL-1β induced migration of trout leukocytes to the site of injection, enhanced phagocytosis of peritoneal cells and increased the systemic expression of IL-1β, COX-2 and lysozyme II [77]. The injection of fish with rtIL-1β also enhanced trout resistance to infection with the fish pathogen *A. salmonicida*. Additionally, the rtIL-1β and the recombinant sea bass IL-1β were demonstrated to induce Ca^{2+} mediated downstream signaling events, abrogated by leukocyte trypsin-treatments and indicating a requirement for receptor engagement [9]. Interestingly, these authors reported that IL-1β of trout, sea bass and humans were highly species specific, which is in contradiction of the early fish IL-1β carp and catfish work, where cross-reactivity was observed [9, 47, 74, 175, 195].

The distinct biological roles of individual sub-domains of the trout IL-1β were also examined by generating appropriate peptides [78, 144, 145]. While a control scrambled peptide (P2) had no effect and the peptide corresponding to the putative trout IL-1β receptor binding region (P1) had little effect on its own, a peptide (P3) corresponding to an alternative receptor binding area of the mammalian IL-1β [5, 196], had chemotactic properties towards head kidney cells alone or in combination with the P1 peptide (at low concentrations) [145]. Intraperitoneal injections of the rtIL-1β P3 peptide induced peritoneal leukocyte migration and enhanced phagocytosis and reactive oxygen production by peritoneal cells. Additionally, the rtIL-1β P3 conferred fish resistance to viral hemorrhagic septicemia virus (VHCV) early after injection. Surprisingly, injection of rtIL-1β P1 caused an increase in *in vivo* expression of the antiviral gene Mx and inhibition of trout TNFα expression, while having no effects on the expression of a panel of other pro-inflammatory genes [78]. The injection of the rtIL-1β P3 peptide resulted in a more robust, widespread pro-inflammatory response with increased expression in trout rtIL-1β, IL-8 and lysozyme. This work suggests that the fish IL-1β utilizes a highly complex receptor-ligand system, possibly unique to that of mammals.

The functional roles of the trout IL-1β have been corroborated in others fish species. For example, injection of carp with a plasmid encoding the carp IL-1β gene caused enhanced PHA-induced proliferation of carp lymphocytes, increased carp macrophage reactive oxygen production, enhanced phagocytosis and improved protection against *A. hydophila* challenge [97]. A sea bass recombinant IL-1β exhibited immuno-adjuvant properties when combined with rsIL-1β in immunization trials against the pathogen, *V. anguillarum*. The sea bass IL-1β also induced the proliferation of the murine IL-1β reporter cell line (D10.G4.1), sea bass thymocytes, enhanced kidney leukocyte phagocytosis and activated peritoneal macrophages when administered i.p. [16, 17, 29]. The orange spotted grouper rIL-1β stimulated the proliferation of grouper head kidney leukocytes and increased the gene expression of IL-1β and COX-2 through p38 MAPK and Jnk signaling pathways [112]. Together, the above findings suggest that the functions of IL-1β have been evolutionarily conserved in vertebrates.

4. Concluding remarks

A successful inflammatory response is defined by the presence and proficient coordination of cytokine networks consisting of hallmark mediators such as TNFα, IFNγ and IL-1β. The synchronized involvement of these pleiotropic yet functionally distinct agents in the recruitment, regulation and functional polarization of inflammatory cells dictates the outcome of the mounted response. Thus, it can be argued that the inflammatory processes are largely defined by the efficacy of the individual and interdependent cytokine pathways.

The regulation of the vertebrate inflammatory response is complex, involving numerous mechanisms, some of which are poorly understood while others remain to be identified. This is particularly true for the teleost model systems, where lack of specific reagents for different fish species hampers our ability to examine different aspects of the regulation of inflammation at a mechanistic level. However, there is growing evidence that the key immune components required for effective inflammatory responses are present in teleosts. Notably, certain fish species possess additional pathways that regulate inflammatory processes (for example IFNγrel and its receptor IFNGR1-1, novel chemokines and PRRs) that are distinct from those reported in mammals. The elucidation of the coordination of inflammatory responses by these factors may shed new light on the evolution of innate host defense mechanisms in lower vertebrates.

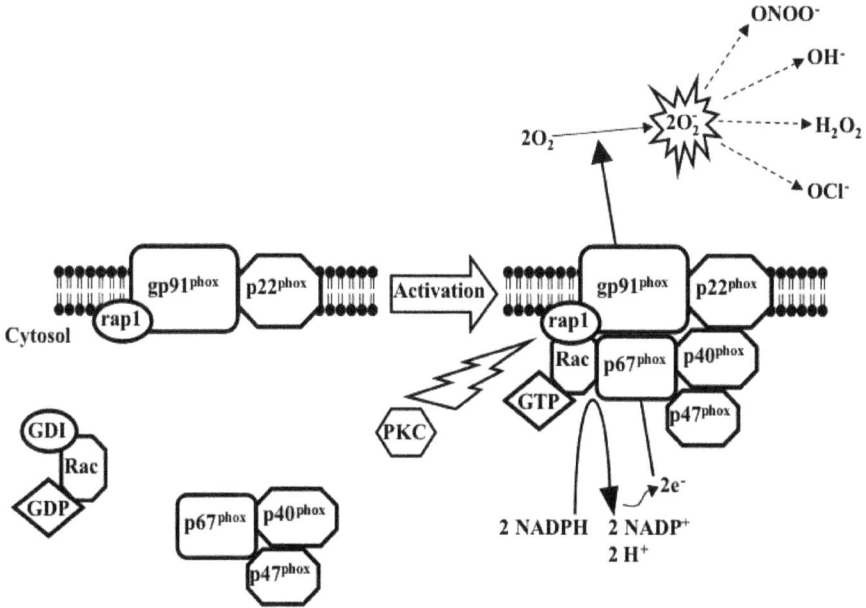

Figure 1. Schematic representation of the NADPH oxidase complex mobilization and reactive oxygen production follow-ing phagocyte activation. In a resting state, the gp91phox and p22 phox components are membrane bound while the p40 phox, p47 phox and p67 phox components are located in the cytosol and the small G-protein Rac is GDP bound. Upon cell acti-vation, Rac is rapidly converted from a GDP- to a GTP-bound state and facilitates the translocation and assembly of the cytosolic NADPH oxidase components at the cell membrane. Following PKC activation, Rap1 is thought to serve as the fi-nal switch in the activation of the NADPH complex. The activated NADPH oxidase complex accepts the electrons form the reduced NADPH and transfers these to molecular oxygen, forming the superoxide anion (O_2^-). The generated O_2^- can sub-sequently be converted to other reactive oxygen species. Reviewed in references [124, 153, 154, 173].

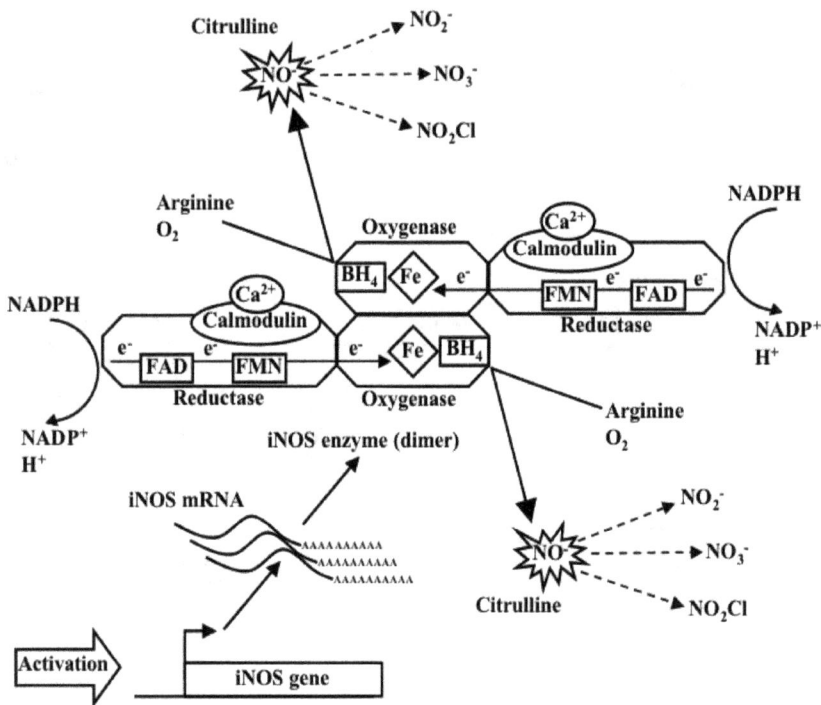

Figure 2. Schematic representation of iNOS gene expression, iNOS protein synthesis and enzymatic production of nitric oxide by the iNOS dimer complex. Upon phagocyte activation, there is a substantial increase in the expression from the iNOS gene and de novo synthesis of the iNOS enzyme. The iNOS enzyme forms a dimer and associates with a Ca^{2+} bound calmodulin, stabilizing the structure and facilitating the function of the enzymatic complex. The iNOS dimer catalyzes the transfer of electrons from a reduced NADPH, through a series of cofactors (FAD, FMN, Fe, BH_4) in the oxidation of L-Arginine to L-Citrulline with a concomitant production of nitric oxide (NO·). The generated NO· can subsequently be converted to other reactive nitrogen species. Reviewed in references [2, 111, 126, 172].

Abbreviations

ConA: concanavalin A; **GAS:** γ-IFN-activated sequence; **Jak:** janus activated kinase; **ICE:** IL-1 cleaving enzyme; **IFN:** interferon; **IFNGR:** interferon gamma receptor; **IL:** interleukin; **IL-1R:** interleukin-1 receptor; **IL-1RAcP:** IL-1R associated protein; **iNOS:** inucible nitric oxides synthase; **IRAK:** IL-1R associated kinase; **IRF:** interferon regulatory factor; **MAF:** macrophage activating factor(s); **NK:** natural killer; **NLS:** nuclear localization signal; **NO:** nitric oxide; **NTR:** neurotropin receptor; **PBL:** peripheral blood leukocyte; **PHA:** phytohemaglutanin; **PKC:** protein kinase C; **PKM:** primary kidney macrophage; **PMA:** phorbol myrystate acitate; **PRR:** pattern recognition receptor; **rg:** recombinant goldfish; **RNI:** reactive nitrogen intermediates; **ROI:** reactive oxygen intermediates; **Stat:** signal transducer of activation transcription factor; **TACE:** TNFα cleaving enzyme; **TLR:** toll-like receptor; **TNF:** tumor necrosis factor; **TNFR:** tumor necrosis factor receptor.

Author details

Leon Grayfer[1] and Miodrag Belosevic[1,2*]

*Address all correspondence to: mike.belosevic@ualberta.ca

1 Department of Biological Sciences, University of Alberta, Canada

2 School of Public Health, University of Alberta, Canada

References

[1] Aggad, D., C. Stein, D. Sieger, M. Mazel, P. Boudinot, P. Herbomel, J. P. Levraud, G. Lutfalla, and M. Leptin. 2010. In vivo analysis of Ifn-gamma1 and Ifn-gamma2 signaling in zebrafish. J Immunol 185:6774-6782.

[2] Andrew, P. J., and B. Mayer. 1999. Enzymatic function of nitric oxide synthases. Cardiovasc Res 43:521-531.

[3] Arts, J. A., E. J. Tijhaar, M. Chadzinska, H. F. Savelkoul, and B. M. Verburg-van Kemenade. 2010. Functional analysis of carp interferon-gamma: evolutionary conservation of classical phagocyte activation. Fish Shellfish Immunol 29:793-802.

[4] Auffray, C., D. Fogg, M. Garfa, G. Elain, O. Join-Lambert, S. Kayal, S. Sarnacki, A. Cumano, G. Lauvau, and F. Geissmann. 2007. Monitoring of blood vessels and tissues by a population of monocytes with patrolling behavior. Science 317:666-670.

[5] Auron, P. E. 1998. The interleukin 1 receptor: ligand interactions and signal transduction. Cytokine Growth Factor Rev 9:221-237.

[6] Auron, P. E., A. C. Webb, L. J. Rosenwasser, S. F. Mucci, A. Rich, S. M. Wolff, and C. A. Dinarello. 1984. Nucleotide sequence of human monocyte interleukin 1 precursor cDNA. Proc Natl Acad Sci U S A 81:7907-7911.

[7] Banner, D. W., A. D'Arcy, W. Janes, R. Gentz, H. J. Schoenfeld, C. Broger, H. Loetscher, and W. Lesslauer. 1993. Crystal structure of the soluble human 55 kd TNF receptor-human TNF beta complex: implications for TNF receptor activation. Cell 73:431-445.

[8] Belosevic, M., C. E. Davis, M. S. Meltzer, and C. A. Nacy. 1988. Regulation of activated macrophage antimicrobial activities. Identification of lymphokines that cooperate with IFN-gamma for induction of resistance to infection. J Immunol 141:890-896.

[9] Benedetti, S., E. Randelli, F. Buonocore, J. Zou, C. J. Secombes, and G. Scapigliati. 2006. Evolution of cytokine responses: IL-1beta directly affects intracellular Ca2+ concentration of teleost fish leukocytes through a receptor-mediated mechanism. Cytokine 34:9-16.

[10] Berton, G., L. Zeni, M. A. Cassatella, and F. Rossi. 1986. Gamma interferon is able to enhance the oxidative metabolism of human neutrophils. Biochem Biophys Res Commun 138:1276-1282.

[11] Bluyssen, H. A., R. Muzaffar, R. J. Vlieststra, A. C. van der Made, S. Leung, G. R. Stark, I. M. Kerr, J. Trapman, and D. E. Levy. 1995. Combinatorial association and abundance of components of interferon-stimulated gene factor 3 dictate the selectivity of interferon responses. Proc Natl Acad Sci U S A 92:5645-5649.

[12] Bobe, J., and F. W. Goetz. 2001. Molecular cloning and expression of a TNF receptor and two TNF ligands in the fish ovary. Comp Biochem Physiol B Biochem Mol Biol 129:475-481.

[13] Boltana, S., C. Donate, F. W. Goetz, S. MacKenzie, and J. C. Balasch. 2009. Characterization and expression of NADPH oxidase in LPS-, poly(I:C)- and zymosan-stimulated trout (Oncorhynchus mykiss W.) macrophages. Fish Shellfish Immunol 26:651-661.

[14] Bosco, M. C., G. L. Gusella, I. Espinoza-Delgado, D. L. Longo, and L. Varesio. 1994. Interferon-gamma upregulates interleukin-8 gene expression in human monocytic cells by a posttranscriptional mechanism. Blood 83:537-542.

[15] Bujak, M., and N. G. Frangogiannis. 2009. The role of IL-1 in the pathogenesis of heart disease. Arch Immunol Ther Exp (Warsz) 57:165-176.

[16] Buonocore, F., M. Forlenza, E. Randelli, S. Benedetti, P. Bossu, S. Meloni, C. J. Secombes, M. Mazzini, and G. Scapigliati. 2005. Biological activity of sea bass (Dicentrarchus labrax L.) recombinant interleukin-1beta. Mar Biotechnol (NY) 7:609-617.

[17] Buonocore, F., M. Mazzini, M. Forlenza, E. Randelli, C. J. Secombes, J. Zou, and G. Scapigliati. 2004. Expression in Escherchia coli and purification of sea bass (Dicen-

trarchus labrax) interleukin 1beta, a possible immunoadjuvant in aquaculture. Mar Biotechnol (NY) 6:53-59.

[18] Cai, Z. H., L. S. Song, C. P. Gao, L. T. Wu, L. H. Qiu, and J. H. Xiang. 2003. [Cloning and expression of tumor necrosis factor (TNFalpha) cDNA from red seabream pagrus major]. Sheng Wu Hua Xue Yu Sheng Wu Wu Li Xue Bao (Shanghai) 35:1111-1116.

[19] Camussi, G., F. Bussolino, G. Salvidio, and C. Baglioni. 1987. Tumor necrosis factor/ cachectin stimulates peritoneal macrophages, polymorphonuclear neutrophils, and vascular endothelial cells to synthesize and release platelet-activating factor. J Exp Med 166:1390-1404.

[20] Carswell, E. A., L. J. Old, R. L. Kassel, S. Green, N. Fiore, and B. Williamson. 1975. An endotoxin-induced serum factor that causes necrosis of tumors. Proc Natl Acad Sci U S A 72:3666-3670.

[21] Cassatella, M. A., F. Bazzoni, R. M. Flynn, S. Dusi, G. Trinchieri, and F. Rossi. 1990. Molecular basis of interferon-gamma and lipopolysaccharide enhancement of phagocyte respiratory burst capability. Studies on the gene expression of several NADPH oxidase components. The Journal of biological chemistry 265:20241-20246.

[22] Cassatella, M. A., V. Della Bianca, G. Berton, and F. Rossi. 1985. Activation by gamma interferon of human macrophage capability to produce toxic oxygen molecules is accompanied by decreased Km of the superoxide-generating NADPH oxidase. Biochem Biophys Res Commun 132:908-914.

[23] Cassatella, M. A., I. Guasparri, M. Ceska, F. Bazzoni, and F. Rossi. 1993. Interferon-gamma inhibits interleukin-8 production by human polymorphonuclear leucocytes. Immunology 78:177-184.

[24] Castro, R., S. A. Martin, S. Bird, J. Lamas, and C. J. Secombes. 2008. Characterisation of gamma-interferon responsive promoters in fish. Molecular immunology 45:3454-3462.

[25] Castro, R., S. A. Martin, J. Zou, and C. J. Secombes. 2010. Establishment of an IFN-gamma specific reporter cell line in fish. Fish Shellfish Immunol 28:312-319.

[26] Chapman, B. S., and I. D. Kuntz. 1995. Modeled structure of the 75-kDa neurotrophin receptor. Protein Sci 4:1696-1707.

[27] Chaves-Pozo, E., P. Pelegrin, J. Garcia-Castillo, A. Garcia-Ayala, V. Mulero, and J. Meseguer. 2004. Acidophilic granulocytes of the marine fish gilthead seabream (Sparus aurata L.) produce interleukin-1beta following infection with Vibrio anguillarum. Cell Tissue Res 316:189-195.

[28] Chen, G., and D. V. Goeddel. 2002. TNF-R1 signaling: a beautiful pathway. Science 296:1634-1635.

[29] Chistiakov, D. A., F. V. Kabanov, O. D. Troepolskaya, and M. M. Tischenko. 2010. A variant of the interleukin-1beta gene in European sea bass, Dicentrarchus labrax L., is

associated with increased resistance against Vibrio anguillarum. J Fish Dis 33:759-767.

[30] Clay, H., H. E. Volkman, and L. Ramakrishnan. 2008. Tumor necrosis factor signaling mediates resistance to mycobacteria by inhibiting bacterial growth and macrophage death. Immunity 29:283-294.

[31] Colotta, F., F. Re, M. Muzio, R. Bertini, N. Polentarutti, M. Sironi, J. G. Giri, S. K. Dower, J. E. Sims, and A. Mantovani. 1993. Interleukin-1 type II receptor: a decoy target for IL-1 that is regulated by IL-4. Science 261:472-475.

[32] Cooper, A. M., D. K. Dalton, T. A. Stewart, J. P. Griffin, D. G. Russell, and I. M. Orme. 1993. Disseminated tuberculosis in interferon gamma gene-disrupted mice. The Journal of experimental medicine 178:2243-2247.

[33] Darnell, J. E., Jr., I. M. Kerr, and G. R. Stark. 1994. Jak-STAT pathways and transcriptional activation in response to IFNs and other extracellular signaling proteins. Science 264:1415-1421.

[34] Debets, R., J. C. Timans, B. Homey, S. Zurawski, T. R. Sana, S. Lo, J. Wagner, G. Edwards, T. Clifford, S. Menon, J. F. Bazan, and R. A. Kastelein. 2001. Two novel IL-1 family members, IL-1 delta and IL-1 epsilon, function as an antagonist and agonist of NF-kappa B activation through the orphan IL-1 receptor-related protein 2. J Immunol 167:1440-1446.

[35] Denicola, A., H. Rubbo, D. Rodriguez, and R. Radi. 1993. Peroxynitrite-mediated cytotoxicity to Trypanosoma cruzi. Arch Biochem Biophys 304:279-286.

[36] Dhawan, S., A. Heredia, R. B. Lal, L. M. Wahl, J. S. Epstein, and I. K. Hewlett. 1994. Interferon-gamma induces resistance in primary monocytes against human immunodeficiency virus type-1 infection. Biochem Biophys Res Commun 201:756-761.

[37] Dinarello, C. A. 1996. Biologic basis for interleukin-1 in disease. Blood 87:2095-2147.

[38] Dinarello, C. A. 1988. Biology of interleukin 1. Faseb J 2:108-115.

[39] Dinarello, C. A. 1998. Interleukin-1 beta, interleukin-18, and the interleukin-1 beta converting enzyme. Ann N Y Acad Sci 856:1-11.

[40] Dinarello, C. A. 1998. Interleukin-1, interleukin-1 receptors and interleukin-1 receptor antagonist. Int Rev Immunol 16:457-499.

[41] Dinarello, C. A. 1987. The biology of interleukin 1 and comparison to tumor necrosis factor. Immunol Lett 16:227-231.

[42] Ding, A. H., C. F. Nathan, and D. J. Stuehr. 1988. Release of reactive nitrogen intermediates and reactive oxygen intermediates from mouse peritoneal macrophages. Comparison of activating cytokines and evidence for independent production. J Immunol 141:2407-2412.

[43] Dranoff, G. 2004. Cytokines in cancer pathogenesis and cancer therapy. Nat Rev Cancer 4:11-22.

[44] Dubravec, D. B., D. R. Spriggs, J. A. Mannick, and M. L. Rodrick. 1990. Circulating human peripheral blood granulocytes synthesize and secrete tumor necrosis factor alpha. Proc Natl Acad Sci U S A 87:6758-6761.

[45] Eck, M. J., B. Beutler, G. Kuo, J. P. Merryweather, and S. R. Sprang. 1988. Crystallization of trimeric recombinant human tumor necrosis factor (cachectin). J Biol Chem 263:12816-12819.

[46] Eder, C. 2009. Mechanisms of interleukin-1beta release. Immunobiology 214:543-553.

[47] Ellsaesser, C. F., and L. W. Clem. 1994. Functionally distinct high and low molecular weight species of channel catfish and mouse IL-1. Cytokine 6:10-20.

[48] Engelsma, M. Y., R. J. Stet, J. P. Saeij, and B. M. Verburg-van Kemenade. 2003. Differential expression and haplotypic variation of two interleukin-1beta genes in the common carp (Cyprinus carpio L.). Cytokine 22:21-32.

[49] Farrar, M. A., J. Fernandez-Luna, and R. D. Schreiber. 1991. Identification of two regions within the cytoplasmic domain of the human interferon-gamma receptor required for function. The Journal of biological chemistry 266:19626-19635.

[50] Ferrari, D., C. Pizzirani, E. Adinolfi, R. M. Lemoli, A. Curti, M. Idzko, E. Panther, and F. Di Virgilio. 2006. The P2X7 receptor: a key player in IL-1 processing and release. J Immunol 176:3877-3883.

[51] Fertsch, D., and S. N. Vogel. 1984. Recombinant interferons increase macrophage Fc receptor capacity. J Immunol 132:2436-2439.

[52] Forlenza, M., S. Magez, J. P. Scharsack, A. Westphal, H. F. Savelkoul, and G. F. Wiegertjes. 2009. Receptor-mediated and lectin-like activities of carp (Cyprinus carpio) TNF-alpha. J Immunol 183:5319-5332.

[53] Fujiki, K., D. H. Shin, M. Nakao, and T. Yano. 2000. Molecular cloning and expression analysis of carp (Cyprinus carpio) interleukin-1 beta, high affinity immunoglobulin E Fc receptor gamma subunit and serum amyloid A. Fish Shellfish Immunol 10:229-242.

[54] Gabbiani, G. 1979. The role of contractile proteins in wound healing and fibrocontractive diseases. Methods Achiev Exp Pathol 9:187-206.

[55] Gao, Q., P. Nie, K. D. Thompson, A. Adams, T. Wang, C. J. Secombes, and J. Zou. 2009. The search for the IFN-gamma receptor in fish: Functional and expression analysis of putative binding and signalling chains in rainbow trout Oncorhynchus mykiss. Developmental and comparative immunology 33:920-931.

[56] Garcia-Castillo, J., E. Chaves-Pozo, P. Olivares, P. Pelegrin, J. Meseguer, and V. Mulero. 2004. The tumor necrosis factor alpha of the bony fish seabream exhibits the in

vivo proinflammatory and proliferative activities of its mammalian counterparts, yet it functions in a species-specific manner. Cell Mol Life Sci 61:1331-1340.

[57] Gong, Y., P. Cao, H. J. Yu, and T. Jiang. 2008. Crystal structure of the neurotrophin-3 and p75NTR symmetrical complex. Nature 454:789-793.

[58] Graham, S., and C. J. Secombes. 1990. Do fish lymphocytes secrete interferon-y? J. Fish Biol. 36:563-573.

[59] Graham, S., and C. J. Secombes. 1988. The production of a macrophage-activating factor from rainbow trout Salmo gairdneri leucocytes. Immunology 65:293-297.

[60] Grayfer, L., and M. Belosevic. 2009. Molecular characterization of novel interferon gamma receptor 1 isoforms in zebrafish (Danio rerio) and goldfish (Carassius auratus L.). Molecular immunology 46:3050-3059.

[61] Grayfer, L., and M. Belosevic. 2009. Molecular characterization of tumor necrosis factor receptors 1 and 2 of the goldfish (Carassius aurutus L.). Molecular immunology 46:2190-2199.

[62] Grayfer, L., and M. Belosevic. 2009. Molecular characterization, expression and functional analysis of goldfish (Carassius aurutus L.) interferon gamma. Developmental and comparative immunology 33:235-246.

[63] Grayfer, L., E. G. Garcia, and M. Belosevic. 2010. Comparison of macrophage antimicrobial responses induced by type II interferons of the goldfish (Carassius auratus L.). J Biol Chem 285:23537-23547.

[64] Grayfer, L., P. C. Hanington, and M. Belosevic. 2009. Macrophage colony-stimulating factor (CSF-1) induces pro-inflammatory gene expression and enhances antimicrobial responses of goldfish (Carassius auratus L.) macrophages. Fish Shellfish Immunol 26:406-413.

[65] Grayfer, L., J. W. Hodgkinson, and M. Belosevic. 2011. Analysis of the antimicrobial responses of primary phagocytes of the goldfish (Carassius auratus L.) against Mycobacterium marinum. Developmental and comparative immunology 35:1146-1158.

[66] Grayfer, L., J. W. Hodgkinson, S. J. Hitchen, and M. Belosevic. 2010. Characterization and functional analysis of goldfish (Carassius auratus L.) interleukin-10. Molecular immunology doi:10.1016/j.molimm.2010.10.013.

[67] Grayfer, L., J. G. Walsh, and M. Belosevic. 2008. Characterization and functional analysis of goldfish (Carassius auratus L.) tumor necrosis factor-alpha. Developmental and comparative immunology 32:532-543.

[68] Greenlund, A. C., M. A. Farrar, B. L. Viviano, and R. D. Schreiber. 1994. Ligand-induced IFN gamma receptor tyrosine phosphorylation couples the receptor to its signal transduction system (p91). The EMBO journal 13:1591-1600.

[69] Greenlund, A. C., M. O. Morales, B. L. Viviano, H. Yan, J. Krolewski, and R. D. Schreiber. 1995. Stat recruitment by tyrosine-phosphorylated cytokine receptors: an ordered reversible affinity-driven process. Immunity 2:677-687.

[70] Grell, M., E. Douni, H. Wajant, M. Lohden, M. Clauss, B. Maxeiner, S. Georgopoulos, W. Lesslauer, G. Kollias, K. Pfizenmaier, and P. Scheurich. 1995. The transmembrane form of tumor necrosis factor is the prime activating ligand of the 80 kDa tumor necrosis factor receptor. Cell 83:793-802.

[71] Griffin, W. S., J. G. Sheng, M. C. Royston, S. M. Gentleman, J. E. McKenzie, D. I. Graham, G. W. Roberts, and R. E. Mrak. 1998. Glial-neuronal interactions in Alzheimer's disease: the potential role of a 'cytokine cycle' in disease progression. Brain Pathol 8:65-72.

[72] Gupta, J. W., M. Kubin, L. Hartman, M. Cassatella, and G. Trinchieri. 1992. Induction of expression of genes encoding components of the respiratory burst oxidase during differentiation of human myeloid cell lines induced by tumor necrosis factor and gamma-interferon. Cancer Res 52:2530-2537.

[73] Hajjar, K. A., D. P. Hajjar, R. L. Silverstein, and R. L. Nachman. 1987. Tumor necrosis factor-mediated release of platelet-derived growth factor from cultured endothelial cells. J Exp Med 166:235-245.

[74] Hamby, B. A., E. M. Huggins, Jr., L. B. Lachman, C. A. Dinarello, and M. M. Sigel. 1986. Fish lymphocytes respond to human IL-1. Lymphokine Res 5:157-162.

[75] Hardie, L. J., L. H. Chappell, and C. J. Secombes. 1994. Human tumor necrosis factor alpha influences rainbow trout Oncorhynchus mykiss leucocyte responses. Vet Immunol Immunopathol 40:73-84.

[76] Hirono, I., B. H. Nam, T. Kurobe, and T. Aoki. 2000. Molecular cloning, characterization, and expression of TNF cDNA and gene from Japanese flounder Paralychthys olivaceus. J Immunol 165:4423-4427.

[77] Hong, S., S. Peddie, J. J. Campos-Perez, J. Zou, and C. J. Secombes. 2003. The effect of intraperitoneally administered recombinant IL-1beta on immune parameters and resistance to Aeromonas salmonicida in the rainbow trout (Oncorhynchus mykiss). Developmental and comparative immunology 27:801-812.

[78] Hong, S., and C. J. Secombes. 2009. Two peptides derived from trout IL-1beta have different stimulatory effects on immune gene expression after intraperitoneal administration. Comp Biochem Physiol B Biochem Mol Biol 153:275-280.

[79] Hong, S., J. Zou, B. Collet, N. C. Bols, and C. J. Secombes. 2004. Analysis and characterisation of IL-1beta processing in rainbow trout, Oncorhynchus mykiss. Fish Shellfish Immunol 16:453-459.

[80] Hong, S., J. Zou, M. Crampe, S. Peddie, G. Scapigliati, N. Bols, C. Cunningham, and C. J. Secombes. 2001. The production and bioactivity of rainbow trout (Oncorhynchus mykiss) recombinant IL-1 beta. Vet Immunol Immunopathol 81:1-14.

[81] Huang, S. H., W., Althage, A., Hemmi, S., Bluethmann, H., Kamijo, R., Vilcek, J., Zin-
      kernagel, RM., Aguet, M.. 1993. Immune response in mice that lack the interferon-γ
      receptor. Science (New York, N.Y 259.

[82] Igawa, D., M. Sakai, and R. Savan. 2006. An unexpected discovery of two interferon
      gamma-like genes along with interleukin (IL)-22 and -26 from teleost: IL-22 and -26
      genes have been described for the first time outside mammals. Molecular immunolo-
      gy 43:999-1009.

[83] Ihle, J. N., and I. M. Kerr. 1995. Jaks and Stats in signaling by the cytokine receptor
      superfamily. Trends Genet 11:69-74.

[84] Inoue, Y., S. Kamota, K. Ito, Y. Yoshiura, M. Ototake, T. Moritomo, and T. Nakanishi.
      2005. Molecular cloning and expression analysis of rainbow trout (Oncorhynchus
      mykiss) interleukin-10 cDNAs. Fish Shellfish Immunol 18:335-344.

[85] Ishibe, K., T. Yamanishi, Y. Wang, K. Osatomi, K. Hara, K. Kanai, K. Yamaguchi, and
      T. Oda. 2009. Comparative analysis of the production of nitric oxide (NO) and tumor
      necrosis factor-alpha (TNF-alpha) from macrophages exposed to high virulent and
      low virulent strains of Edwardsiella tarda. Fish Shellfish Immunol 27:386-389.

[86] Iyengar, R., D. J. Stuehr, and M. A. Marletta. 1987. Macrophage synthesis of nitrite,
      nitrate, and N-nitrosamines: precursors and role of the respiratory burst. Proc Natl
      Acad Sci U S A 84:6369-6373.

[87] Jang, S. I., L. J. Hardie, and C. J. Secombes. 1995. Elevation of rainbow trout Onco-
      rhynchus mykiss macrophage respiratory burst activity with macrophage-derived
      supernatants. Journal of leukocyte biology 57:943-947.

[88] Jurecka, P., I. Irnazarow, J. L. Stafford, A. Ruszczyk, N. Taverne, M. Belosevic, H. F.
      Savelkoul, and G. F. Wiegertjes. 2009. The induction of nitric oxide response of carp
      macrophages by transferrin is influenced by the allelic diversity of the molecule. Fish
      Shellfish Immunol 26:632-638.

[89] Kadowaki, T., H. Harada, Y. Sawada, C. Kohchi, G. Soma, Y. Takahashi, and H. Ina-
      gawa. 2009. Two types of tumor necrosis factor-alpha in bluefin tuna (Thunnus ori-
      entalis) genes: Molecular cloning and expression profile in response to several
      immunological stimulants. Fish Shellfish Immunol 27:585-594.

[90] Kasama, T., R. M. Strieter, N. W. Lukacs, P. M. Lincoln, M. D. Burdick, and S. L. Kun-
      kel. 1995. Interferon gamma modulates the expression of neutrophil-derived chemo-
      kines. J Investig Med 43:58-67.

[91] Kerr, I. M., and G. R. Stark. 1992. The antiviral effects of the interferons and their in-
      hibition. Journal of interferon research 12:237-240.

[92] Khan, S., J. D. Greenberg, and N. Bhardwaj. 2009. Dendritic cells as targets for thera-
      py in rheumatoid arthritis. Nat Rev Rheumatol 5:566-571.

[93] Kim, E. Y., J. J. Priatel, S. J. Teh, and H. S. Teh. 2006. TNF receptor type 2 (p75) functions as a costimulator for antigen-driven T cell responses in vivo. J Immunol 176:1026-1035.

[94] Kim, E. Y., and H. S. Teh. 2004. Critical role of TNF receptor type-2 (p75) as a costimulator for IL-2 induction and T cell survival: a functional link to CD28. J Immunol 173:4500-4509.

[95] Klebanoff, S. J., M. A. Vadas, J. M. Harlan, L. H. Sparks, J. R. Gamble, J. M. Agosti, and A. M. Waltersdorph. 1986. Stimulation of neutrophils by tumor necrosis factor. J Immunol 136:4220-4225.

[96] Kobayashi, Y., K. Yamamoto, T. Saido, H. Kawasaki, J. J. Oppenheim, and K. Matsushima. 1990. Identification of calcium-activated neutral protease as a processing enzyme of human interleukin 1 alpha. Proc Natl Acad Sci U S A 87:5548-5552.

[97] Kono, T., K. Fujiki, M. Nakao, T. Yano, M. Endo, and M. Sakai. 2002. The immune responses of common carp, Cyprinus carpio L., injected with carp interleukin-1beta gene. J Interferon Cytokine Res 22:413-419.

[98] Kriegler, M., C. Perez, K. DeFay, I. Albert, and S. D. Lu. 1988. A novel form of TNF/cachectin is a cell surface cytotoxic transmembrane protein: ramifications for the complex physiology of TNF. Cell 53:45-53.

[99] Krishnan, L., L. J. Guilbert, T. G. Wegmann, M. Belosevic, and T. R. Mosmann. 1996. T helper 1 response against Leishmania major in pregnant C57BL/6 mice increases implantation failure and fetal resorptions. Correlation with increased IFN-gamma and TNF and reduced IL-10 production by placental cells. J Immunol 156:653-662.

[100] Kurt-Jones, E. A., D. I. Beller, S. B. Mizel, and E. R. Unanue. 1985. Identification of a membrane-associated interleukin 1 in macrophages. Proc Natl Acad Sci U S A 82:1204-1208.

[101] Laing, K. J., P. S. Grabowski, M. Belosevic, and C. J. Secombes. 1996. A partial sequence for nitric oxide synthase from a goldfish (Carassius auratus) macrophage cell line. Immunol Cell Biol 74:374-379.

[102] Laing, K. J., L. J. Hardie, W. Aartsen, P. S. Grabowski, and C. J. Secombes. 1999. Expression of an inducible nitric oxide synthase gene in rainbow trout Oncorhynchus mykiss. Developmental and comparative immunology 23:71-85.

[103] Laing, K. J., T. Wang, J. Zou, J. Holland, S. Hong, N. Bols, I. Hirono, T. Aoki, and C. J. Secombes. 2001. Cloning and expression analysis of rainbow trout Oncorhynchus mykiss tumour necrosis factor-alpha. Eur J Biochem 268:1315-1322.

[104] Li, J., M. Zhang, and Y. C. Rui. 1997. Tumor necrosis factor mediated release of platelet-derived growth factor from bovine cerebral microvascular endothelial cells. Zhongguo Yao Li Xue Bao 18:133-136.

[105] Lohmann-Matthes, M. L., B. Luttig, and S. Hockertz. 1991. Involvement of membrane-associated TNF in the killing of Leishmania donovani parasites by macrophages. Behring Inst Mitt:125-132.

[106] Lomedico, P. T., U. Gubler, C. P. Hellmann, M. Dukovich, J. G. Giri, Y. C. Pan, K. Collier, R. Semionow, A. O. Chua, and S. B. Mizel. 1984. Cloning and expression of murine interleukin-1 cDNA in Escherichia coli. Nature 312:458-462.

[107] Lopez-Castejon, G., M. P. Sepulcre, I. Mulero, P. Pelegrin, J. Meseguer, and V. Mulero. 2008. Molecular and functional characterization of gilthead seabream Sparus aurata caspase-1: the first identification of an inflammatory caspase in fish. Molecular immunology 45:49-57.

[108] Lopez-Castejon, G., M. P. Sepulcre, F. J. Roca, B. Castellana, J. V. Planas, J. Meseguer, and V. Mulero. 2007. The type II interleukin-1 receptor (IL-1RII) of the bony fish gilthead seabream Sparus aurata is strongly induced after infection and tightly regulated at transcriptional and post-transcriptional levels. Molecular immunology 44:2772-2780.

[109] Lopez-Castejon, G., M. T. Young, J. Meseguer, A. Surprenant, and V. Mulero. 2007. Characterization of ATP-gated P2X7 receptors in fish provides new insights into the mechanism of release of the leaderless cytokine interleukin-1 beta. Molecular immunology 44:1286-1299.

[110] Lopez-Munoz, A., F. J. Roca, M. P. Sepulcre, J. Meseguer, and V. Mulero. 2011. Zebrafish larvae are unable to mount a protective antiviral response against waterborne infection by spring viremia of carp virus. Developmental and comparative immunology 34:546-552.

[111] Lowenstein, C. J., and E. Padalko. 2004. iNOS (NOS2) at a glance. J Cell Sci 117:2865-2867.

[112] Lu, D. Q., J. X. Bei, L. N. Feng, Y. Zhang, X. C. Liu, L. Wang, J. L. Chen, and H. R. Lin. 2008. Interleukin-1beta gene in orange-spotted grouper, Epinephelus coioides: molecular cloning, expression, biological activities and signal transduction. Molecular immunology 45:857-867.

[113] MacKenzie, A., H. L. Wilson, E. Kiss-Toth, S. K. Dower, R. A. North, and A. Surprenant. 2001. Rapid secretion of interleukin-1beta by microvesicle shedding. Immunity 15:825-835.

[114] MacKenzie, S., J. V. Planas, and F. W. Goetz. 2003. LPS-stimulated expression of a tumor necrosis factor-alpha mRNA in primary trout monocytes and in vitro differentiated macrophages. Developmental and comparative immunology 27:393-400.

[115] Martin, E., C. Nathan, and Q. W. Xie. 1994. Role of interferon regulatory factor 1 in induction of nitric oxide synthase. J Exp Med 180:977-984.

[116] Martin, J. H., and S. W. Edwards. 1993. Changes in mechanisms of monocyte/macrophage-mediated cytotoxicity during culture. Reactive oxygen intermediates are in-

volved in monocyte-mediated cytotoxicity, whereas reactive nitrogen intermediates are employed by macrophages in tumor cell killing. J Immunol 150:3478-3486.

[117] Mathew, J. A., Y. X. Guo, K. P. Goh, J. Chan, B. M. Verburg-van Kemenade, and J. Kwang. 2002. Characterisation of a monoclonal antibody to carp IL-1beta and the development of a sensitive capture ELISA. Fish Shellfish Immunol 13:85-95.

[118] Matsumoto, M., N. Tanaka, H. Harada, T. Kimura, T. Yokochi, M. Kitagawa, C. Schindler, and T. Taniguchi. 1999. Activation of the transcription factor ISGF3 by interferon-gamma. Biol Chem 380:699-703.

[119] Mayumi, M., Y. Takeda, M. Hoshiko, K. Serada, M. Murata, T. Moritomo, F. Takizawa, I. Kobayashi, K. Araki, T. Nakanishi, and H. Sumimoto. 2008. Characterization of teleost phagocyte NADPH oxidase: molecular cloning and expression analysis of carp (Cyprinus carpio) phagocyte NADPH oxidase. Molecular immunology 45:1720-1731.

[120] Meda, L., S. Gasperini, M. Ceska, and M. A. Cassatella. 1994. Modulation of proinflammatory cytokine release from human polymorphonuclear leukocytes by gamma interferon. Cell Immunol 157:448-461.

[121] Metz, J. R., M. O. Huising, K. Leon, B. M. Verburg-van Kemenade, and G. Flik. 2006. Central and peripheral interleukin-1beta and interleukin-1 receptor I expression and their role in the acute stress response of common carp, Cyprinus carpio L. J Endocrinol 191:25-35.

[122] Milev-Milovanovic, I., S. Long, M. Wilson, E. Bengten, N. W. Miller, and V. G. Chinchar. 2006. Identification and expression analysis of interferon gamma genes in channel catfish. Immunogenetics 58:70-80.

[123] Ming, W. J., L. Bersani, and A. Mantovani. 1987. Tumor necrosis factor is chemotactic for monocytes and polymorphonuclear leukocytes. J Immunol 138:1469-1474.

[124] Mizrahi, A., Y. Berdichevsky, Y. Ugolev, S. Molshanski-Mor, Y. Nakash, I. Dahan, N. Alloul, Y. Gorzalczany, R. Sarfstein, M. Hirshberg, and E. Pick. 2006. Assembly of the phagocyte NADPH oxidase complex: chimeric constructs derived from the cytosolic components as tools for exploring structure-function relationships. J Leukoc Biol 79:881-895.

[125] Mizutani, H., R. Black, and T. S. Kupper. 1991. Human keratinocytes produce but do not process pro-interleukin-1 (IL-1) beta. Different strategies of IL-1 production and processing in monocytes and keratinocytes. J Clin Invest 87:1066-1071.

[126] Mori, M. 2007. Regulation of nitric oxide synthesis and apoptosis by arginase and arginine recycling. J Nutr 137:1616S-1620S.

[127] Mosmann, T. R., and R. L. Coffman. 1989. TH1 and TH2 cells: different patterns of lymphokine secretion lead to different functional properties. Annu Rev Immunol 7:145-173.

[128] Moss, M. L., S. L. Jin, J. D. Becherer, D. M. Bickett, W. Burkhart, W. J. Chen, D. Hassler, M. T. Leesnitzer, G. McGeehan, M. Milla, M. Moyer, W. Rocque, T. Seaton, F. Schoenen, J. Warner, and D. Willard. 1997. Structural features and biochemical properties of TNF-alpha converting enzyme (TACE). J Neuroimmunol 72:127-129.

[129] Mukhopadhyay, A., J. Suttles, R. D. Stout, and B. B. Aggarwal. 2001. Genetic deletion of the tumor necrosis factor receptor p60 or p80 abrogates ligand-mediated activation of nuclear factor-kappa B and of mitogen-activated protein kinases in macrophages. J Biol Chem 276:31906-31912.

[130] Nahrendorf, M., F. K. Swirski, E. Aikawa, L. Stangenberg, T. Wurdinger, J. L. Figueiredo, P. Libby, R. Weissleder, and M. J. Pittet. 2007. The healing myocardium sequentially mobilizes two monocyte subsets with divergent and complementary functions. J Exp Med 204:3037-3047.

[131] Naismith, J. H., B. J. Brandhuber, T. Q. Devine, and S. R. Sprang. 1996. Seeing double: crystal structures of the type I TNF receptor. J Mol Recognit 9:113-117.

[132] Naismith, J. H., T. Q. Devine, T. Kohno, and S. R. Sprang. 1996. Structures of the extracellular domain of the type I tumor necrosis factor receptor. Structure 4:1251-1262.

[133] Nascimento, D. S., P. J. Pereira, M. I. Reis, A. do Vale, J. Zou, M. T. Silva, C. J. Secombes, and N. M. dos Santos. 2007. Molecular cloning and expression analysis of sea bass (Dicentrarchus labrax L.) tumor necrosis factor-alpha (TNF-alpha). Fish Shellfish Immunol 23:701-710.

[134] Nathan, C., and Q. W. Xie. 1994. Nitric oxide synthases: roles, tolls, and controls. Cell 78:915-918.

[135] Neumann, M., and E. Kownatzki. 1989. The effect of adherence on the generation of reactive oxygen species by human neutrophilic granulocytes. Agents Actions 26:183-185.

[136] Neumann, N. F., D. R. Barreda, and M. Belosevic. 2000. Generation and functional analysis of distinct macrophage sub-populations from goldfish (Carassius auratus L.) kidney leukocyte cultures. Fish Shellfish Immunol 10:1-20.

[137] Neumann, N. F., D. Fagan, and M. Belosevic. 1995. Macrophage activating factor(s) secreted by mitogen stimulated goldfish kidney leukocytes synergize with bacterial lipopolysaccharide to induce nitric oxide production in teleost macrophages. Developmental and comparative immunology 19:473-482.

[138] Neumann, N. F., J. L. Stafford, D. Barreda, A. J. Ainsworth, and M. Belosevic. 2001. Antimicrobial mechanisms of fish phagocytes and their role in host defense. Developmental and comparative immunology 25:807-825.

[139] O'Neill, L. A. 2000. The interleukin-1 receptor/Toll-like receptor superfamily: signal transduction during inflammation and host defense. Sci STKE 2000:re1.

[140] Oladiran, A., and M. Belosevic. 2009. Trypanosoma carassii hsp70 increases expression of inflammatory cytokines and chemokines in macrophages of the goldfish (Carassius auratus L.). Developmental and comparative immunology 33:1128-1136.

[141] Olavarria, V. H., L. Gallardo, J. E. Figueroa, and V. Mulero. Lipopolysaccharide primes the respiratory burst of Atlantic salmon SHK-1 cells through protein kinase C-mediated phosphorylation of p47phox. Developmental and comparative immunology 34:1242-1253.

[142] Ordas, M. C., M. M. Costa, F. J. Roca, G. Lopez-Castejon, V. Mulero, J. Meseguer, A. Figueras, and B. Novoa. 2007. Turbot TNFalpha gene: molecular characterization and biological activity of the recombinant protein. Molecular immunology 44:389-400.

[143] Parameswaran, N., and S. Patial. 2010. Tumor necrosis factor-alpha signaling in macrophages. Crit Rev Eukaryot Gene Expr 20:87-103.

[144] Peddie, S., P. E. McLauchlan, A. E. Ellis, and C. J. Secombes. 2003. Effect of intraperitoneally administered IL-1beta-derived peptides on resistance to viral haemorrhagic septicaemia in rainbow trout Oncorhynchus mykiss. Dis Aquat Organ 56:195-200.

[145] Peddie, S., J. Zou, C. Cunningham, and C. J. Secombes. 2001. Rainbow trout (Oncorhynchus mykiss) recombinant IL-1beta and derived peptides induce migration of head-kidney leucocytes in vitro. Fish Shellfish Immunol 11:697-709.

[146] Pelegrin, P., E. Chaves-Pozo, V. Mulero, and J. Meseguer. 2004. Production and mechanism of secretion of interleukin-1beta from the marine fish gilthead seabream. Developmental and comparative immunology 28:229-237.

[147] Pennica, D., W. J. Kohr, B. M. Fendly, S. J. Shire, H. E. Raab, P. E. Borchardt, M. Lewis, and D. V. Goeddel. 1992. Characterization of a recombinant extracellular domain of the type 1 tumor necrosis factor receptor: evidence for tumor necrosis factor-alpha induced receptor aggregation. Biochemistry 31:1134-1141.

[148] Perez, C., I. Albert, K. DeFay, N. Zachariades, L. Gooding, and M. Kriegler. 1990. A nonsecretable cell surface mutant of tumor necrosis factor (TNF) kills by cell-to-cell contact. Cell 63:251-258.

[149] Perussia, B. 1991. Lymphokine-activated killer cells, natural killer cells and cytokines. Curr Opin Immunol 3:49-55.

[150] Pleguezuelos, O., J. Zou, C. Cunningham, and C. J. Secombes. 2000. Cloning, sequencing, and analysis of expression of a second IL-1beta gene in rainbow trout (Oncorhynchus mykiss). Immunogenetics 51:1002-1011.

[151] Praveen, K., D. L. Evans, and L. Jaso-Friedmann. 2006. Constitutive expression of tumor necrosis factor-alpha in cytotoxic cells of teleosts and its role in regulation of cell-mediated cytotoxicity. Molecular immunology 43:279-291.

[152] Rieger, A. M., B. E. Hall, and D. R. Barreda. 2010. Macrophage activation differentially modulates particle binding, phagocytosis and downstream antimicrobial mechanisms. Developmental and comparative immunology 34:1144-1159.

[153] Robinson, J. M. 2009. Phagocytic leukocytes and reactive oxygen species. Histochem
Cell Biol 131:465-469.

[154] Robinson, J. M. 2008. Reactive oxygen species in phagocytic leukocytes. Histochem
Cell Biol 130:281-297.

[155] Roca, F. J., I. Mulero, A. Lopez-Munoz, M. P. Sepulcre, S. A. Renshaw, J. Meseguer,
and V. Mulero. 2008. Evolution of the inflammatory response in vertebrates: fish
TNF-alpha is a powerful activator of endothelial cells but hardly activates phagocytes. J Immunol 181:5071-5081.

[156] Rubbo, H., A. Denicola, and R. Radi. 1994. Peroxynitrite inactivates thiol-containing
enzymes of Trypanosoma cruzi energetic metabolism and inhibits cell respiration.
Arch Biochem Biophys 308:96-102.

[157] Ruddle, N. H. 1986. Activation of human polymorphonuclear neutrophil functions
by interferon-gamma and tumor necrosis factors. J Immunol 136:2335-2336.

[158] Russo, R., C. A. Shoemaker, V. S. Panangala, and P. H. Klesius. 2009. In vitro and in
vivo interaction of macrophages from vaccinated and non-vaccinated channel catfish
(Ictalurus punctatus) to Edwardsiella ictaluri. Fish Shellfish Immunol 26:543-552.

[159] Ryckaert, J., P. Bossier, K. D'Herde, A. Diez-Fraile, P. Sorgeloos, F. Haesebrouck, and
F. Pasmans. 2010. Persistence of Yersinia ruckeri in trout macrophages. Fish Shellfish
Immunol 29:648-655.

[160] Rycyzyn, M. A., M. R. Wilson, E. Bengten, G. W. Warr, L. W. Clem, and N. W. Miller.
1998. Mitogen and growth factor-induced activation of a STAT-like molecule in channel catfish lymphoid cells. Molecular immunology 35:127-136.

[161] Sad, S., R. Marcotte, and T. R. Mosmann. 1995. Cytokine-induced differentiation of
precursor mouse CD8+ T cells into cytotoxic CD8+ T cells secreting Th1 or Th2 cytokines. Immunity 2:271-279.

[162] Saeij, J. P., R. J. Stet, B. J. de Vries, W. B. van Muiswinkel, and G. F. Wiegertjes. 2003.
Molecular and functional characterization of carp TNF: a link between TNF polymorphism and trypanotolerance? Developmental and comparative immunology
27:29-41.

[163] Saeij, J. P., R. J. Stet, A. Groeneveld, L. B. Verburg-van Kemenade, W. B. van Muiswinkel, and G. F. Wiegertjes. 2000. Molecular and functional characterization of a
fish inducible-type nitric oxide synthase. Immunogenetics 51:339-346.

[164] Sangrador-Vegas, A., S. A. Martin, P. G. O'Dea, and T. J. Smith. 2000. Cloning and
characterization of the rainbow trout (Oncorhynchus mykiss) type II interleukin-1 receptor cDNA. Eur J Biochem 267:7031-7037.

[165] Savan, R., T. Kono, D. Igawa, and M. Sakai. 2005. A novel tumor necrosis factor (TNF) gene present in tandem with theTNF-alpha gene on the same chromosome in teleosts. Immunogenetics 57:140-150.

[166] Savan, R., S. Ravichandran, J. R. Collins, M. Sakai, and H. A. Young. 2009. Structural conservation of interferon gamma among vertebrates. Cytokine Growth Factor Rev 20:115-124.

[167] Savan, R., and M. Sakai. 2004. Presence of multiple isoforms of TNF alpha in carp (Cyprinus carpio L.): genomic and expression analysis. Fish Shellfish Immunol 17:87-94.

[168] Schirren, C. G., K. Scharffetter, R. Hein, O. Braun-Falco, and T. Krieg. 1990. Tumor necrosis factor alpha induces invasiveness of human skin fibroblasts in vitro. J Invest Dermatol 94:706-710.

[169] Schroder, K., P. J. Hertzog, T. Ravasi, and D. A. Hume. 2004. Interferon-gamma: an overview of signals, mechanisms and functions. J Leukoc Biol 75:163-189.

[170] Secombes, C., J. Zou, G. Daniels, C. Cunningham, A. Koussounadis, and G. Kemp. 1998. Rainbow trout cytokine and cytokine receptor genes. Immunol Rev 166:333-340.

[171] Secombes, C. J. 1987. Lymphokine-release from rainbow trout leucocytes stimulated with concanavalin A. Effects upon macrophage spreading and adherence. Developmental and comparative immunology 11:513-520.

[172] Shen, Y., M. Naujokas, M. Park, and K. Ireton. 2000. InlB-dependent internalization of Listeria is mediated by the Met receptor tyrosine kinase. Cell 103:501-510.

[173] Sheppard, F. R., M. R. Kelher, E. E. Moore, N. J. McLaughlin, A. Banerjee, and C. C. Silliman. 2005. Structural organization of the neutrophil NADPH oxidase: phosphorylation and translocation during priming and activation. J Leukoc Biol 78:1025-1042.

[174] Sieger, D., C. Stein, D. Neifer, A. M. van der Sar, and M. Leptin. 2009. The role of gamma interferon in innate immunity in the zebrafish embryo. Dis Model Mech 2:571-581.

[175] Sigel, M. M., B. A. Hamby, and E. M. Huggins, Jr. 1986. Phylogenetic studies on lymphokines. Fish lymphocytes respond to human IL-1 and epithelial cells produce an IL-1 like factor. Vet Immunol Immunopathol 12:47-58.

[176] Sims, J. E. 2002. IL-1 and IL-18 receptors, and their extended family. Curr Opin Immunol 14:117-122.

[177] Sims, J. E., M. J. Nicklin, J. F. Bazan, J. L. Barton, S. J. Busfield, J. E. Ford, R. A. Kastelein, S. Kumar, H. Lin, J. J. Mulero, J. Pan, Y. Pan, D. E. Smith, and P. R. Young. 2001. A new nomenclature for IL-1-family genes. Trends Immunol 22:536-537.

[178] Smith, D. E., B. R. Renshaw, R. R. Ketchem, M. Kubin, K. E. Garka, and J. E. Sims. 2000. Four new members expand the interleukin-1 superfamily. J Biol Chem 275:1169-1175.

[179] Staeheli, P. 1990. Interferon-induced proteins and the antiviral state. Adv Virus Res 38:147-200.

[180] Stafford, J. L., E. C. Wilson, and M. Belosevic. 2004. Recombinant transferrin induces nitric oxide response in goldfish and murine macrophages. Fish Shellfish Immunol 17:171-185.

[181] Stevenson, F. T., S. L. Bursten, C. Fanton, R. M. Locksley, and D. H. Lovett. 1993. The 31-kDa precursor of interleukin 1 alpha is myristoylated on specific lysines within the 16-kDa N-terminal propiece. Proc Natl Acad Sci U S A 90:7245-7249.

[182] Stevenson, M. M., M. F. Tam, M. Belosevic, P. H. van der Meide, and J. E. Podoba. 1990. Role of endogenous gamma interferon in host response to infection with blood-stage Plasmodium chabaudi AS. Infection and immunity 58:3225-3232.

[183] Stolte, E. H., H. F. Savelkoul, G. Wiegertjes, G. Flik, and B. M. Lidy Verburg-van Kemenade. 2008. Differential expression of two interferon-gamma genes in common carp (Cyprinus carpio L.). Developmental and comparative immunology 32:1467-1481.

[184] Stuehr, D. J., and C. F. Nathan. 1989. Nitric oxide. A macrophage product responsible for cytostasis and respiratory inhibition in tumor target cells. J Exp Med 169:1543-1555.

[185] Subramaniam, P. S., B. A. Torres, and H. M. Johnson. 2001. So many ligands, so few transcription factors: a new paradigm for signaling through the STAT transcription factors. Cytokine 15:175-187.

[186] Subramaniam, S., C. Stansberg, L. Olsen, J. Zou, C. J. Secombes, and C. Cunningham. 2002. Cloning of a Salmo salar interleukin-1 receptor-like cDNA. Developmental and comparative immunology 26:415-431.

[187] Takaoka, A., Y. Mitani, H. Suemori, M. Sato, T. Yokochi, S. Noguchi, N. Tanaka, and T. Taniguchi. 2000. Cross talk between interferon-gamma and -alpha/beta signaling components in caveolar membrane domains. Science 288:2357-2360.

[188] Takaoka, A., and H. Yanai. 2006. Interferon signalling network in innate defence. Cell Microbiol 8:907-922.

[189] Tartaglia, L. A., T. M. Ayres, G. H. Wong, and D. V. Goeddel. 1993. A novel domain within the 55 kd TNF receptor signals cell death. Cell 74:845-853.

[190] Theiss, A. L., J. G. Simmons, C. Jobin, and P. K. Lund. 2005. Tumor necrosis factor (TNF) alpha increases collagen accumulation and proliferation in intestinal myofibroblasts via TNF receptor 2. J Biol Chem 280:36099-36109.

[191] Thomson, L., A. Denicola, and R. Radi. 2003. The trypanothione-thiol system in Trypanosoma cruzi as a key antioxidant mechanism against peroxynitrite-mediated cytotoxicity. Arch Biochem Biophys 412:55-64.

[192] Towne, J. E., K. E. Garka, B. R. Renshaw, G. D. Virca, and J. E. Sims. 2004. Interleukin (IL)-1F6, IL-1F8, and IL-1F9 signal through IL-1Rrp2 and IL-1RAcP to activate the pathway leading to NF-kappaB and MAPKs. J Biol Chem 279:13677-13688.

[193] Uenobe, M., C. Kohchi, N. Yoshioka, A. Yuasa, H. Inagawa, K. Morii, T. Nishizawa, Y. Takahashi, and G. Soma. 2007. Cloning and characterization of a TNF-like protein of Plecoglossus altivelis (ayu fish). Molecular immunology 44:1115-1122.

[194] van Strijp, J. A., M. E. van der Tol, L. A. Miltenburg, K. P. van Kessel, and J. Verhoef. 1991. Tumour necrosis factor triggers granulocytes to internalize complement-coated virus particles. Immunology 73:77-82.

[195] Verburg-van Kemenade, B. M., F. A. Weyts, R. Debets, and G. Flik. 1995. Carp macrophages and neutrophilic granulocytes secrete an interleukin-1-like factor. Developmental and comparative immunology 19:59-70.

[196] Vigers, G. P., L. J. Anderson, P. Caffes, and B. J. Brandhuber. 1997. Crystal structure of the type-I interleukin-1 receptor complexed with interleukin-1beta. Nature 386:190-194.

[197] Wallach, D., E. E. Varfolomeev, N. L. Malinin, Y. V. Goltsev, A. V. Kovalenko, and M. P. Boldin. 1999. Tumor necrosis factor receptor and Fas signaling mechanisms. Annu Rev Immunol 17:331-367.

[198] Wang, T., S. Bird, A. Koussounadis, J. W. Holland, A. Carrington, J. Zou, and C. J. Secombes. 2009. Identification of a novel IL-1 cytokine family member in teleost fish. J Immunol 183:962-974.

[199] Wang, T., J. W. Holland, A. Carrington, J. Zou, and C. J. Secombes. 2007. Molecular and functional characterization of IL-15 in rainbow trout Oncorhynchus mykiss: a potent inducer of IFN-gamma expression in spleen leukocytes. J Immunol 179:1475-1488.

[200] Wang, T., M. Ward, P. Grabowski, and C. J. Secombes. 2001. Molecular cloning, gene organization and expression of rainbow trout (Oncorhynchus mykiss) inducible nitric oxide synthase (iNOS) gene. Biochem J 358:747-755.

[201] Wang, Y., Q. Wang, P. Baoprasertkul, E. Peatman, and Z. Liu. 2006. Genomic organization, gene duplication, and expression analysis of interleukin-1beta in channel catfish (Ictalurus punctatus). Molecular immunology 43:1653-1664.

[202] Wang, Z. E., S. L. Reiner, S. Zheng, D. K. Dalton, and R. M. Locksley. 1994. CD4+ effector cells default to the Th2 pathway in interferon gamma-deficient mice infected with Leishmania major. J Exp Med 179:1367-1371.

[203] Warner, S. J., and P. Libby. 1989. Human vascular smooth muscle cells. Target for and source of tumor necrosis factor. J Immunol 142:100-109.

[204] Weiss, T., M. Grell, K. Siemienski, F. Muhlenbeck, H. Durkop, K. Pfizenmaier, P. Scheurich, and H. Wajant. 1998. TNFR80-dependent enhancement of TNFR60-induced cell death is mediated by TNFR-associated factor 2 and is specific for TNFR60. J Immunol 161:3136-3142.

[205] Wheelock, E. F. 1965. Interferon-Like Virus-Inhibitor Induced in Human Leukocytes by Phytohemagglutinin. Science 149:310-311.

[206] Yin, Z., and J. Kwang. 2000. Carp interleukin-1 beta in the role of an immuno-adjuvant. Fish Shellfish Immunol 10:375-378.

[207] Yonemaru, M., K. E. Stephens, A. Ishizaka, H. Zheng, R. S. Hogue, J. J. Crowley, J. R. Hatherill, and T. A. Raffin. 1989. Effects of tumor necrosis factor on PMN chemotaxis, chemiluminescence, and elastase activity. J Lab Clin Med 114:674-681.

[208] Young, H. A., and K. J. Hardy. 1995. Role of interferon-gamma in immune cell regulation. J Leukoc Biol 58:373-381.

[209] Young, J. D., C. C. Liu, G. Butler, Z. A. Cohn, and S. J. Galli. 1987. Identification, purification, and characterization of a mast cell-associated cytolytic factor related to tumor necrosis factor. Proc Natl Acad Sci U S A 84:9175-9179.

[210] Zhang, M., Z. Z. Xiao, and L. Sun. 2011. Suppressor of cytokine signaling 3 inhibits head kidney macrophage activation and cytokine expression in Scophthalmus maximus. Developmental and comparative immunology 35:174-181.

[211] Zhao, C., H. Zhang, W. C. Wong, X. Sem, H. Han, S. M. Ong, Y. C. Tan, W. H. Yeap, C. S. Gan, K. Q. Ng, M. B. Koh, P. Kourilsky, S. K. Sze, and S. C. Wong. 2009. Identification of novel functional differences in monocyte subsets using proteomic and transcriptomic methods. J Proteome Res 8:4028-4038.

[212] Zhao, X., M. Mohaupt, J. Jiang, S. Liu, B. Li, and Z. Qin. 2007. Tumor necrosis factor receptor 2-mediated tumor suppression is nitric oxide dependent and involves angiostasis. Cancer Res 67:4443-4450.

[213] Zhu, L., C. Gunn, and J. S. Beckman. 1992. Bactericidal activity of peroxynitrite. Arch Biochem Biophys 298:452-457.

[214] Ziegler-Heitbrock, L. 2007. The CD14+ CD16+ blood monocytes: their role in infection and inflammation. J Leukoc Biol 81:584-592.

[215] Zou, J., A. Carrington, B. Collet, J. M. Dijkstra, Y. Yoshiura, N. Bols, and C. Secombes. 2005. Identification and bioactivities of IFN-gamma in rainbow trout Oncorhynchus mykiss: the first Th1-type cytokine characterized functionally in fish. J Immunol 175:2484-2494.

[216] Zou, J., P. S. Grabowski, C. Cunningham, and C. J. Secombes. 1999. Molecular cloning of interleukin 1beta from rainbow trout Oncorhynchus mykiss reveals no evidence of an ice cut site. Cytokine 11:552-560.

[217] Zou, J., S. Peddie, G. Scapigliati, Y. Zhang, N. C. Bols, A. E. Ellis, and C. J. Secombes. 2003. Functional characterisation of the recombinant tumor necrosis factors in rainbow trout, Oncorhynchus mykiss. Developmental and comparative immunology 27:813-822.

[218] Zou, J., C. J. Secombes, S. Long, N. Miller, L. W. Clem, and V. G. Chinchar. 2003. Molecular identification and expression analysis of tumor necrosis factor in channel catfish (Ictalurus punctatus). Developmental and comparative immunology 27:845-858.

[219] Zou, J., T. Wang, I. Hirono, T. Aoki, H. Inagawa, T. Honda, G. I. Soma, M. Ototake, T. Nakanishi, A. E. Ellis, and C. J. Secombes. 2002. Differential expression of two tumor necrosis factor genes in rainbow trout, Oncorhynchus mykiss. Developmental and comparative immunology 26:161-172.

[220] Zou, J., Y. Yoshiura, J. M. Dijkstra, M. Sakai, M. Ototake, and C. Secombes. 2004. Identification of an interferon gamma homologue in Fugu, Takifugu rubripes. Fish Shellfish Immunol 17:403-409.

Chapter 4

Freshwater Fish as Sentinel Organisms: From the Molecular to the Population Level, a Review

Jacinto Elías Sedeño-Díaz and Eugenia López-López

Additional information is available at the end of the chapter

1. Introduction

Fish around the world are found occupying almost any aquatic habitat. In particular, freshwater fish are severely threatened as the freshwater ecosystems are considered the most endangered of the world [1]. The ultimate destination of most contaminants is water; rivers, lakes, aquifers, or sea, are receptors of wastewaters with a complex mixture of xenobiotics. The variety of contaminants and their mixtures that daily reach the water bodies coupled with a multitude of irresponsible water management practices and destructive land uses, are currently threatening freshwater ecosystems [2], such is the case of discharge of municipal and industrial wastewaters, deforestation, increase of land crops, and water extraction from water bodies to human consumption and other uses. The impact of contaminants in an aquatic ecosystem is complex, therefore has increased the need for determining the ambient status in order to provide an indication of changes induced by anthropogenic activities and their influence on aquatic organisms. As physicochemical analyses shed no light on the biological status of ecosystems, a biological approach is needed to evaluate environmental health; moreover, the biological effects of contaminant interactions cannot be expressed by physicochemical investigations [3].The aquatic ecosystem health is often reflected by the health of organisms that reside in that system. Fish in their natural environments are typically exposed to numerous stressors including unfavorable or fluctuating temperatures, high water velocities and sediment loads, low dissolved oxygen concentrations, limited food availability, and among other types of natural episodic variables. In addition, anthropogenic stressors such as contaminant loading can add to the insults that fish may already experience in many systems. All these factors, individually or together, can impose considerable stress on physiological systems of fish and impair their health [4, 5]. Environmental contaminants are known to induce measurable biochemical changes in exposed aquatic organisms [6]. Likewise, stressors can load or limit physiological systems, reduce growth, impair reproduction, predispose fish to disease, and reduce the capacity of fish to tolerate additional stressors. Many species of fish, in particular those species near the top of the food chain, are generally regarded as integrators of environmental conditions and may reflect, therefore, the health of aquatic ecosystems [4] and therefore, they are excellent indicators of the relative health of aquatic ecosystems and their surrounding watersheds [7]. Thus, effects of contaminants on aquatic organisms may be manifested at all levels of biological organization (in a hierarchical scale that can be at cellular level, organisms, populations, communities, and ecosystems). In this way, the measuring of a suite of indicators across such levels of organization is often necessary to assess ecological integrity; these indicators also should include molecular, biochemical, physiological, population, community, and ecosystem responses.

The indicators allow us to isolate key aspects of the environment from an overwhelming array of signals [8]. Ecological indicators have been defined as measurable characteristics of the structure (e.g. genetic, population, habitat, and landscape pattern), composition (e.g., genes, species, populations, communities, and landscape types), or function (e.g., genetic, demographic/life history, ecosystem, and landscape disturbance processes) of ecological systems [9]. On the other hand, other authors [10], established that bioindicators are organisms or communities of organisms, which reactions are observed representatively to evaluate a situation, giving clues for the condition of the whole ecosystem; Gerhardt also indicate that bioindicators are species reacting to anthropogenical effects on the environment, concluding that a biological indicator would be: a species or group of species that readily reflects the abiotic or biotic state of an environment, represents the impact of environmental change on a habitat, community or ecosystem or is indicative of the diversity of a subset of taxa or the whole diversity within an area. In this sense, the primary role of ecological indicators is to measure the response of the ecosystem to anthropogenic disturbances [9]. A sentinel species can be defined as any domestic or wild microorganism, plant or animal, that can be used as an indicator of exposure to and toxicity of a xenobiotic that can be used in assessing the impact on human and/or environmental health because of the organism's sensitivity, position in a community, likelihood of exposure, geographic and ecological distribution or abundance [11].

The specific objective of this review is to provide a short framework of effects of xenobiotics on the responses of freshwater fish across molecular to population level when have been exposed to environmental stressors. Likewise, the present review considers the use of fish as sentinel organisms to assess the anthropogenic impacts over the freshwater ecosystems. The review asks whether fish can be able to reflect the environmental damage from molecular to population levels. Also, the present review offers a selection of examples of studies employing fish as sentinel organisms in ecological, toxicological and environmental risk assessments.

2. Suborganismal responses

When an organism is exposed to stressors like contaminants or a mixture of them, energy is demanded to deal with that stress [4]; stressors tend to impact ecosystems at lower levels of organization first [12]. One of the methods to quantify the exposure to xenobiotics and its potential impact on living organisms is the monitoring by the use of the so-called biomarkers [13]. Biomarkers have been defined by several authors, all of them, in reference to biological responses to contaminants exposure, as a) measurements in body fluids, cells or tissues indicating biochemical or cellular modifications due to the presence and magnitude of toxicants, or of host response [8]; b) a change in a biological response (ranging from molecular through cellular and physiological responses to behavioral changes) which can be related to exposure to or toxic effects of environmental chemicals [14]; c) any biological response to an environmental chemical at the subindividual level, measured inside an organism or in its products (urine, faeces, hair, feathers, etc.), indicating a deviation from the normal status that cannot be detected in the intact organism [15]; d) a xenobiotically induced variation in cellular or biochemical components or processes, structures, or functions that is measurable in a biological system or samples [16]; e) contaminant-induced physiological, biochemical, or histological response of an organism, and f) as functional measures of exposure to stressors expressed at the sub-organismal, physiological or behavioural level. Considering these definitions of biomarkers, we could adopt our own definition: *"any biological measurable response from an organism, induced by the exposure to a xenobiotic or complex mixture of them"*. Biomarkers can provide valuable information in field or semifield testing and be used to measure a wide range of physiological responses to chemicals at the biochemical, cellular, or tissular level [17].

In concordance with [18] and other authors [19, 20], biomarkers have been classified in three different categories: a) biomarkers of exposure, which represent responses such as induction or inhibition of specific enzymes involved in biotransformation and detoxification as a consequence of chemical exposure [21], b) biomarkers of effect, are any changes in a biological system that reflects qualitative or quantitative impairment resulting from exposure [20], including responses measurable at level biochemical, physiological or some other alterations within tissues or body fluids of an organism that can be recognized as associated with an established or possible health impairment or disease [19], and c) biomarkers of susceptibility, which serve as indicators of a particular sensitivity of individuals to respond to the challenge of exposure to a effect of a xenobiotic or to the effects of a group of such compounds, in this case, individual changes included genetic factors and changes in receptors which alter the susceptibility of an organism to that exposure [19]. However, other authors have been subdivided the biomarkers in exposure biomarkers, effects biomarkers and predictive biomarkers.

Responses of fish at suborganismal level to xenobiotic exposure are complex and varied and depending of type of contaminant and time of exposure. The most general effect of xenobiotics on fish is oxidative stress, which is experienced when antioxidant defenses are overcome by prooxidant compounds. Oxidative stress include a variety of oxidative reactions, usually started by free radicals and propagated by molecular oxygen, which results in the oxidation of lipids, proteins, and nucleic acids [22]. Free radicals are atoms, molecules, or ions with unpaired electrons on an otherwise open shell configuration. These unpaired electrons are usually highly reactive due to which radicals are likely to take part in chemical reactions. Very often free radicals are confused with reactive oxygen species (ROS) such as molecular and singlet oxygen, superoxide anion, hydroxyl radical and some their derivatives; however, hydrogen peroxide is not a radical, but it is a reactive species because has higher activity than molecular oxygen [23]. Hydroxyl radical is the most important free radical of biological importance, because of its potent oxidative potential and indiscriminate reactivity with cellular components of enzymes and DNA [24, 25]; likewise, being oxidant, all ROS are agents which at high concentrations are toxic to cells. Oxidative stress is a risky condition in which increases in free radical production, and/or decreases in antioxidant levels can lead to potential damage. The antioxidant system in aerobic organisms includes several biochemical safety mechanisms such as antioxidant enzymes and other compounds like vitamins, glutathione, matallothioneins, and others.

Antioxidant defense enzymes are induced by various environmental pollutants under pro-oxydant conditions among these enzymes we can found superoxide dismutase (SOD), Catalase (Cat) and Glutation Peroxidase (GPx), Glutation reductase and Glutation S Transferase (GST, catalyze the nucleophilic conjugation of different biologically and potentially carcinogenic compounds [20]). The role of SOD is to catalyze the reaction of superoxide radical (O_2^-) to peroxide (H_2O_2). CAT detoxifies H_2O_2 to H_2O and O_2. GPx detoxifies mainly organic peroxides. CAT is an enzyme with high biological relevance because reduce the concentration of peroxide, a precursor of OH-, which is a highly reactive toxic form of ROS [26]. Cytochrome P450 monooxygenases (CYPs) are a multi-gene family of enzymes that play a key role in the biotransformation of pollutants, such as dioxins, pesticides, polychlorinated biphenyls (PCBs) and polycyclic aromatic hydrocarbons (PAHs). One of the most common and highly conserved is the CYP1A subfamily. The CYP1A biomarker is widely used as a biomarker of effect both in vertebrates and invertebrates for environmental biomonitoring, especially in marine bivalves and fish. The induction of CYP1A is triggered via the cytosolic aryl hydrocarbon (Ah) receptor due to exposure to pollutants, such as PCBs, dioxins, and numerous PAHs. CYP1A activity is typically measured using the substrate ethoxyresorufin, which is o-deethylated by ethoxyresorufin-O-deethylase (EROD) to a fluorescent product, resorufin, which can be easily measured. Because EROD activities are generally measured using liver homogenates that also tend to accumulate numerous CYP1A substrates, activity may be inhibited by residual substrates or metals [20].

3. Individual responses

Morphological alterations is one of the most individual level parameters that are measure to identified damage in sentinel organisms [27, 28]. In reference [28] identified severe histological damage in gills and liver of *Goodea atripinnis* a goodeid fish from Central Mexico, after the chronic exposure to Yerbimat, an herbicide with glyphosate. After 75 days of exposure to pesticide, they found lamellar hypertrophy and leukocyte infiltration in gills, and hepatocytes with vacuolization in the cytoplasm and piknotic nuclei in liver, concluding that Yerbimat induces histological alterations in the gills and liver that might impair normal organ function-ing that could lead to health damage in fish because of the important physiological roll of these organs.

Also, haematological parameters, such as erythrocyte and leucocyte count, erythrocyte indices and thrombocyte number vis-a-vis coagulation of blood has been considered bioindicators of toxicosis in fish following exposure to xenobiotics [29]. Any alteration in normal cellular com-ponents (morphology and number), and level of fluids of blood is named dyscrasia. Some authors [29] carry out an extensive review over dyscrasia in fish. They show several examples over alterations in morphological changes in blood cells, total count, haemoglobin content, thrombocytes and clotting time, concluding that haematological parameters are not specific to faced acute and chronic exposure of fish to xenobiotics, mainly organochlorinated, organo-phosphonate, pyrethroid and carbamate pesticides.

4. Population responses

The status of a fish population is a reflection of the overall condition of the aquatic environment in which that population resides. As such, fish population characteristics can be used as indi-cators of environmental health. Although changes in population structure may act as a sensi-tive indicator of changing environmental conditions, the timing, degree and nature of the feedback response to altered conditions will vary with the intensity, identity and the number of stressors, as well as the availability of energy [30].

Bioindicators are responses to environmental effects that occur at higher levels of biological organization than sub-organism (biomarkers); This kind of responses can be measured at dif-ferent high levels of biological organization, from individual, through population (reproduc-tive success, mortality, size distribution, reduction in abundance and biomass), community (primary production, disruption of the nutrient cycle) to ecosystem levels [31, 32], whose main characteristic is that the measure change with exposure to negative environmental factors. Organosomatic indices are common approaches for assessing fish health [4]. In this review we consider three organosomatic indices: Condition factor (CF), hepatosomatic index (HSI) and gonadosomatic index (GSI).

Since the nutritional status of fish can change according to different factors, such as season, food availability, and among they the exposure to xenobiotics, is important to have a measure of corporal condition of fish. The chronic exposure to contaminants may cause changes in feeding behavior, leading to a deterioration of the health. The CF is a frequently used index for fish biology study, as it furnishes important information related to fish physiological state [33]. It is a measure of corporal condition since measuring the body mass associated with the body length. A fish that is heavier for a given length (higher CF) is considered to be a healthier fish, because extra weight means extra energy reserves. While a lighter fish lack energy reserves and therefore, tend to be more susceptible to environmental stressors. A low body condition may also suggest muscle wasting (proteolysis) indicating a starvation response [34]. It has also been suggested that females with a lower body condition reduce reproductive investment yet still have an increased risk of mortality.

The liver plays a major role in the metabolism of xenobiotic compounds with biochemical alterations occurring under some toxic conditions; likewise, the liver is a primary detoxification organ in fish [35]. Therefore, this strong activity can lead to an increase in liver size, from hypertrophy (an increase in size) to hyperplasia (an increase in number) of hepatocytes [35], or both. Studies evaluating the relative liver size of fishes from contaminated and reference sites often utilize the Hepatosomatic Index (HIS), which expresses the ratio of liver weight to body weight as a percentage.

Gonadosomatic index (GSI) is also a percentage relationship between the gonad weight and fish weight. Depending on the severity of exposure to xenobiotics, the sublethal effects can be to limit physiological capacity, reduce growth, and impair reproduction, therefore GSI is a convenient organosomaitc index [36].

5. Specific case studies about fish as sentinel organisms

In México some studies have analyzed the use of freshwater fish as sentinels. These studies include the use of biomarkers and sentinel fish and the use of the whole fish population as indicators of environmental change. Furthermore, these studies are in areas of contrasting environmental conditions, in the case of biomarkers studies are in: the course of a river in the Atlantic slope (Río Champotón), a Lake in the Central Plateau (Yuriria Lake), a spring and a reservoir in the upper portion of a river of the Pacific slope (Ameca river). These studies make evident the utility of freshwater fish as sentinels and are briefly exposed.

5.1. Case of study of *Astyanax aeneus* in the Champotón river

The Champotón river, located in the humid subtropics of southeastern Mexico in terrain with a high content of karstic material, is the main surface stream in the Yucatán Peninsula; is a coastal river with 48 km in length to its outlet with a drainage basin surface area of 650 km2. The fish studied inhabit the fresh water zone of the river with salinity up to 1.2 practical salinity units [37]. The climatic regime is hot subhumid with summer rain (June to September) and occasional winter precipitation as a result of the windy (northerly) and hurricane seasons. The main anthropogenic activity in the basin is agriculture and livestock raising [38]. During the study period (2007-2008) the region was affected by several hurricanes (mostly from August to October), that caused the river overflow. This study assessed the effects of the environmental conditions along the freshwater portion of the Champotón river on the native fish *A. aeneus*, analyzing responses between lower and higher levels of organization, and linking spatial and seasonal fish responses with water quality features. A Water Quality Index (WQI) was employed as an indicator of environmental conditions, a set of sub-organismal biomarkers in *A. aeneus* lipid peroxidation (LPO), GST, EROD, and lactate dehydrogenase (LDH)) was monitored to determine the Integrated biomarker response (IBR), and organosomatic indices: GSI, HSI and CF were characterized. Three study sites were analyzed: San Juan Carpizo (SJC) in the upper portion of the river, San Antonio del Río (SAR) in the middle portion of the river (where there are a rustic swimming spot lacking of sanitary facilities), and downstream Ulumal (U); the study periods were: April, July, and November 2007 and February 2008.

The WQI scores exhibited spatial and temporal variations (from 53.21 to 78.49) on a scale of 0 to 100. The lowest value was recorded in July at site SAR, where the lack of sanitary facilities provoke fecal materials are swept away by runoff during the rainy season (July), increasing coliform numbers and lowering the WQI. In addition, this river flows through a region in which calcareous substrates predominate; with a high content of calcium carbonate and increase in conductivity both provoked decreases WQI scores, particularly during the drought. However, WQI scores were higher in November and February, following the hurricane season that brought large amounts of precipitation and increased river flow favoring dilution. As a result, the values of several WQI parameters (including hardness and conductivity) decreased while the river was in flood. Several studies reported similar WQI fluctuation patterns. WQI scores in the Lerma-Chapala Basin, Mexico, indicated severe degradation of the basin, particularly during the dry season, when its rating ranged from contaminated to highly contaminated; however, WQI improved during the wet season, [39, 40].

Although the WQI scores indicated that the Champotón river had acceptable water quality, some pollutants (residues of persistent organic compounds (POCs), such as PCBs, hexachlorocyclohexanes, aldrin-related pesticides, heptachlor, and dichlorodiphenyltrichloroethane) have been detected in the Champotón river [41]. In reference [42] reported that sediments from several Champotón river sites contained two or more of the 16 PAHs considered by the Environmental Protection Agency United States as priority pollutants that represent a potential threat to exposed organisms. Seasonal variations in POCs were found by [41]; PCBs and hexachlorocyclohexanes reached their highest values during the rainy season, while dichlorodiphenyltrichloroethane, drines, and heptachlor peaked during the dry season. Additionally, high episodic loadings of contaminants have been detected in aquatic ecosystems following flooding events [43].

Regarding biomarkers in the sentinel fish *A. aeneus*, the highest LPO values were detected in November and February (post-hurricane and windy seasons), while the lowest values were detected in July at all study sites. GST activity was highest in November at site U and lowest in July and April at all study sites. EROD activity in general was highest in April, while the lowest means occurred in July at all sites. Spatial analysis revealed that SAR recorded the highest mean EROD activity, while the lowest occurred at sites SJC and U. LDH values peaked in November and February following the hurricane and windy seasons, while minimum LDH levels occurred in April during the dry season. Spatial variability occurred; sites in the middle reaches (U, SAR) had the highest LDH values, while the lowest were recorded at headwater site SJC [44].

The increased LPO values in the post-hurricane season (November) and in the windy season (February) was associated with the hurricane season in the Champotón river which provokes flooding of adjacent areas where field crops are treated with agrochemicals that, along with the POCs and PAHs detected in sediments by [41], may be incorporated into the aquatic system. Similar results were found by [45] that detected the mobilization of agriculture-related xenobiotics, during the flooding of the river Elbe. Oxidative stress in fish after extensive flooding was also detected by [43] in the Pamlico Sound estuary. The observed increase in LPO levels in *A. aeneus* may be related to climate-induced stress and by exposure to the mixture of xenobiotics that may be mobilized during that period.

Produced ROS are detoxified by antioxidant defense mechanisms, which are essential for protection of cellular systems against xenobiotic-induced oxidative stress [46]. GST is particularly interesting, since it is involved in elimination of reactive compounds and is the transport system for glutathione [22]. GST activity in *A. aeneus* peaked during the same period that LPO increased, following the hurricane season, which may reflect a defense of the fish against oxidative stress, as reported by [47] in *Prochilodus lineatus* in Argentina and by [48] in three subspecies of *Salmo truta* in Turkey. The liver usually contains high levels of antioxidant enzyme activity, which may be due to high rates of free radical generation in this tissue [49].

A. aeneus exhibited seasonal fluctuations in EROD activity, peaking in April at the SAR site. Several authors [43, 50] point out that POCs (including PAHs) typically cause elevated levels of EROD activity, and fossil fuel spills and infiltrations are a major source of PAH input [51]. In *A. aeneus* EROD activity was highest at the SAR site, a recreational spot reached only by unpaved road where motor vehicles park on the riverbank. EROD induction may thus be due to surface runoff carrying PAHs.

LDHs are cytoplasmic enzymes that catalyze the reversible reduction of pyruvate to lactate [52], an important step in the energy processes of many animal groups. Its use as a biomarker is based on the assumption that organisms subjected to chemical stress must obtain additional energy rapidly, thus increasing anaerobic glycolysis. The LDH response is apparently time-dependent and may vary with the pollutant and organism involved [41]. LDH in *A. aeneus* increased at all sites in November and February, coinciding with the period of maximum LPO as well as the post-hurricane season and highest river flow.

The IBR enables evaluation of the global variations of biomarkers, taking into account the contributions and variations in the biomarkers assessed [17]. IBR data revealed seasonal fluctuations, the maximum total IBR values occurred in April and November (April, 15.73; November, 14.73) and the minimum values in February (2.53). Response in April and November suggested that *A. aeneus* was exposed to greater stress during this period, one at the end of the dry season and the other coinciding with the post-hurricane season that also led to high LPO values (oxidative stress), high GST levels (antioxidant responses), and high levels of LDH (high energetic need). Multiple stressors are involved during flooding, such as altered habitat, changes in hydrological regime, and mobilization of pollutants. Responses of biota to environmental stressors are the integrated result of natural and anthropogenic stressors that can be ultimately manifested in biotic changes at several levels of organization [53]. IBR differences among the study periods may reflect a compensatory mechanism by which the fish regains homeostasis following the period of highest stress. Fish display a large variety of physiological stress responses that manifest as an increase in certain biomarkers after a stressful event [54]. These responses are considered adaptations of the fish to adjust itself to the disturbance and regain homeostasis [55]. Fish have also been found to exhibit recovery responses dependent on stress duration and magnitude. If the stressors are too severe or persistent and the fish is unable to regain homeostasis, the responses may become maladaptive and may pose a risk to the health and wellbeing of the fish [54].

Regarding somatic indices, GSI displayed the reproductive period of *A. aeneus* occurred from April (end of the dry season) to July (early rainy season), with the reproductive peak in July. HSI was highest in *A. aeneus* during periods of reproductive inactivity, which is interpreted as an increase in liver-stored reserve materials for later use during gametogenesis. Furthermore, low HSI values prior to and during reproduction may result from the transfer of liver energy reserves toward gonadal maturation and the reproductive event, with consequent depletion of these reserves and a decrease in HSI values [56]. CF remained constant among sites and study periods, with maximum values during July. CF reflects the interactions between abiotic and biotic factors in the physiological conditions of the fish [57]. Observations demonstrate that *A. aeneus* maintains a stable, robust condition. CF and HSI trends may reflect physiological conditions in *A. aeneus* that enable oocyte maturation and release, suggesting that oocyte production relies on energy stored in the liver and not on energy stored in the musculature [58, 59]. GSI, HSI, and CF revealed that the reproductive success of the sentinel species had not been affected, since the GSI values concurred with those reported for other species in the genus *Astyanax*. HSI values showed the transfer of energy during the reproductive period, and CF remained stable throughout the study, evidencing no effects on the general condition of the fish during periods of higher stress and indicating that the stress was temporal and the fish were able to compensate for it.

Despite WQI scores suggest that the Champotón river water is not highly polluted, the set of *A. aeneus* biomarkers constitutes a more sensitive and effective tool for identifying periods of environmental conditions adverse to fish health. Markers of oxidative damage (LPO), energy processes (LDH), detoxification (EROD), and antioxidant activity (GST) suggested that two stress periods affected the health condition of *A. aeneus* in different ways. These biomarkers can be used as early warning signals of environmental change prior to the onset of irreversible damage at the population level. Indeed, the IBR values highlighted the two periods of high biomarker response. Overall, this study provides evidence supporting the use of a biomarker set in assessing the health of aquatic systems, corroborating the suitability of *Astyanas aeneus* as a sentinel species.

5.2. Study case of *Chirostoma jordani* in Yurira Lake

Chirostoma jordani is an Atherinopsid fish endemic of Central México. Data presented are from a population living in Yuriria Lake, one of the most important lakes in México, located in the Central Plateau (hydrologic region Lerma-Chapala-Santiago 20 ° 20'24" -19 ° 04'48" N and 101 ° 55'48" -100 ° 48 '36" W). It is an artificial, small and shallow lake (area= 66 km^2 and maximum depth of 3.2 m) feed by a diversion of Lerma river (their main tributary), that carries wastes from mining activities, livestock, industrial, urban and rural areas [60]. In the western end the lake receives water from two small an intermittent tributaries. Yuriria Lake supplies water for the surrounding farming areas, harbors migratory bird populations, and supports fisheries and tourism of several human settlements on the littoral zone.

This study shows an assessment of water quality by means of a WQI and a battery of oxidative stress: LPO, and the activity of antioxidant enzymes, SOD, CAT, and GPx along with somatic indices, such as the GSI, HSI, and CF were analyzed to assess the health condition of *C. jordani* in Yuriria Lake. The study was carried on in a period with scarce pluvial precipitation and prolonged drought (May, August, November 2009 and February and May 2010).

Yuriria Lake is characterized by a high deterioration in their water quality, WQI values ranged from 55 to 70, with a global mean of 65.85. The lake has spatial differences in water quality, the limnetic zone has higher scores (63 to 70), and the tributaries have the lower quality (55 to 58). Yuriria Lake being located in the Central Plateau, one of the most highly populated areas in México, displays the general problem of water quality of the basin (the Lerma-Chapala basin), where urban and industrial wastewater discharges, and leachates of agrochemicals are the main pollutants that diminish the water quality [61]. Particularly the middle Lerma (where Yuriria Lake is located) is recognized as the most affected area, with WQI scores between 41.1 and 54.2 in 1999 [40, 61, 62]. Furthermore, previous studies have recognized the entry of pollutants in Yuriria Lake [61]; however, the effect of the mixtures of these pollutants on the aquatic biota inhabiting the lake had not been analyzed.

The biomarker assessment suggests that the lake conditions exert stress on the fish *C. jordani*. The biomarker response showed pronounced seasonal variations. The gills presented higher values of LPO. May 2009 displayed the highest levels and November the lowest. In the liver the higher levels of LPO were detected during November and February and May 2010 (the end of the rainy season and the dry season). In muscle, the highest level of LPO was observed during February and May 2010. Gills being the first organ of contact with water are exposed directly to any xenobiotic in the aquatic environment and their biomarker responses are the result of the exposure to stressors. In addition, toxics can also enter via the intake of water and food and be absorbed and transported by the portal system to the liver before entering the general circulatory system; in consequence, the liver is one of the most sensitive organs to environmental stressors [63].

The activity of the antioxidant enzymes in liver and gills also showed a marked seasonal variation. SOD and GPx significantly increased during November, mainly in the gills, compared to the rest of the seasons. CAT also showed higher values in activity during November; however, its highest value was found in gills during February. In general, the activity of antioxidant enzymes decreased from November to May 2010.

Exposure to various xenobiotics, such as metals and organic compounds that enter water bodies, can promote the formation of ROS and induce oxidative stress [12]. The increase in the level of LPO in liver observed in November 2009 and February and May 2010, suggests the existence of pro-oxidant agents in Yuriria Lake and indicates increased oxidative stress in these seasons. Seasonal variation in LPO values could be related to the rainy and the dry seasons; the rainy season can promote dilution of xenobiotics that induce less stress during this season. Rainfall also increases leaching and runoff that enhance the entry of xenobiotics (chemicals), in this study results show that the damage generated in the fish liver became evident from November (the end of the rainy season) until February and May 2010 (dry season), when the processes of evaporation and consequent concentration of xenobiotics could be higher. Furthermore, in May 2009, there were higher levels of LPO in gills, which may indicate that water in the lake at the beginning of the rainy season provoke oxidative stress in gills.

An increase in LPO levels in fish can trigger an antioxidant response as a defense mechanism to prevent cell damage caused by pro-oxidant agents [19, 64] and could be expressed as increased or depleted CAT, SOD, and GPx activities [12]; in both cases, the result is damage to the antioxidant system. In *C. jordani* inhabiting Yuriria Lake, both responses were detected: 1) stimulation of the activity, when LPO levels in liver and muscle were higher (November and February) and antioxidant activity showed the highest values; and 2) depletion in antioxidant activity during May 2009 with high levels of LPO in the gills, and May 2010 with increased LPO in liver and muscle. These highly variable responses of the antioxidant system depend on the type and concentration of contaminant to which fish are exposed, as well as on the intensity and duration of exposure [65]. The clear decrease in the activities of CAT, SOD, and GPx in *C. jordani* is extremely important because it suggests severe damage to the antioxidant system of fish. These damages have been previously documented by [66], who recorded a collapse of the antioxidant defense system of *Liza aurata* in the Ria de Aveiro, Portugal, with exposure to mercury. In addition, [67] reported damage to the antioxidant system of *Oreochromis niloticus* from acute and chronic exposure to Cd, Cr, Cu, Zn, and Fe.

At the population level the assessment can reveal changes in the fish biology and ecology resulting from natural fluctuations and/or ecosystem changes caused by environmental degradation. The standard length of *C. jordani* ranges from 21.16 to 77.61 mm. Three size classes were determined: Class I with a mean size of 26 mm, represented only in the month of May 2010; Class II with a mean size of 56 mm; and class III with a mean size of 62 mm. There was a gap in the size frequencies in the class of 32–50 mm from May to August 2009, whose abundances were not sufficient to form this cohort. The lack of smaller sizes from May to November 2009 can be interpreted as a possible reproductive failure during that year, resulting in low recruitment and consequently precluding estimation of the cohort [68]. These gaps or missing cohorts have been previously documented for other fish as a result of overfishing and/or environmental degradation [69]. Several causes could explain this event. According to [70], the hydrological cycle plays an important role in the development of different biological attributes such as gonadal maturation, migration, spawning, larval development, growth, and feeding. Prolonged periods of drought are also associated with failures in the recruitment and subsequently reduced adult stocks and serious effects on fisheries.

The somatic indices revealed that K displayed small variation between sites and between periods. HSI showed significant differences between sites in August, and between periods in May, values were significantly lower than those of other periods. The GSI showed the greatest variation, with a clear reproductive peak during May 2010. There was a positive correlation between the GSI and HSI. By size class, only the GSI showed variations between seasons. Class I was significantly lower than the rest of the classes. Classes II and III the GSI showed a reproductive peak in May 2010.

In fish, the cost of reproduction may be considerable; thus, fish can express different patterns of energy storage and depletion in relation to reproductive cycles, with an alternation in energy storage (56). The comparison of K and the HSI with the GSI could therefore be useful for estimating the possible balances or energy transfer between the reproductive period and nutritional status [71]. *C. jordani* in Yuriria Lake show that K was maintained at stable levels. Furthermore, there was no alternation of energy storage between the liver and gonads; GSI and HSI correlated positively, indicating that the reproductive period did not compromise energy reserves or the liver or the soma.

According to [72], fish living in waters contaminated with domestic sewage exhibit higher K and GSI, these authors suggest it is because these sites have more available food for fish, enabling them to compensate for the environmental impairments. This scenario is likely occurring in Lake Yuriria, where *C. jordani*, having enough food, can complete reproductive cycles and reach larger sizes despite the presence of stressors in the lake.

According to the results critical periods in the health of *C. jordani* occur from November, February to May 2010 (end of the rainy season and the dry season) because in this period, the higher LPO, the lowest antioxidant response, and the lowest K were observed in contrast to the higher WQI scores detected in the same period. This result indicates that fish health assessed by biological indicators as oxidative stress biomarkers and the lack of a cohort, are highly sensitive to environmental conditions imposed by the dry season. The drought has been recognized as one of the critical periods in fish health because during this period, the dilution capacity of aquatic ecosystems is low, which increases the risk of exposure to high concentrations of pollutants [73].

The findings suggest that *C. jordani* faces oxidative stress resulting from the presence of pro-oxidant agents in Yuriria Lake. At the population level, *C. jordani* has adapted to the conditions in Yuriria Lake, with mean values of HSI and GSI greater than those in other sites and K values stable throughout the year. Changes were observed in recruitment and reproductive success associated with low water levels in the lake in 2009 which shows that the fish population is highly dependent on water levels in the lake and climate changes. Previous studies indicate that Yuriria Lake receives various xenobiotics with levels that vary both spatially and temporally. Biomarkers of oxidative stress, somatic indices, and monitoring of the size classes in the fish *C. jordani* are appropriate indicators of Yuriria Lake conditions.

5.3. Study case of *Ameca splendens* and *Goodea atripinnis* in Ameca river

The Ameca River is located in the western slope of Mexico which drains to the Pacific Ocean. It is a river characterized by their great fish biodiversity, however, the upper portion of Ameca River, is affected by several environmental disturbers: the construction of a reservoir, the inputs of wastewater from a sugar-processing facility and water extraction, which have resulted in a drastic reduction in fish biodiversity [74, 75]. In addition, some endemic fish species such as *Ameca splendens*, have suffered a reduction in their range, and have become more prone to extinction (NOM-059, 2002) than those with a broad distribution (like *Goodea atripinnis*).

The authors of this paper [32] analyze biomarkers and bioindicators of two viviparous fish species, *A. splendens* and *G. atripinnis* living in a reservoir of the upper portion of the Ameca River, which receives wastewater, and in a spring of the same river that is free from such polluting water. In this study a comparison of the biomarker responses and bioindicators in two fish species were assessed, according to the main objectives: a) to assess water quality of a spring (ER reference site) and De LaVega reservoir (LV impacted site) where *G. atripinnis* and *A. splendens* coexist; b) to examine the health of both fish species in the reference and impacted sites by means of a battery of biomarkers; and c) to analyze physiological condition indices and population level assessment by mean of bioindicators (population measurements). A WQI was assessed. The set of biomarkers was composed by: enzyme activities of gamma-glutamyl-transpeptidase (γ-GTP), acetylcholinesterase (AChE), EROD and the LPO were determined. Additionally, somatic index were analyzed: CF, HSI and GSI.

Data of WQI scores showed spatial and temporal variations, the spring ER achieved the highest scores in all months over the course of the study and the highest values occurred in March in both sites.

Regarding biomarkers highest values of LPO were found in September (rainy season); the highest values were detected for female livers and gills of *A. splendens* in the impacted site LV and for female livers and male gills of *G. atripinnis* in LV. The LPO activity in the upper Ameca River displayed the stress to which organisms are subjected, since in the reservoir LPO showed the highest values, being *A. splendens* the most affected species. Many environmental pollutants and their metabolites have shown to exert toxic effects associated to oxidative stress, producing free radicals that initiate the LPO and cause damage to membrane proteins [76].

Results of γ-GTP showed less marked seasonal differences than LPO. In this study, in most comparisons between sites, γ-GTP activity was slightly higher at the spring; the inhibitory effect of this activity at LV could indicate a diminution in the amount of membrane proteins caused by LPO [77]. On the other hand, an increase of γ-GTP activity towards March in LV could be to prevent increases in LPO [76]; this increase coincides with the rainy season. Several authors have found seasonal variations in the response of this enzyme as a result of exposition to alkylphenols, the final degradation products of pesticides, detergents and other formulated products [78].

There was a seasonal variation in AchE activity. Organophosphates and carbamates, as well as PAHs, have been widely recognized for causing AchE inhibition, through their reaction with the serine at active site of the enzyme [79]. In nervous tissue AchE is responsible for the breakdown of acetylcholine (Ach) during transmission of an impulse; if the enzyme is inhibited, Ach is accumulated and thus a prolonged transmission of impulses could result in tetani and often in respiratory failure and death. In September in *A. splendens* living at LV, AchE exhibited lower values related to reference site that could have resulted from fish exposure to diverse pesticides used in the adjacent agricultural lands that run off in the rainy season. In March during the dry season, in *G. atripinnis*, living at LV, AchE also exhibited lower values related to ER, may be the effect of evaporation and consequently the concentration of total solid dissolved at LV, including pesticides and other xenobiotics.

Regarding EROD, the hydrocarbons discharged by the sugar-processing facility and other effluents into LV from December to June, could be responsible for activating EROD detoxification mechanism in fishes living there [80]. This mechanism is considered as the main measure of the CYP1A activity, which in turn constitutes a part of the enzyme complex of the Mixed-Function Oxidase (MFO). Since MFO facilitates the excretion of aromatic contaminants from the body induction of this complex is an effective biomarker of exposure [81, 82]; there are several studies reporting elevated levels of MFO activity in liver fishes as a result of exposure to organic contaminants, such as PAHs, dioxins, PCBs and agricultural and urban wastewater [83]. Also, there are other factors, such as UV radiation, that causes increase in MFO activity [84], moreover, damaged livers, like those of LV organisms, are less capable to MFO induction [85].

Responses to environmental stress also were reflected in bioindicators in both species studied. The major HSI values in LV concurs with [86] who reported major HSI values, related with higher EROD induction, at contaminated sites in comparison with a reference site. Moreover, in our study, higher LSI in LV concurs with higher IBR values. High values of HSI could have resulted from exposition to hydrocarbons which cause hypertrophia in liver [86].

Species may differ in the nature of their physiological response and reproductive consequences to stressors [87]. Tolerant species to environmental stress, like *G. atripinnis*, are more abundant in more disturbed environments, like LV [74, 88]; on the other hand, *A. splendens* is more abundant in a more stable environment, like ER. In LV females/male ratio for *A. splendens* could be affected by environmental estrogens, like pesticides [89]; these could act by merging receptor binding properties of estradiol, alteration of estradiol/testosterone ratios or estrogen receptor levels [12].

The higher SL, weight and CF values in LV are in concordance with [71], that found higher CF and GSI in fishes living in waters polluted with untreated domestic sewage; they suggested that fishes in these sites could find abundant food availability, and they are able to compensate for environmental changes caused by untreated domestic sewage discharges. Fishes could have major GSI values, higher fecundity and lower maturity age even under conditions of high pH [89]; in the present study, *A splendens* in LV presented these features; only in *G. atripinnis* organisms GSI was higher in the ER. [90] found a decrease in GSI as a result of the exposition of chubs to effluents carrying out organic pollutants and metals. [71] revealed a negative relation between CF and HSI with GSI, but this relation was observed only in *G. atripinnis*. The larger size, higher growth, longevity and reproductive success of organisms living at LV suggest a tactic to compensate for the stress to which the populations of both species studied are subjected; however, offspring is smaller and has a lower weight.

Throughout this study, water quality was higher in ER than in LV due to human activities; but in both sites there were different spatial and temporal factors that produced stress on fishes living there. Therefore, fishes had responses at biomarker and population levels of biological organization. Every biomarker and IBR in this study showed seasonal variation and they were useful environmental tools to demonstrate that, as consequence of pollution, LV is a more stressing place to organisms living them in comparison with ER. In general, organisms in LV presented oxidative stress by the LPO levels, and then neurotoxic impacts by the AchE and some detoxification mechanisms were evident by the γ-GTP and EROD activities. Bioindicators showed evidences of physiological changes due to contaminants exposure and make evident the plasticity of the organisms to survive in this site, in turn the responses should be considered as tactics to survive under stress condition. Moreover, both biomarkers and bioindicators revealed that *A. splendens* is a less tolerant species than *G. atripinnis* to environmental stress. Differences in biological response could be attributed to different physiological status of each fish species during the wet and dry season as well as to differences in the type and quantity of the xenobiotics that input at LV due to the period of maximum and minimal activity of the sugar industry; the lixiviation of the agrochemicals from the adjacent lands to the water bodies and also to the complexity of the mixtures of pollutants that are conform at LV that provoke several biological responses.

6. Holistic approach

Environmental stressors can cause several and different damages over aquatic organisms. These damages could be from molecular to population levels, likewise community and eco-system levels. Through the biomarkers such as defined in this document (any biological measurable response from an organism, induced by the exposure to a xenobiotic or complex mixture of them) we can determine only some of possible causal relationships. Therefore, it is necessary always, measure a set of biomarkers to identify different stressors or damage on sentinel organisms. Several indices have been proposed to try to integrate the multi responses of different biomarkers in a single number that is indicative of the severity of the damage or stress. Such is the case of the IBR proposed by [17]. In this index, the biomarker data must first be normalized and standardized; then the score is represented by the area of a star plot. IBR considers the responses of activation or inhibition of the biomarkers assessed. IBR is an exploratory tool and should be appropriate only if an a priori justification exists for each biomarker used and if the physiological significance of the changes to each biomarker is well known [17].

An other case of index based on a battery of biomarkers was proposed by [91]. This biomarker index was obtained by summing the biomarker values expressed in term of classes. Classes were determined by a distribution-free approach derived from the theory of rough sets. No synergistic or antagonistic assumptions were incorporated into this index.

In [4] the authors proposed a quantitative health assessment index for rapid evaluation of fish condition in the field named Health Assessment Index (HAI). This index is not based on a battery of biomarkers; however, it is a quantitative index that allows statistical comparisons of fish health among data sets. Index variables are assigned numerical values based on the degree of severity or damage incurred by an organ or tissue from environmental stressors.

The Bioeffect Assessment Index (BAI), is based on the integration of several pathological endpoints measured in the liver of fish [21]. The BAI represents a modification of the HAI since it includes solely validated biomarkers reflecting toxically induced alterations at different levels of biological organisation in order to quantify the effects of environmental pollution. BAI is able to reflect deleterious effects of several classes of xenobiotics such as heavy metals, organochlorines, pesticides, PAHs, and therefore is also considered as an integrative index of health in aquatic ecosystems.

When we use sentinel organisms, a key point is the study of baseline or natural variation of responses of the sentinel organism selected, or characterizing the response of the same sentinel organism in reference sites.

The use of fish as sentinel organism is feasible for pollution monitoring in aquatic systems; however, the survey should consider the application of a suite of measurable responses (bio-markers and bioindicators) to identify potential sources of stress and damage to which organisms are exposed, as shown in the case studies presented above. The set of biomarkers or bioindicators should also, consider several levels of biological organization in order to identify effects of environmental stressors, spatio temporal trends in environmental conditions and to identify early warning signals to prevent that damage continue from low biological organizing levels to higher levels of organization.

Author details

Jacinto Elías Sedeño-Díaz[1*] and Eugenia López-López[2]

*Address all correspondence to: jsedeno@ipn.mx

1 Polytechnic Coordination for Sustainability, National Polytechnic Institute. Mexico, Distrito Federal, México

2 Ichtiology and Limnology Laboratory, National School of Biological Sciences, México, Distrito Federal, México

References

[1] Dudgeon D, Arthington AH, Gessner O, Kawabata ZI, Knowler DJ, Lévêque C, Naiman RJ, Prieur-Richard AH, Soto D, Stiassny MLJ, and Sullivan CA. Freshwater biodiversity: importance, threats, status and conservation challenges. Biological Reviews 2005; 81: 163–182.

[2] Strobl RO, Robillard PD. Network design for water quality monitoring of surface freshwaters: A review. Journal of Environmental Management 2008; 87: 639–648.

[3] Vasseur P, Cossu Leguille C. Biomarkers and community indices as complementary tools for environmental safety. Environment International 2003; 28: 711 – 717.

[4] Adams SM, Brown AM, Goede RW. A Quantitative Health Assessment Index for Rapid Evaluation of Fish Condition in the Field. Transactions of the American Fisheries Society 1993; 122: 63-73.

[5] Wedemeyer GA, McLeay DJ, Goodyear CP. Assessing the tolerance of fish and fish populations to environmental stress: the problems and methods of monitoring. In:

Cairns P, Hodson V, Nriagu JO. (ed) Contaminated effects on fisheries. Wiley, New York; 1984. p163-195.

[6] Eufemia NA, Collier TK, Stein JE, Watson DE, and Di Giulio RT. Biochemical responses to sediment-associated contaminants in brown bullhead (Ameriurus nebulosus) from the Niagara River ecosystem. Ecotoxicology 1997; 6: 13–34.

[7] Fausch KD, Lyons J, Karr JR and Angermeier PL. Fish Communities as Indicators of Environmental Degradation. American Fisheries Society Symposium 1990; 8:123-144.

[8] NRC: Committee on Biological Markers of the National Research Council. Biological markers in environmental health research. Environmental Health Perspectives 1987; 74: 3-9.

[9] Niemi GJ, McDonald ME. Application of Ecological Indicators. Annual Review of Ecology, Evolution, and Systematics 2004; 35: 89–111.

[10] Gerhardt A. Bioindicator species and their use in biomonitoring. Environmental Monitoring I. EOLSS.

[11] US Goverment Printing Office. http://www.gpo.gov/fdsys/pkg/CZIC-z5863-m65-p33-1995/xml/CZIC-z5863-m65-p33-1995.xml (accesed 12 july 2012).

[12] Sherry JP. The role of biomarkers in the health assessment of aquatic ecosystems. Aquatic Ecosystems Health and Management 2003; 6: 443-440.

[13] Livingstone d. Biotechnology and pollution monitoring: use of molecular biomarkers in the aquaitc environment. Journal of Chemistry, Technology and Biotechnology 1993; 57: 195-211.

[14] Peakall DB, Walker CH. The role of biomarkers in environmental risk assessment. 3: Vertebrates. Ecotoxicology 1994; 3: 173-179.

[15] Van Gestel CAM, Van Brummelen TC. Incorporation of the biomarker concept in ecotoxicology calls for a redefinition of terms. Ecotoxicology 1996; 5: 217-225.

[16] Shugart LR, McCarthy JF, Halbrook RS. Biological markers of environmental and ecological contamination: An overview. Risk Analysis 1992; 12: 353–360.

[17] Beliaeff B, Burgeot T. Integrated biomarker response: a useful tool for ecological risk assessment. Environmental Toxicology and Chemistry 2002; 21: 1316-1322.

[18] WHO International Programme on Chemical Safety (IPCS). Biomarkers and risk assessment: concepts and principles. Environmental Health Criteria 155, World Health Organization, Geneva 1993.

[19] van der Oost R, Beyer J, Vermeulen NPE. Fish bioaccumulation and biomarkers in environmental risk. Environmental Toxicology and Pharmacology 2003; 13: 57-149.

[20] Rojas-García AE, Medina-Díaz IM, Robledo-Marenco ML, Barrón-Vivanco BS, Pérez-Herrera N. Pesticide Biomarkers. In Margarita Stoytcheva (ed.) Pesticides in the mod-

ern world-Pest Control and pesticides exposure and toxicity assessment. InTech; 161-190.

[21] Broeg K, Westernhagen HV, Zander S, Körting W, Koehler A. The "bioeffect assessment index" (BAI) A concept for the quantification of effects of marine pollution by an integrated biomarker approach. Marine Pollution Bulletin 2005; 50: 495–503.

[22] Turrens JF. Topical Review Mitochondrial formation of reactive oxygen species. Journal of Physiology 2003; 552(2): 335-344.

[23] Lushchak VI. Environmentally induced oxidative stress in aquatic animals. Aquatic Toxicology 2011; 101: 13-30.

[24] Valavanidis A, Vlahogianni T, Dassenakis M, Scoullos M. Molecular biomarkers of oxidative stress in aquatic organisms in relation to toxic environmental pollutants. Ecotoxicology Environmental Safety 2006; 64: 178-189.

[25] Stadtman ER, Levine RL. Protein oxidation. Annals of the New York Academy of Sciences 2000; 899: 191–208.

[26] Vlahogianni T, Dassenakis M, Scoullos MJ, Valavanidis A. Integrated use of biomarkers (superoxide dismutase, catalase and lipid peroxidation) in mussels Mytilus galloprovincialis for assessing heavy metals' pollution in coastal areas from the Saronikos Gulf of Greece. Marine Pollution Bulletin 2007; 54: 1361 – 1371.

[27] Prieto AI, Jos A, Pichardo S, Moreno I, Cameán AM. Differential oxidative stress responses to microcystins LR and RR in intraperitoneally exposed tilapia fish (Oreochromis sp.). Aquatic Toxicology 2006; 77: 314-321.

[28] Ortiz-Ordoñez E, Uría-Galicia E, Ruiz-Picos RA, Sánchez Duran AG, Hernández Trejo Y, Sedeño-Díaz JE, López-López E. Effect of Yerbimat Herbicide on Lipid Peroxidation, Catalase Activity, and Histological Damage in Gills and Liver of the Freshwater Fish Goodea Atripinnis. Archives of Environmental Contamination and Toxicology 2011; 61: 443–452.

[29] Nath Singh N, Srivastava AK. Haematological parameters as bioindicators of insecticide exposure in teleosts. Ecotoxicology 2010; 19: 838–854.

[30] Munkittrick KR, Dixon DG. A holistic approach to ecosystem health assessment using fish population characteristics. Hydrobiologia 1989; 188-189: 123-135.

[31] Oertel N, Salánki J. Biomonitoring and bioindicators in aquatic ecosystems. In: Ambasht RS, Ambasht NK (ed) Modern trends in applied aquatic ecology. Nueva York, USA: Kluwer Academic 2003 p219–246.

[32] Tejeda-Vera R, López-López E, Sedeño-Díaz JE. Biomarkers and bioindicators of the health condition of Ameca splendens and Goodea atripinnis (Pisces: Goodeaidae) in the Ameca River, Mexico. Environment International 2007; 33: 521–531.

[33] Lima-Junior SE, Braz Cardone I, Goitein R. Determination of a method for calculation of Allometric Condition Factor of fish. Maringá 2002; 24(2): 397-400.

[34] Barton BA, Morgan JD, Vijayan MM. Physiological and condition-related indicators of environmental stress in fish. In Biological Indicators of Aquatic Ecosystem Stress, ed. S. Marshall Adams, Maryland: American Fisheries Society. 2002; 111–148.

[35] Hinton DF, Laurén DJ. Liver structural alterations accompanying chronic toxicity in fishes: potential biomarkers of exposure. In: McCarthy JF, Shugart LR (eds) Biomarkers of environmental contamination. Lewis, Boca Raton, 1990. pp17–57

[36] Courtney J, Klinkmann T, Courtney A, Torano J, Courtney M. Relative Condition Factors of Fish as Bioindicators One Year after the Deepwater Horizon Oil Spill. www.arXiv.org > q-bio > arXiv:1208.5095.

[37] López-López E, Sedeño-Díaz JE, López-Romero F, Trujillo-Jiménez P. Spatial and seasonal distribution patterns of fish assemblages in the Río Champotón, southeastern México. Reviews in Fish Biology and Fisheries 2009; 21: 127-142.

[38] CONABIO (Comisión Nacional para el Conocimiento y Uso de la Biodiversidad). Uso del suelo y vegetación modificado por CONABIO. Escala 1:1,000,000. Comisión Nacional para el Conocimiento y Uso de la Biodiversidad. 1999. Ciudad de México, México.

[39] Mestre, R.J.E., 2002. Integrated approach to River Basin management: Lerma-Chapala case study- attribution and experiences in water management in Mexico. Water Int. 22, 140-152.

[40] Sedeño-Díaz JE, López-López E. Water quality in the Río Lerma, Mexico: An overview of the Last Quarter of the Twentieth Century. Water Resources Management 2007; 21: 1797-1812.

[41] Rendón von Osten J. Uso de biomarcadores en ecosistemas acuáticos. In: Botello AV, Rendon von Osten J, Gold-Bouchot G, Agraz-Hernández C. (Eds.) Golfo de México contaminación e impacto ambiental: Diagnóstico y tendencias. México: Univ Autón Campeche, Univ Nal México, Instituto Nacional de Ecología. 2005 pp121-140.

[42] Quetz L, Memije M, Benítez J, Rendón von Osten J. Hidrocarburos aromáticos policíclicos en sedimentos del río y costa de Champotón, Campeche. Jaina 2009; 20, 27-34.

[43] Adams SM, Greeley Jr MS, McHugh JL, Noga EJ, Zelikoff JT. Application of multiple sublethal stress indicators to assess the health of fish in Pamlico Sound following extensive flooding. Estuaries 2003; 26, 1365-1382.

[44] Trujillo-Jiménez P, Sedeño-Díaz JE, Camargo JA, López-López E. Assessing environmental conditions of the Río Champotón (México) using diverse indices and biomarkers in the fish Astyanax aeneus (Günther, 1860). Ecological Indicators 2011; 636-1646.

[45] Einsporn S, Broeg K, Koehler A. The Elbe flood 2002—toxic effects of transported contaminants in flatfish and mussels of the Wadden Sea. Marine. Pollution Bulletin 2005; 50: 423–429.

[46] Ochoa DM, González JF. Estrés oxidativo en peces inducido por contaminantes ambientales. Rev. Med. Vet. Zoo. 2008; 55: 115-126.

[47] Cazenave J, Bacchetta C, Parma JM, Scarabotti AP, Wunderlin AD. Multiple biomarkers responses in Prochilodus lineatus allowed assessing changes in the water quality of Salado River basin (Santa Fe, Argentina). Environmental Pollution 2009; 157, 3025–3033.

[48] Aras NM, Bayir A, Sirkecioglu AN, Bayir M, Aksakal E, Haliloglu HI. Seasonal changes in antioxidant defence system of liver and gills of Salmo trutta caspius, Salmo trutta labraz and Salmo trutta macrostigma. Journal of Fish Biology 2009; 74, 842-856.

[49] Mohammed SAA, Al-Qahtani A. The distribution of glutathione and glutathione-S-transferase activity in the oprgans of dhub (The agamid Lizard; Uromastyx aegyptius). Journal of Biological Sciences 2007; 7: 558-561.

[50] Havelkova M, Blahova J, Kroupova H, Randak T, Slatinska I, Leontovycova D, Grabic R, Pospisil R, Svobodova Z. Biomarkers of contaminant exposure in Chub (Leuciscus cephalus L.) – Biomonitoring of major rivers in the Czech Republic. Sensors. 2008; 8: 2589–2603.

[51] Hartl MGL. Benthic fish as sentinel organisms of estuarine sediment toxicity, in: Bright M, Dworschak PC, Stachowitsch M. (Eds.) The Vienna School of Marine Biology: A Tribute to Jörg Ott. Facultas Universitätsverlag, Wien. 2002 pp89–100.

[52] López JE, Marcano Torres M, López Salazar JE, López Salazar Y, Fasanella H, Michelangelli L. Lactato deshidrogenasa total y sus isoenzimas LDH_1, LDH_2, LDH_3, LDH_4, LDH_5, en el diagnóstico diferencial del síndrome ascítico. Gaceta Médica de Caracas 1996; 104(4): 345-364.

[53] Adams SM. Assessing cause and effect of multiple stressors on marine systems. Marine Pollution Bulletin 2005; 51, 649-657.

[54] Barton AB. Stress in fishes: A diversity of responses with particular reference to changes in circulating corticosteroids. Integrative and Comparative Biology 2002; 42: 513-525.

[55] Wendelaar BSE. The stress response in fish. Physiological Reviews 1997; 77: 591-625.

[56] Dethloff MG, Christopher J, Schmitt JC. Condition factor and organosomatic indices, in: Schmitt CJ, Dethloff MG (ed) Biomonitoring of Environmental Status and Trends (BEST) Program: selected methods for monitoring chemical contaminants and their effects in aquatic ecosystems. U.S. Geological Survey, Biological Resources Division, Columbia, (MO): Information and Technology Report USGS/BRD-2000--2005. 2000; pp13-18.

[57] Lizama MAP, Ambrósio AM. Condition factor in nine species of fish of the Characidae family in the upper Paraná river floodplain, Brazil. Brazilian Journal of Biology 2002; 62: 113-124.

[58]   Zimmerman M. Maturity and fecundity of arrow tooth flounder, Atheresthes stomias, from the Gulf of Alaska. Fishery Bulletin 1997; 95: 598-611.

[59]   Yoneda M, Tokomura M, Fujita H, Takeshita N, Takeshita K, Matsuyama M, Matsuura S. Reproductive cycle and sexual maturity of the anglerfish Lophiomus setigerus in the East China Sea with a note on specialized spermatogenesis. Journal of Fish Biology 1998; 53: 164-178.

[60]   López-Hernández M, Ramos-Espinosa MG, Carranza-Fraser J. Análisis multimétrico para evaluar contaminación en el Río Lerma y Lago de Chapala, México. Hidrobiológica 2007; 17: 17-30.

[61]   Jiménez-Cisneros BE. Sustentabilidad, un debate a fondo: Información y calidad del agua en México. Trayectorias 2007; 9: 45-56.

[62]   Torres de la O, M. Salud de veinte embalses de la cuenca Lerma-Chapala. Instituto Mexicano de Tecnología del Agua. 2007; México.

[63]   Peña CE, Carter DE, Ayala-Fierro F. Toxicología ambiental: Evaluación de riesgos y restauración ambiental. Southwest hazardous waste program. University of Arizona. USA. 2007.

[64]   Cheung CCC, Zheng GJ, Li AMY, Richardson BJ, Lam PKS. Relationships between issue concentrations of polycyclic aromatic hydrocarbons and antioxidative responses of marine mussels, Perna viridis. Aquat. Toxicol. 2001; 52: 189-203.

[65]   Mieiro CM, Ahmad I, Pereira ME, Duarte AC, Pacheco M. Antioxidant system breakdown in brain of feral Golden grey mullet (Liza aurata) as an effect of mercury exposure. Ecotoxicology 2010; 19: 1034-1045.

[66]   Atili G, Canli M. Response of antioxidant system of freshwater fish Oreochromis niloticus to acute and chronic metal (Cd, Cu, Cr, Zn, Fe) exposures. Ecotoxicology and Environmental Safety 2010; 73: 1884-1889.

[67]   Houde ED. Recruitment variability, in: Jacobsen, T., Fogarty, M., Megrey, B., Mocksness, E. (Eds), Fish reproductive biology, Implications for assessment and management, Wiley-Blackwell, USA, 2009 pp91-171.

[68]   Secor DH. Longevity and resilience of Chesapeake Bay striped bass. ICES J. Mar. Sci. 2000; 57: 808-815.

[69]   Agostinho A, Gomes LC, Veríssimo S, Okada EK. Flood regime, dam regulation and fish in the Upper Paraná River: effects on assemblage attributes, reproduction and recruitment. Rev. Fish. Biol. Fish. 2004; 14: 11-19.

[70]   González P, Oyarzun C. Variabilidad de índices biológicos en Pinguipes chilensis valenciennes 1833 (Perciformes, Pinguipedidae): ¿están realmente correlacionados?. Gayana (Concep.). 2002; 66, 249-253.

[71] Alberto A, Camargo A,Verani JR, Costa OFT, Fernándes MN. Health variables and gill morphology in the tropical fish Astyanax fasciatus from a sewage-contaminated river. Ecotoxicol. Environ. 2005; 61: 247-255.

[72] Adedeji OB. Acute Efects of Diazinon on Blood Paramters in the african Catfish (Clarias Gariepinus). Journal of Clinical Medicine Research 2010; 2: 1-6. http://www.academicjournals.org/jcmr/PDF/PDF2010/Jan/Adedeji.pdf. Accessed 7 November 2011

[73] López-López E, Paulo-Maya J. Changes in the fish assemblages in the upper Río Ameca, México. Journal of Freshwater Ecology 2001; 16: 179–87.

[74] López-López E, Paulo-Maya J, Carvajal AL, Ortiz-Ordóñez E, Uría-Galicia E, Mendoza-Reynosa E. Populations of the butterfly goodeid (Ameca splendens) in the upper Rio Ameca Basin, Mexico. Journal of Freshwater Ecology 2004; 19(4): 575–80.

[75] Oruç EÖ, Üner. Combined effects of 2,4-D and azinphosmethyl on antioxidant enzymes and lipid peroxidation in liver of Oreochromis niloticus. Comp Biochem Physiol C 2000; 127: 291–6.

[76] Gutteridge JMC. Lipid peroxidation and antioxidants as biomarkers of tissue damage. Clin Chem 1995; 41(12): 1819–28.

[77] Kinnberg K, Korsgaard B, Bjerregaard P, Jespersen Å. Effects of nonylphenol and 17β-estradiol on vitellogenin síntesis and testis morphology in male platyfish Xiphophorus maculatus. J Exp Biol 2000; 203: 171–81.

[78] Bocquené G, Chantereau S, Clérendeau C, Beausir E, Ménard D, Raffin B, et al. Biological effects of the "Erika" oil spill on the common mussel (Mytilus edulis). Aquat Living Resour 2004; 17: 309–16.

[79] Arinç E, Sen A, Bozcaarmutlu A. Cytochrome P4501A and associated mixed function oxidase induction in fish as a biomarker for toxic carcinogenic pollutants in the aquatic environment. Pure Appl Chem 2000; 72(6): 985–94.

[80] Cash KJ, Gibbons WN, Munkittrick KR, Brown SB, John Carey J. Fish health in the Peace, Athabasca and Slave river systems. J Aquat Ecosyst Stress Recovery 2000; 8: 77–86.

[81] Fent K. Ecotoxicological effects at contaminated sites. Toxicology 2004; 205: 223–40.

[82] Sarkar A, Ray D, Shrivastava AN, Sarker S. Molecular biomarkers: their significance and application in marine pollution monitoring. Ecotoxicology 2006; 15: 333–40.

[83] Whyte JJ, Jung RE, Schmitt DJ, Tillin DE. Ethoxyresorufin-O-deethylase (EROD) activity in fish as a biomarker of chemical exposure. Crit Rev Toxicol 2000; 30: 348–495.

[84] Webb D, Gagnon MM. Biomarkers of exposure in fish inhabiting the Swan-Canning Estuary, Western Australia: a preliminary study. J Aquat Ecosyst Stress Recovery 2002; 9: 259–69.

[85] Corsi I, Mariottini M, Sensini C, Lancini L, Focardi S. Fish as bioindicators of brackish ecosystem health: integrating biomarker responses and target pollutant concentrations. Oceanol Acta 2003; 26: 129–38.

[86] Schreck CB, Contreras-Sanchez W, Fitzpatrick MS. Effects of stress on fish reproduction, gamete quality, and progeny. Aquaculture 2001; 197 (1–4): 3–24.

[87] Varela-Romero A, Ruiz-Campos G, Yépiz-Velázquez LM, Alaníz-García J. Distribution, habitat and conservation status of desert pupfish (Cyprinodon macularius) in the Lower Colorado River Basin, Mexico. Rev Fish Biol Fish 2002; 12: 157–65.

[88] Toppari J, Larsen JC, Christiansen P, Giwercman A, Grandjean P, Guillette Jr LJ, et al. Male reproductive health and environmental xenoestrogens. Environmental Health Perspectives Supplement l 1996; 104(S4): 741–803.

[89] Heibo E, Vøllestad LA. Life-history variation in perch (Perca fluviatilis L.) in five neighbouring Norwegian lakes. Ecol Freshw Fish 2002; 11: 270–80.

[90] Flammarion P, Devaux A, Nehls S, Migeon B, Noury P, Garric J. Multibiomarker responses in fish from the Moselle River (France). Ecotoxicol Environ Saf 2002; 51: 145–53.

[91] Chèvre N, Gagné F, Blaise C. Development of a biomarker-based index for assessing the ecotoxic potential of aquatic sites. Biomarkers 2003; 8: 287–98.

Chapter 5

Evaluation of Toxicity in Silver Catfish

Cláudia Turra Pimpão, Ênio Moura,
Ana Carolina Fredianelli, Luciana G. Galeb,
Rita Maria V. Mangrich Rocha and
Francisco P. Montanha

Additional information is available at the end of the chapter

1. Introduction

The global population growth observed in the recent decades coupled with continual tech-nological advancement and increase in the generation of new industrial products, including the manufacture of chemicals such as fertilizers and pesticides, has led to an expansion in the levels of xenobiotic compounds in aquatic ecosystems (Jesus; Carvalho, 2008).

The pollution of rivers and lakes with chemicals of anthropogenic origin may have adverse consequences: the waters become unsuitable for drinking and other household purposes, ir-rigation, and fish cultivation, and the animal communities living in them may suffer serious-ly. Massive fish kills are recorded rather frequently, and changes in the population of the fauna as a consequence of sublethal effects on ecologically important species have also been described (Koprocu; Aydin, 2004).

Insecticides are used extensively in agriculture and industry because it is easy to apply, cost effective, and in some situations, it is only a practical method of control. However, benefits of pesticides are not derived without consequences. They are one of the most potentially harmful chemicals and are released into the environment by direct applica-tions, spraying, atmospheric deposition, and surface runoff. Given the fact that, insecti-cides are not selective and affect non target species as readily as target organisms, it isn't surprising that a chemical that acts on the insect's different systems will elicit similar ef-fects in higher forms of life (Dogan; Can, 2011).

Levels of insecticides in superficial waters generally range far below lethal concentra-tions for aquatic organisms. However, sublethal adverse effects may result from exposure of aquatic organisms to insecticides at environmentally relevant concentrations (Das; Mukherjee, 2003).

Pesticides in the environment may be used as a model for the study of ecotoxicology, because they contaminate air, land and water, causing adverse effects that affect from bacteria to humans. It is well proven that these chemicals are toxic to aquatic arthropods, bees and fish (Santos et al., 2007). The effects of the use of pesticides are recognized worldwide and aggravated by misuse since part of this material is accumulated in plants and soil and much of it is transported to the rivers by rain (Tsuda et al., 1995; Wilson; Tisdell, 2001).

As a result of a great variety of human activities, the aquatic environment is becoming increasingly threatened by an alarming number of foreign chemicals or xenobiotics. Fish populations living in highly polluted areas often have high incidences of gross pathological lesions (Malins et al., 1988), associated with elevated levels of toxic contaminants in the sediments. However, pesticides applied to the land may be washed into surface waters and may kill or at least adversely influence the life of aquatic organisms.

Contamination of water with large amounts of pesticides leads to fish mortality or starvation by destruction of food organism, many toxicants have been shown to affect growth rate, reproduction and behavior, with evidence of tissue damage (Van Der Oost et al., 2003; Srivastav et al., 2002).

The poisoning of fish by pesticides can be acute or chronic and in general acute poisoning causes mass mortality. However, pollution is an often chronic process, apparently without any visible damage but sometimes producing several sublethal effects (Rodrigues, 2003).

The aim of this chapter was present a review based on some aspects of silver catfish's toxicology.

2. The fish

Silfvergrip (1996) conducted extensive taxonomic revision of the genus based on characters of internal morphology, and concluded that the genre *Rhamdia* consists of only 11 species among 100 previously described. According to the same author, quelen taxonomic division belongs to the following: Class: Osteichthues, Series: Teleostei, Order: Siluriformes, Family: Pimelodidae. Genre: Rhamdia, Species: quelen.

Rhamdia quelen, popularly known, as silver catfish is a species of fish found from southern Mexico to Argentina that displays the absence of teeth and scales with variable-length cylindrical wattles. This is an omnivorous species with an eating preference for fish, crustaceans, insects, plants and organic debris (Silfvergrip, 1996).

The silver catfish (*Rhamdia quelen*, Quoy & Gaimard) is an endemic South American fish species that with stands cold winters and presents fast growth rate in summer. These characteristics make catfish a suitable species for fish production in southern South America or any region with a temperate or subtropical climate. In aquaculture systems, at a density of two to four fish/m^2 catfish reach a 600-800 g body weight in eight months. Our unpublished observations in experimental field trials and at fish farms have shown that this weight is easily reached, but high mortality rates (40-50%) might occur if small fish (1-3 g) are used to initiate the culture. However, when beginning with heavier juveniles (30-60 g), the final weight will still range from 600 to 800 g, but with mortality rates not exceeding 5-10%. Thus, the more reasonable order in silver catfish culture is: hatchery, larviculture (1-6 g), nursery (from 5-6 g to 30-60 g), and termination (from 30-60 g to 600-800 g) (Barcellos et al., 2001). This species can be considered eurytermal because the fry acclimated to 31°C withstand temperatures from 15 to 34°C (Barcellos et al., 2003).

3. Chemical contaminants in aquatic systems

Toxic compounds or natural anthropogenic are called xenobiotics. With the onset of the epidemiological investigations was to confirm the hypothesis of many xenobiotics to be dangerous to living things, as well as their respective offspring, exerting toxic effects in the short, medium or long term (Reys, 2001). These substances are persistent in the environment eventually absorbed and accumulated by living organisms, toxic effects on various organs and systems. Thus, it was noted that the use of xenobiotics without evaluation of risks to the ecosystem, constituted a potential threat to the health of people, animals and plants (Sanches, 2006).

The introduction of toxic substances in the aquatic environment causes local and immediate effects, but can also lead to contamination of watersheds and commitment of underground reservoirs by infiltration through the soil. Numerous compounds have been detected in surface water, groundwater and water supply relating to agricultural activities and human cases of environmental contamination (Sanches, 2006).

3.1. Insecticides

The growing use of synthetic insecticides is intensifying global pollution risks. Insecticides are toxic and were designed to repel or kill unwanted organisms and when used for their different purposes they may be brought to water bodies killing or influencing the lives of aquatic organisms (El Sayed et al., 2007). The effects of the use of insecticides are recognized worldwide and compounded by their improper use (Tsuda et al., 1995; Wilson; Tisdell, 2001).

Organophosphates comprise a group of chemical compounds extensively used in farming as insecticides, which cause accidental poisoning in animals and men. The toxicity of these compounds is due especially to the respiratory and cardiac impairment in consequence of autonomic nervous system disorders.

The primary effect of organophosphates (Ops) on vertebrate and invertebrate organisms is the inhibition of the enzyme Acetylcholinesterase (AChE), which is responsible for terminating the transmission of the nerve impulse. OPs block the hydrolysis of the neurotransmitter acetylcholine (ACh) at the central and peripheral neuronal synapses, leading to excessive accumulation of ACh and activation of ACh receptors (Peña-Llopis et al., 2003). The overstimulation of cholinergic neurones initiates a process of hyperexcitation and convulsive activity that progresses rapidly to *status epilecticus*, leading to profound structural brain damage, respiratory distress, coma, and ultimately the death of the organism if the muscarinic ACh receptor antagonist atropine is not rapidly administered (Shih; Mcdonough, 1997).

Melo (2004) showed that *Rhamdia quelen* (silver catfish) juveniles exposed for 96 hours to a sublethal dose (0.01 mL/L) of Folidol® 600, was target for the toxicant action and some alterations became evident after 4 hours of exposure. The alterations observed in the liver were reduction in the density of melanomacrophages, focuses of necrosis, enhancement in the density of hepatocytes, loss of the cellular contour of the hepatocytes, cytoplasmic granulation, reduction of the cytoplasmic vacuolization, mithocondrial disruption, disorganization of the rough endoplasmic reticulum, nuclear heterochromatization and decharacterization of the endothelium. These alterations could diminish the liver metabolism, and as a consequence, they could cause damages to the health of the *R. quelen* juveniles.

Deltamethrin (DM) and other pyrethroids have proven to be toxic to aquatic organisms, mainly to fish. Due to its lipophilic characteristics it can be highly absorbed by the fish gills, which partially explains the high sensitivity of these animals to DM exposure in concentrations up to a thousand times lower than in mammals (Rodrigues, 2003).

Galeb et al. (2010) and Montanha (2010) studied the behavior of silver catfish exposed to sublethal concentrations of DM and CM, respectively, and they have presented loss of balance, swimming alteration, dyspnea (they kept their mouths and opercula open). *Post-mortem* signs observed in the animals exposed to DM and CM were mainly darkening of the surface of the body, tail and wattles erosion and hemorrhagic spots on the body surface.

The main behavioral changes observed are represented by respiratory and neurological manifestations. Such results corroborate with Polat et al. (2002) and Ylmaz et al. (2004), who have tested Cypermethrin in guppies (*Poecilia reticulata*); Borges (2007), who have tested CM in silver catfish (*Rhamdia quelen*). These changes can be attributed to the neurotoxic effect of DM/CM by blocking sodium channels and inhibiting the GABA receptors in the nervous filaments which results in an excessive stimulation of the central nervous system that sometimes can lead to brain hypoxia (El- Sayed et al., 2007).

Galeb et al. (2010) studied the haematological response of silver catfish exposed by DM, it was a significant increase in the total leukocyte counts. Similar results were reported by El-Sayed et al. (2007) in Nile tilapia (*Oreochromis niloticus*) and Pimpão et al. (2007) in catfish (*Ancistrus multispinis*). Montanha (2010) showed similar results in silver catfish exposed to CM. The leukocytosis showed that these pesticides can generate inflammatory or stress responses.

It was observed decrease of serum levels of ALT (alanine aminotransferase), AST (aspartate aminotransferase) and FA (alkaline phosphatase) were observed in silver catfish exposed to DM (Galeb, 2010) and CM (Montanha, 2010).

Fipronil is a non-systemic, chiral phenylpyrazole insecticide registered for use to control ants, beetles, cockroaches, fleas, mole crickets, ticks, termites, thrips, and other insects in a variety of agricultural and residential uses. Its mechanism of action involves non-competitive binding to the GABA receptor, effectively blocking the chloride channel and resulting in paralysis (USEPA, 2011).

Their toxicity to fish varies with species and it is highly toxic to *Lepomis macrochirus* (LC_{50} = 85 g/L), the *Oncorhynchus mykiss* (LC_{50} = 248 g/L), the *O. niloticus* (LC_{50} = 42 mg/L), *Poecilia reticulata* (LC_{50} less than 100 g/L) and *Cyprinus carpio* (LC_{50} = 430 g/L). Low concentrations of fipronil are lethal to most species of fish that were tested, and is especially toxic to young fish. Studies indicate that the dose 15 mg/L reduced the growth of trout. In addition, this compound is also bioaccumulated fish.

3.2. Drugs

The occurrence of pharmaceuticals for human and veterinary use has been detected in surface waters, sediments and drain in worldwide. Various substances such as anti-inflammatories, analgesics, antibiotics, hormones and antidepressants represent them. Although they have been subjected to pharmacokinetic studies, little information exists about the environmental fate and toxic effects in several organisms of aquatic fauna and flora, certainly affected (Stumpf et al., 1999; Fent et al. 2006).

Chemical toxicity distribution approaches have been employed for identifying Thresholds of Toxicological Concern for many industrial chemicals (Kroes et al., 2005), comparing the sensitivities of in vitro and fish models for estrogenicity (Dobbins et al., 2008), and predicting aquatic concentrations of ecotoxicological concern for chemical with common MOAs in plant models and invertebrates and fish (Dobbins et al., 2008).

4. Assessment of toxicity in fish by biomarkers

The biological response of an organism to xenobiotics following absorption and distribution starts with toxicant induced changes at the cellular and biochemical levels, leading to changes in the structure and function of the cells, tissues, physiology and behavior of the organism. These changes can perhaps ultimately affect the integrity of the population and ecosystem. For the biomonitoring and management of the aquatic ecosystems, these biological responses (biomarkers) have been proposed to complement and enhance the reliability of the chemical analysis data (Parvéz; Raisuddin, 2005).

There are very few pollutants that have been confirmed to cause adverse effects. In most cases, casual relationships have not been established to a large group of persistent pollutants, due the complex chemical contamination of environmental compartments, which difficulty to attribute harmful effects to any particular pollutant or category of pollutants.

Several definitions have been given for the term 'biomarker', which is generally used in a broad sense to include almost any measurement reflecting an interaction between a biological system and a potential hazard, which may be chemical, physical or biological (WHO, 1993). A biomarker is defined as a change in a biological response (ranging from molecular through cellular and physiological responses to behavioral changes) that can be related to exposure to or toxic effects of environmental chemicals (Van Der Oost, R et al., 2003).

A bioindicator is defined as organism giving information on the environmental conditions of its habitat by its presence or absence or by its behavior, and an ecological indicator is an ecosystem parameter, describing the structure and functioning of ecosystems.

According to the WHO (1993), biomarkers can be subdivided into three classes:

- **Biomarkers of exposure:** covering the detection and measurement of an exogenous substance or its metabolite or the product of an interaction between a xenobiotic agent and some target molecule or cell that is measured in a compartment within an organism;

- **Biomarkers of effect:** including measurable biochemical, physiological or other alterations within tissues or body fluids of an organism that can be recognized as associated with an established or possible health impairment or disease;

- **Biomarkers of susceptibility:** indicating the inherent or acquired ability of an organism to respond to the challenge of exposure to a specific xenobiotic substance, including genetic factors and changes in receptors which alter the susceptibility of an organism to that exposure.

4.1. Histopathology

Histopathological characteristics of specific organs express condition and represent time-integrated endogenous and exogenous impacts on the organism stemming from alterations at lower levels of biological organization. Therefore, histological changes occur earlier than reproductive changes and are more sensitive than growth or reproductive parameters and, as an integrative parameter, provide a better evaluation of organism health than a single biochemical parameter. Histopathological biomarkers signal the effects of exposures both acute and chronic to toxic agents in high or low levels. The pathological changes may be adaptive or degenerative and will determine the survival or death of the organism. The organs most commonly damaged by toxic agents are gills, liver and kidneys, which may be also affected by bacteria, viruses and parasites. That is why a health assessment of the animal is important to differentiate damage induced by toxic agents and diseases. Pharmaceuticals of many categories may be detected on the environment contaminating water flows. In a study for the evaluation of the effect of the analgesic dipyron in surface waters, Pamplona (2009) observed important histological disorders in the kidneys of silver catfish. The fish were exposed to three concentrations of dipyrone (0.5, 5 and 50 µg/L) in the water for 15 days and then evaluated. According to the author, the histological parameters of the intoxicated animals showed important tissue damage in the silver catfish kidneys, such as necrosis, fibrous deposition, vacuolization and constriction of blood vessels of the renal parenchyma, hyperplasia, and even necrosis of great renal vessels and glomeruli in the higher concentration of dipyron. Such alterations were not observed in the control group.

Pesticides are the substances most extensively researched in the aquatic ecotoxicology due to the large amounts used in agriculture and livestock in the whole world. Melo (2004) exposed the silver catfish to the organophosphate methyl parathion at the sublethal concentration of 6 mg/L for 96 hours with the aim of histological and ultrastructural analysis of liver tissue. Melo (2004) suggested that the histological changes are related to the intoxication by methyl parathion that is likely to bring the individuals metabolic, cellular and subcellular problems. It was observed in *Rhamdia quelen* loss of the contour of the hepatocytes and endothelial cells, as well as changes in the nuclear chromatin. According to Melo (2004), changes in the organization of rough endoplasmic reticulum and mitochondrial disruption within four hours of intoxication and strongly eosin stained granulation in the cytosol within 24 hours of exposure could be noticed. Ultrastructurally, a high incidence of lipid droplets was observed in the cytosol of hepatocytes when compared with control within 48 hours of exposure to fipronil. Increased foci of necrosis in the liver of silver catfish after 48 and 72 hours of exposure in relation to the control group, which presents only occasional foci of necrosis, were noted, among other structural changes. After 96 hours of exposure, Melo (2004) describes cells indistinguishable contour, presence necrosis focus, in addition of damaged blood vessels, vacuolization of the cytosol and the presence of an unknown material strongly eosin stained in the cytoplasm of the hepatocytes. Blood leukocyte infiltration and congestion were observed in all animals analyzed. Melanomacrophage were found in various locations between the hepatocytes as well as pyknotic nuclei cells.

Ghisi (2010) studied the effects of the phenyl pyrazolefipronil in the gills of the silver catfish after 60 days of intoxication in the sublethal concentrations 0.05; 0.10 and 0.23 µg/L. The author reports hyperplasia, lamellar fusion and aneurysms in all groups treated, including the control group that can impair the gill function. However, Ghisi (2010) considers the injuries of low severity and possible regression if the source of stress is eliminated, since the concentration of fipronil used was very low.

Cattaneo (2009) studied the effects of 2,4 - dichlorophenoxiacetic acid (2,4-D) herbicide in *Rhamdia quelen* fingerlings in an acute toxicity assay. After acclimation, silver catfish were intoxicated with concentrations of 0, 400, 600 and 700 mg/mL of 2,4 – D for 96 hours. The author states that the histological analysis showed alterations in the liver of silver catfish after exposure to 2,4-D herbicide, such as abnormal arrangement of hepatocytes cords, cell membrane rupture and hepatocytes vacuolation.

The effects of the herbicide clomazone in teleost fish *Rhamdia quelen* were studied by Crestani et al. (2007). The silver catfish were exposed to the concentrations 0.5 and 1.0 mg/L of clomazone for 12, 24, 48, 96 and 192 h. Histological analysis showed vacuolation in the liver after herbicide exposure, some of that were completely restored after a recovery period. Ferreira (2010) studied the sublethal effects of glyphosate (1.21 mg/L), methyl parathion (0.8 mg/L) and tebuconazole (0.88 mg/L) herbicides to silver catfish. The fingerlings were intoxicated during 96 hours and then sampled after anesthesia. The mapathologicalcal changes in liver histology were: diffuse hepatocyte degeneration, bile stagnation, granules on the cytoplasm, hyperemia and vacuoles in the cytoplasm and nucleus of fingerlings exposed to methyl parathion and tebuconazole. The liver of fish exposed to glyphosate did not show any visible histological changes.

4.2. Hematological and biochemical analyses

Hematological analysis in fish is not routinely used for fish diseases diagnosis. Hematology of fish lags behind that of other classes of vertebrates, but analysis of blood still can be informative about disease processes in teleosts and elasmobranchs (Clauss et al., 2008). The most important factor limiting accurate hematological analysis in different fish species is the variation in types, numbers, and appearance of leukocytes. In addition, the considerable variation in reported leukocyte values from healthy fish, even within a species, is also partly caused by differences in the methodology used as well as by subjective interpretation of cell types by the investigator. Nevertheless, some effort should be made to try to surpass these difficulties since these parameters can be a useful tool for evaluation of the effects of pesticides in the cellular components of blood and even in the immune system. Therefore, the analysis of hematological and biochemical parameters in fish can contribute to the assessment of the animal's health and also the habitat conditions (Pimpão et al., 2007).

Table 1 provides an overview of some hematological and blood chemistry results in catfish (*Rhamdia quelen*) exposed to toxic chemicals.

4.3. Enzymatic analyses

One of the more sensitive effect biomarkers is alteration in levels and activities of biotransformation enzymes. The activity of these enzymes may be induced or inhibited upon in fish exposure to xenobiotic. Enzyme induction is an increase in the amount or activity of these enzymes, or both and inhibition is the opposite of induction. In this case, enzymatic activity is blocked, possibly due to a strong binding or complex formation between the enzyme and the inhibitors.

The according by Van Der Oost, R et al. (2003) enzymatic analyzes to fish are used more:

- Phase I enzymes: Total cytochrome P450 (cyt P450); Cytochrome P450 1A ; Cytochrome b5 (cyt b5); Ethoxyresorufin O-deethylase (EROD) and arylhydrocarbon hydroxylase (AHH); NADPH cytochrome P450 reductase (P450 RED);

- Phase II enzymes and cofactors: Reduced and oxidized glutathione (GSH and GSSG); Glutathione S-transferases (GSTs); UDP-glucuronyltransferases (UDPGTs);

- Oxidative stress parameters: Superoxide dismutase (SOD); Catalase (CAT); Glutathione peroxidase (GPOX); Glutathione reductase (GRED); Non-enzymatic antioxidants

- Biochemical indices of oxidative damage: Lipid peroxidation; DNA oxidation;

Table 2 provides an overview of some enzymatic results in catfish exposed to toxic chemicals.

Chemical substance	Diets containing aflatoxins *				Treatment with diflubenzuron in 24-hour immersion baths **				***
	Control	41ppb/ aflatoxin/ kg	90ppb/ aflatoxin/ kg	204ppb/ aflatoxin/ kg	Control	0,01 mg/ l⁻¹diflubenzuron	0,1 mg/ l⁻¹diflubenzuron	1,0 mg/ l⁻¹diflubenzuron	None
Erythrocytes (x 10⁶/μL)					2,71 ± 0,3	2,67 ± 0,4	2,56 ± 0,3	2,6 ± 0,4	1,95 ± 0,40
Hemoglobin (g/dL)	8,2 ± 0,5	5,5 ± 0,3	4,0 ± 0,3	3,5 ± 0,2					6,73 ± 1,15
Hematocrit (%)	24,51 ± 1,51	13,16 ± 1,47	11,75 ± 0,41	11,16 ± 0,88	32,7 ± 3,62	32,9 ± 2,61	31,1 ± 2,8	30,1 ± 1,7	26,5 ± 5,3
Leucocytes (x 10³/μL)					13,4 ± 2,7	13,2 ± 2,9	12,6 ± 2,1	12,2 ± 2,4	
Lymphocytes (%)					62,1 ± 5,7	61,9 ± 6,8	62,4 ± 6,3	62,6 ± 6,4	11,61 ± 6,57
Neutrophils (%)					25,3 ± 5,5	26,9 ± 5,8	25,9 ± 6,7	25,9 ± 6,7	6,02 ± 3,33
Granulocytic especial cell (%)					2,7 ± 1,2	2,0 ± 0	2,2 ± 0	4,0 ± 2,0	1,24 ± 1,96
Monocytes (%)					11,6 ± 2,9	10,9 ± 4,5	11,1 ± 3,5	10,6 ± 4,3	1,13 ± 1,09
Eosinophils (%)					2,0 ± 0	2,0 ± 0	2,0 ± 0	2,0 ± 0	0,02 ± 0,14
Glucose (mg/dL)	45,00 ± 6,28	36,40 ± 2,42	24,40 ± 6,69	23,20 ± 5,93					

* Vieira, et al. (2006).

**Mabilia; Souza (2006)

***Tavares-Dias et al. (2002)

Table 1. Some hematological and blood chemistry results in catfish (*Rhamdia quelen*) exposed to toxic chemicals.

Species	Pollutants	cyt P450	CYP1A	AHH	EROD	cyt b5	P450 RED	References
Channel catfish	PCB Aroclor1254	=		++	++	=	=	Ankley et al., 1986
Ictaluruspuncta-	BKME	=			++			Mather-Mihaich;DiGiulio, 1991
tus	PAH (BNF)				++			Hasspieler et al., 1994
	2,4D+ Picloram							Gallagher and Di Giulio, 1991
	2 Aminocanthracene (AA)				-			Watson et al.,1995
	PAH (BaP)		+	+				Ploch et al., 1998
	PCB126		++		++			Rice and Roszell, 1998
Species	Pollutants	SOD	GPOX	GSSG	CAT	LPOX	GSH	References
Catfish*Ictalurus-nebulosus*	OPP (dichlorvos)	=	=		=	+	=	Hai et al., 1995
Channel catfish *Ictaluruspuncta-tus*	BKME	=	=		+	=	-	Mather-Mihaich and Di Giulio, 1991
	PAH			+	+		+	Di Giulio et al., 1989

Van Der Oost, R et al. (2003)

Symbols: --, strong inhibition (<20% of control); -, inhibition; =, no (significant) response; +, induction; ++, strong induction (>500% of control);

Table 2. Laboratory studies on responses of organic trace pollutants on fish enzymes

4.4. Genetic analyses

Genotoxic substances produce chemical or physical changes inDNA, commonly measured as breaks or DNA adducts, respectively (Nacci et al., 1996). The exposure of an organism to genotoxic substances can induce a cascade of events such as changes in the structure of DNA and the expression of damaged DNA using modified products. Thus, biological consequences may be triggered in cells, organs, body and finally in community and population (Lee; Steinert, 2003; Van Der Oost et al., 2003).

Many biomarkers have been used as tools for the detection of exposure and to evaluate the genotoxic effects of pollution. Among the major genetic biomarkers, we can mention the evaluation of the frequency of chromosomal aberrations, analysis of the frequency of sister chromatid exchanges, formation of DNA adducts, DNA breakage assessed by measuring the comet assay and micronucleus frequency and nuclear morphological abnormalities (Bombail et al., 2001).

According to the studies of Pamplona (2009) by the Comet Assay, widely used to measure DNA damage in aquatic organisms, after the intoxication of silver catfish by dipyrone, the genetic damages occurred in the lowest concentration of dipyron (0.5 µg/L). DNA fragmentation may be triggered by the release of free radicals in response to a stressor such as a xenobionte that may damage the nucleus chromatin, or by lipoperoxidation that causes the loss of cellular membrane integrity, subjecting the attack of the nucleic acid by oxygen radicals.

In a study of genotoxicity by copper sulphate in four different doses (5, 30, 50 and 500 mg/Kg) using the catfish, Costa (2011) induced DNA breakage triggered by comet assay. The fish were intoxicated by trophic copper sulphate during 60 days and suffered genetic damage in brain tissue, kidney, liver and blood tissues. Blood was less sensitive tissue to levels of copper used in this study. The piscine micronucleus test developed by Costa (2011), however, showed no significant changes in relation to the negative control, as well as the polychromatic erythrocytes frequency in relation to the total number of erythrocytes. There was an increase in the frequency of viable cells in blood tissue, according to the author.

Piancini (2011) performed the piscine micronucleus test and the comet assay in erythrocytes and gill cells of the silver catfish to evaluate the genotoxic effects of atrazine in concentrations 2, 10 and 100 µg/L for 96 hours. The author observed an increased frequency of nuclear morphological changes at all concentrations, a dose-dependent effect of atrazine on DNA trough the comet assay in erythrocytes: damage was significantly higher on 10 and 100 µg/L concentrations compared to control. In gills, there was no difference between treatments. Piancini (2011) observed a dose-dependent effect of atrazine in relation to genotoxicity and its capacity of damaging the DNA of the silver catfish in concentration considered safe by the Brazilian legislation. The same author evaluated the genotoxic effects of copper on *Rhamdia quelen* through the piscine micronucleus test and the comet assay in erythrocytes and gills. The silver catfish were exposed to concentrations of 2.0; 9.0 and 18 µg/L of cuprous chloride for 96 hours. Dose-dependent breaks to the DNA were detected by the comet assays in all treatments, as well as in gills.

Ghisi (2010) did not observe micronuclei in the catfish cells, but did confirm nuclear morphologic changes, after 60 days of intoxication in the sublethal concentrations 0.05; 0.10 and 0.23 µg/L of fipronil. The control group was not significantly different from the group intoxicated by the lower concentration of the insecticide. The higher concentrations were showed significantly higher among of micronuclei compared to the control group, not intoxicated by fipronil, suggesting that the pesticide in higher concentrations (0.10 e 0.23 µg/L) may induce damages to the DNA of *Rhamdia quelen*. The comet assay, also performed by Ghisi (2010) in the same study, showed no significant difference between the intoxicated groups and the control, corroborating the results of histopathological analysis.

The aim of the study performed by Salvagni et al. (2011) was to determine, by micronucleus testing on silver catfish, the risk of genotoxic impact by the use of pesticides such as glyphosate, cyhalothrin, atrazine, simazine and azoxystrobinon in farms located in the Lambedor River watershed in Guatambu, State of Santa Catarina. Blood samples were collected from several species of fish, including *Rhamdia quelen*, captured in 10 different dams existing in agricultural rural properties. Micronucleate erythrocytes were found in blood samples of silver catfish close to the overall average of the other species. Salvagniet at. (2011) believe through *in vivo* piscine micronucleus testing, that water from the Lambedor watershed can be considered genotoxic, with emphasis on the degree of genotoxicity from pollution in the area, since spontaneous formation of micronuclei in fish is normally very rare. Ferraro (2009) also exposed the silver catfish to different molecules of pesticides and the combinations of them in the aim to determine the bioindicator potential of this teleost species. After the exposure of the fish to the pesticides glyphosate (1.58 e 3.16 mg/L), tebuconazole (0.4 e 0.8 mg/L) and a mixture of them (3.16 e 0.8 mg/L) for 5, 10 and 15 days, were proceeded the micronucleus test and the comet assay in blood samples collected from the fish. DNA damage was confirmed by both techniques in all treatments, pointing the suitability of the species for biomonioring. In a similar way, Ramsdorf (2011) studied the genotoxic effects of fipronil (0.05; 0.10 and 0.23 µg/L), lead nitrate (0.01; 0.03 and 0.10 mg/L) and naphthalene (0.005; 0.06 and 3 mg/L) in the water for *Rhamdia quelen* during 60, 30 and 28 days, respectively. In silver catfish the concentrations 0.10 and 0.23 µg/L of fipronil increased the frequency of micronucleus, nuclear morphological changes, and damages to DNA observed after comet assay. After the intoxication by lead nitrate, it was observed that the concentrations 0.03 e 0.1 mg/L increased the levels of DNA breaks, as well as in all concentration of naphthalene tested for the species.

5. Impact of toxic agents in the reproductive system

The use of fish embryos is gaining popularity for research in the area of toxicology and teratology. Particularly embryos of the zebrafish, catfish and rainbow trout offer an array of different applications ranging from regulatory testing to mechanistic research.

Recently, much attention has been given to possible adverse effects resulting from exposure of aquatic animals to chemicals during the prenatal and perinatal (Tramujas et al., 2006; Montanha; Pimpão, 2012).

The impairment of reproductive function in humans and animal species has been of particular concern in recent years. Many factors can interact with the components and the reproductive function and cause infertility and other functional and structural changes. Illness, stress, hormonal changes and exposure to chemicals, are some factors that contribute to the emergence of reproductive system disorders (Sanchez, 2006).

Fish are one of the most thoroughly studied organisms in terms of effect of substance with estrogenic activity in the development of abnormalities in the reproductive system. According to Sumpter (1998), research on how estrogenic substances affect the sexual system of fish began in the 1980s (Bila; Dezotti, 2003).

The fecundity, the spawning period and type-specific characteristics are essential for the maintenance of any species of fish (Gomiero et al., 2007; Montanha; Pimpão, 2012). Reproduction in fish, as in other vertebrates, is affected by environmental factors, social and nutritional (Parra et al., 2008; Montanha, 2010). Reproductive parameters are more complex indicators of exposure and accumulation of chemicals, hampered for several reasons, the two main effects of pollutants on reproduction caused directly and indirectly and physiological process (Bernardi et al., 2008).

Few data exist regarding the potential sublethal effects of pesticides on reproduction and long-term viability of fish populations (Moore; Waring, 2001).

In 60 years, Rachel Carson published her famous book considered a landmark in the history of environmental pollution, Silent Spring, calling attention to the reproductive failure in birds and fish caused by bioaccumulation of persistent organochlorinepesticides such as diclorofodifeniltricloroetano (DDT). This work brought out the fears of modern society in relation to the introduction of synthetic substances in the environment and renewed public interest in science and government toxicology (SANCHEZ, 2006).

In the mid-70s, researchers found that other chemicals, such as the Kepone and PCBs (polychlorinated biphenyls), also had hormonal effects. Thereafter the toxicological effects of the mixture of individual congeners have been studied mainly in fish, mammalian cells and even humans (Sanchez, 2006).

The structural integrity of the gonads can be altered by xenobiotics (Mayon et al., 2006). Chemical agents that can affect the endocrine system are called endocrine-active chemicals (ECAs), expression adapted to Portuguese as "endocrine disruptors" (Sanchez, 2006).

The Environmental Protection Agency of the United States (USEPA) alternatively defines endocrine disrupters as chemicals that lead to toxic results as various types of cancer and a wide range of adverse effects on the reproductive system (Sanchez, 2006).

The pollution in the aquatic environment can affect the reproductive potential, thus reducing the spread of fish species. This can occur by the possible interaction between the gametes and water pollutants, such as the blockage of the micropyle, which prevents the entry of sperm in the fertilization process (Hilbig et al., 2008). In addition, solutions containing certain levels of pollutants can directly interfere with sperm motility and morphology, and subsequently at fertilization (Hilbig et al., 2008; Witeck et al., 2011).

In teleost fish, such as quelen (catfish), with external fertilization, which occurs when spawning, the gametes are released into the environment for fertilization to occur. At that moment, the gametes are exposed to various contaminants in the water, including heavy metals mercury, zinc, lead, copper and cadmus. These trace elements, at certain levels, affect the motility of spermatozoa and fertilization of oocytes (Witeck et al., 2011).

The gonadotropins stimulate gonadal maturation and release of steroid hormones from the gonads. The steroid hormones and the pituitary determine the development of sexual characteristics and various influencing courtship and parental care (SANCHEZ, 2006).

In teleosts, there are two gonadotropin: R gonadotropin (GTH I) and II gonadotropin (GtH II). In females, the GTH I stimulates gonadal growth, gametogenesis and the entry of vitellogenin in the oocyte. GTH II is important for the final maturation of oocytes and spawning. In males, GTH II acts in the testis, acting on testosterone production (Sanchez, 2006).

Endocrine disrupters are chemicals, natural or synthetic compounds also contained in fungicides, pesticides and insecticides, which have estrogenic and antiandrogenic action. Some of these compounds can maintain their chemical nature for many years contaminating the waters, being accumulated in fish. Many are potentially harmful to health since they are difficult excretion, may accumulate in the body and cause changes in all systems, especially in the endocrine system (Lara et al., 2011).

Recently, the monitoring of drug residues in the environment has been gaining great interest due to the fact that many of these substances are frequently found in effluents from sewage treatment plants (STPs) and natural waters. Some groups of residual drugs deserve special attention, among them are estrogen, because of its potential to adversely affect the reproductive system of aquatic organisms such as the feminization of male fish found in rivers contaminated with effluent disposal TEE. This was observed in fish species such as *Cyprinus carpio* and *Rutilus rutilus*. Similar effects (induction of hermaphroditism or complete feminization) were also observed when fish species *Oryzias latipes* were exposed to estrogen 17 β-estradiol (Bila; Dezotti, 2003).

Currently, there is growing concern about the presence of pharmaceuticals in aquatic environments and their possible environmental impacts. Ternes et al. (1999) identified the presence of various estrogens in sewage effluent and wastwater treatment plant (WWTP) in Germany, Brazil and Canada (Bila; Dezotti, 2003).

These compounds are characterized by mimicking the structure and function of natural sex steroids leading to endocrine disorders, which can result in abnormal sexual and reproductive function in animals and humans (LARA et al., 2011).

Deltamethrin, a pyrethroid pesticide widely used, is listed by the USEPA as endocrine disruptor, interfering with the reproductive system (Tramujas et al., 2006). Concentrations of 0.01 and 0.1 mg/L of deltamethrin presented toxic to fertilization of eggs of silver catfish (*Rhamdia quelen*). Since the concentrations 0.1, 0.5 and 1.0 mg/L proved to be toxic in the hatching and survival rates of 12 and 24 hours for eggs and larvae of silver catfish (Montanha, 2010). Tramujas et al. (2006) reported histological changes in levels in the gonads of aquatic organisms subjected to different concentrations of deltamethrin.

In recent years, monitoring the impacts of pesticides and other pollutants on sperm production of various kinds of research has been aimed (Mabilia et al., 2008). According to Moore and Waring (2001), even at low levels, cypermethrin (synthetic pyrethroid) in water caused a negative effect on reproductive functions in Atlantic salmon (*Salmo salar*) (Begum, 2005; Mabilia et al., 2008; Montanha, 2010; Montanha; Pimpão, 2012), causing a reduction in fertilization rate (Moore; Waring, 2001). Tramujas et al. (2006) reported a reduction in egg hatching rate of zebrafish (*Danio rerio*) exposed to deltamethrin. Rodrigues (2007) reported hatching in early zebrafish (*Danio rerio*) exposed to dichlorodiphenyltrichloroethane (DDT), with consequent reduction in survival rate. Aydin et al. (2005) and Rodrigues (2007) reported mortality of larvae of carp (*Cyprinus carpio*) exposed to cypermethrin and zebrafish (*Danio rerio*) exposed to DDT, respectively after 12 hours of exposure.

DDT at concentrations of 50 µg/L, and cause premature death hatching of larvae of *Danio rerio*, as well as parathion, from which 20 µg/L cause a decrease in the growth rates of fish of the same species (Rodrigues, 2007).

Cypermethrin proved toxic to the catfish during embryonic development in concentrations of 0.001, 0.01 and 0.1 mg/L relative to the fertilization rate and hatching. As regards the survival rate at 12 hours of exposure to cypermethrin after hatching, the concentration 10 mg/L performed toxic to larvae of catfish and 24 hours, concentrations of 0.001 to 10 mg/L (Montanha, 2010).

Some studies have reported the influence of heavy metals in fish breeding, referring to the influences of contaminants on the quality of the gametes on the fertilization of the oocytes and the quality of the larvae (Hilbig et al., 2008).

The water contamination by lead, the moment of fertilization to the first 8 hours of incubation eggs artificially catfish gray (quelen) at a concentration greater than 0.25 mg/L cause a deleterious effect on the percentage of fertilized eggs (Hilbig et al., 2008).

Water contamination fertilization with cadmus at levels up to 28.65 mg/L does not reduce rates of fertilization and hatching. However, concentrations above 20 mg/L cause reduction in sperm motility catfish (quelen) (Witeck et al., 2011).

In the study by Routledge et al. (1998), two fish species, *Oncorhynchus mykiss* and *Rutilus rutilus*, were exposed for 21 days at concentrations of 17 β-estradiol and estrone environmentally relevant. According to these researchers, the results confirmed that estrogens identified in effluents are present in sufficient amounts to cause the synthesis of vitellogenin, a protein that plays an important role in the reproductive system of female oviparous vertebrates, in fish species. Kang et al. (2002) clearly show that exposure to estrogen concentration of 17 β-estradiol environmentally relevant (in the range of 30-500 ng/L), for three weeks, induces high levels of vitellogenin hermaphrodite and the incidence of male fish species *Oryzias latipes* (Bila; Dezotti, 2003).

According to studies by Panter et al. (1998), low concentrations of 17 β-estradiol and estrone caused effects in male fish species *Pimephales promelas*, as vitellogenin synthesis and testicular inhibition (Bila; Dezotti, 2003).

Exposure to pesticides and other toxic substances during the prenatal and perinatal, can alter the reproductive system of fish without compromising growth and viability of offspring, but cause functional changes that become apparent later in adulthood (Tramujas et al., 2006; Montanha; Pimpão2012). Many xenobiotics are known to affect reproduction in many organisms (Mayonet al., 2006; Montanha, 2010).

6. The fish and the fish embryo toxicity test as an animal alternative method in hazard and risk assessment and scientific research

Biological and ecological responses to contaminant stressors may range from changes at the molecular level, where genetic integrity and subcellular processes are evaluated, to population and community levels where dynamics and structure of entire foodwebs can be affected. Indicators of stress at several levels of biological organization have been used to evaluate effects of contaminants at the organism level.

A variety of *in vitro* and *in vivo* assays with fish are being used as model systems for toxicological, biochemical and developmental studies (Powers, 1989). Fish species such as rainbow trout *(Oncorhynchus mykiss)*, carpas, catfish and Japanese medaka *(Oryzias latipes)* have been used extensively as test organisms for studies of carcinogenesis. A variety of teleost species have been used to study the development of early life stages of fish and the teratogenic effects of environmental contaminants (Wisk; Cooper, 1990). Fish respond in a manner similar to mammalian test species to chemicals that induce peroxisome proliferation in hepatocytes (Yang et al., 1990), and oxidative damage in hepatocytes (Washburn; Di Giulio, 1989). Hepatic cytochrome P-450 monooxygenases in fish also metabolize many carcinogens in a manner analogous to mammalian organisms (Stegeman; Lech, 1991). The advantages of using fish as model organisms include the ease with which teleosts, especially small aquarium species, can be held in the laboratory and exposed to toxic chemicals. Since fish often respond to toxicants in a manner similar to higher vertebrates, they can be used to screen for chemicals that have the potential to cause teratogenic and carcinogenic effects in humans. However, the main application for model systems with fish is to determine the distribution and toxic effects of chemical contaminants in the aquatic environment. Industrial and urban discharges are responsible for high concentrations of toxic substances in the aquatic environment, although in many countries, regulatory activity is beginning to limit point-source discharges of toxic chemicals to our water resources.

Animal alternatives research has historically focused on human safety assessments and has only recently been extended to environmental testing. This is particularly for those assays that involve the use of fish. A number of alternatives are being pursued by the scientific community including the fish embryo toxicity (FET) test, a proposed replacement alternative to the acute fish test. Discussion of the FET methodology and its application in environmental assessments on a global level was needed. With this emerging issue in mind, the ILSI Health and Environmental Sciences Institute (HESI) and the European Centre for Ecotoxicology and Toxicology of Chemicals (ECETOC) held an International Workshop on the Application of the Fish Embryo Test as an Animal Alternative Method in Hazard and Risk Assessment and Scientific Research in March, 2008.

7. Conclusion

Environmental pollution is of current environmental concern. It is therefore necessary to find solutions to determine environmental contamination, because the organisms are continuously exposed to a variety of anthropogenic toxicants daily released to the environmental.

The silver catfish have been used for bioindicator of environmental contamination for many researches and can be used to aquatic biological systems. Moreover, a considerable expansion of knowledge in using various model species which are easily cultured and have lesion induction times of relatively short duration, such silver catfish.

Author details

Cláudia Turra Pimpão[1*], Ênio Moura[2], Ana Carolina Fredianelli[3], Luciana G. Galeb[3], Rita Maria V. Mangrich Rocha[4] and Francisco P. Montanha[5]

1 Laboratory of Pharmacology and Toxicology/PUCPR, College of Agricultural, Environmental Sciences and Veterinary Medicine,Pontifícia Universidade Católica do Paraná, Curitiba, Brazil

2 Service of Medical Genetics/PUCPR, College of Agricultural, Environmental Sciences and Veterinary Medicine,Pontifícia Universidade Católica do Paraná, Curitiba, Brazil

3 College of Agricultural, Environmental Sciences and Veterinary Medicine,Pontifícia Universidade Católica do Paraná, Curitiba, Brazil

4 Service of Veterinary Clinical Pathology/PUCPR, College of Agricultural, Environmental Sciences and Veterinary Medicine,Pontifícia Universidade Católica do Paraná, Curitiba, Brazil

5 Laboratory of Pharmacology and Toxicology/FAMED, College of Agricultural, Environmental Sciences and Veterinary Medicine,Pontifícia Universidade Católica do Paraná, Curitiba, Brazil

References

[1] Ankley, G. T., Blazer, V. S., Reinert, R. E., Agosin, M. (1986). Effects of Aroclor 1254 on cytochrome P450-dependent monooxygenase, glutathione S-transferase, and UDP-glucuronosyltransferase activities in channel catfish liver. Aquat. Toxicol. 9, 91-103.

[2] Aydin, R. ; Koprucu, K. ; Dorucu, M. ; Koprucu, S. S. ; Pala, M. (2005) Acute toxicity of synthetic pyrethroidcypermethrin on the common carp (*Cyprinus carpio* L.) embryos and larvae. Aquaculture International, v. 13, p. 451-458.

[3] Barcellos, L. J. G., Kreutz, L. C., Rodrigues, L. B., Fioreze, I., Quevedo, R. M., Cericato, L., Conrad, J., Soso, A. B., Fagundes, M., Lacerda, L. A. & Terra, S. (2003) Haematological and biochemical characteristics of male jundiá (*Rhamdia quelen* Quoy & *Gaimardpimelodidae*): changes after acute stress. Aquaculture Research, *34*: 1465-1469.

[4] Barcellos, L. J. G., Woehl, V. M., Wassermann, G. F., Krieger, M. H., Quevedo, R. M. & Lulhier, F. (2001) Plasma levels of cortisol and glucose in response to capture and tank transference in *Rhamdia quelen* (Quoy & Gaimard), a South American catfish. Aquaculture Research,32(3):123-125.

[5] Begum, G. In vivo biochemical changes in liver and gill of *Clarias batrachus* during cypermethrin exposure and following cessation of exposure. (2005) Pesticide Biochemical and Physiology,v. 82, p. 185-196.

[6] Bernardi, M. M. ; Moraes, R. C. ; Varoli, F. M. F. ; Osti, S. C. Ecotoxicologia. In: Spinosa, H. de S. ; Gómiak, S. L. ; Palermo-Neto, J. Toxicologia aplicada à medicina veterinária. São Paulo: Manole, 2008. 942 p.

[7] Bila, D. M. ; Dezotti, M. (2003) Fármacos no meio ambiente. Quim. Nova. v. 26, n. 4, p. 523-530.

[8] Bombail, V. ; AW, D. ; Gordon, E. & Bartty, J. (2001). Aplication of the comet and micronucleus assays to butterfish (*Pholis gunnellus*) erythrocyte from the Firth of Forth, cotland, Chemosphere, 44: 383–392.

[9] Borges, A. (2007) Changes in hematological and serum biochemical values in Jundiá *Rhamdia quelen* due to sublethal toxicity of cipermethrin. Chemosphere, v. 69, p. 920-926.

[10] Cattaneo, R. (2009) Parâmetros metabólicos e histológicos de jundiás (*Rhamdia quelen*) expostos à formulação comercial do herbicida 2,4 – diclorofenoxiacético (2,4 – D). Santa Maria. *Thesis*, Programa de Pós Graduação em Ciências Biológicas, Universidade Federal de Santa Maria, Santa Maria, 2009.

[11] Clauss, T. M., Dove, A. D. M., Arnold, J. E. (2008) Hematologic Disorders of Fish. Vet Clin Exotic Anim. 11, 445-462.

[12] Costa, P. M. (2011) Avaliação do efeito tóxico de sulfato de alumínio e sulfato de cobre em bioensaio de contaminação subcrônica via trófica no bioindicador *Rhamdia quelen* (siluriforme). Curitiba. Thesis. Área de Concentração em Genética, Universidade Federal do Paraná, Curitiba, 2011.

[13] Crestani, M. ; Menezes, C. ; Glusczak, L. ; Miron, D. D. ; Spanevello, R. ; Silveira, A. ; Gonçalves, F. F. ; Zanella, R. ; LORO, V. L. (2007) Effect of clomazone herbicide on biochemical and histological aspects of silver catfish (*Rhamdia quelen*) and recovery pattern. Chemosphere, 67: 2305-2311.

[14] Das, B. K. ; Mukherjee, S. C. (2003) Toxicity of cypermethrin in *Labeo rohita* fingerlings: biochemical, enzymatic and hematological consequences. Comparative Biochemistry and Physiology, n. 134, p. 109-121.

[15] Di Giulio, R. T., Washburn, P. C., Wenning, R. J., Winston, G. W., Jewell, C. S. (1989). Biochemical responses in aquatic animals: a review of determinants of oxidative stress. Environ. Toxicological Chemistry. 8, 1103-1123.

[16] Dobbins, L. L., Brain, R. A., Brooks, B. W. (2008) Comparison of the sensitivities of common *in vitro* and *in vivo* assays of estrogenic activity: application of chemical toxicity distributions. Environmental Toxicological Chemistry, 27(12):2608-16.

[17] Dogan D. ; Can, C. (2011) Hematological, biochemical, and behavioral responses of *Oncorhynchus mykiss* to dimethoate. Fish Physiology Biochemistry, 37:951–958.

[18] El-Sayed, Y. S. ; Saad, T. T. ; El-Bahr, S. M. (2007) Acute intoxication of deltamethrin in monosex Nile Tilapia, *Oreochromis niloticus* with special reference to the clinical, biochemical and hematological effects. Environmental Toxicology and Pharmacology, n. 24, p. 212-217.

[19] Fent, K. ; Weston, A. A. ; Caminada, D. (2006) Ecotoxicology of human pharmaceuticals. Aquatic Toxicology, v. 76, p. 122 - 159.

[20] Ferraro, M. V. M. (2009) Avaliação de três espécies de peixes – *Rhamdia quelen, Cyprinus carpio* e *Astyanax bimaculatus*, como potenciais bioindicadores em sistemas hídricos através dos ensaios: cometa e dos micronúcleos. Curitiba, 2009. Thesis, Curso de Pós-Graduação em Genética, Setor de Ciências Biológicas, Universidade Federal do Paraná, Curitiba.

[21] Ferreira, D. (2010) Parâmetros de estresse oxidativo e estudo de lesões histopatológicas em jundiás (*Rhamdia quelen*) expostos a agroquímicos. Thesis, Programa de Pós-Graduação em Farmacologia, Universidade Federal de Santa Maria, Santa Maria, RS.

[22] Gallagher, E. P., Di Giulio, R. T. (1991) Effects of 2,4-dichlorophenoxyacetic acid and picloram on biotransformation, peroxisomal and serum enzyme activities in channel catfish (*Ictalurus punctatus*). Toxicological Letters. 57, 65-72.

[23] Galeb, L. A. G. (2010) Avaliação dos efeitos toxicológicos da deltametrina em uma espécie de peixe fluvial nativo jundiá (*Rhamdia quelen*). Curitiba, 2010. 70 f. Thesis, Programa de Pós-Graduação em Ciência Animal, Pontifícia Universidade Católica do Paraná, 2010.

[24] Galeb, L. A. G., Fredianelli, A. C. ; Montanha, F. P.,Mikos, J. D., Fam, A. L. D.,Rocha, R. M. V. M., Pimpão, C. T. (2011) Determination of lethal and sublethal concentrations of deltamethrin in Jundia (*Rhamdia quelen*) during 96 hours of exposure. Rev. Acad. (in press).

[25] Ghisi, N. C. (2010) Avaliação genotóxica em *Rhamdia quelen* após contaminação subcrônica com fipronil. Thesis, Pós-Graduação em Genética, Setor de Ciências Biológicas, Universidade Federal do Paraná, Curitiba, 2010.

[26] Gomiero, L. M. ; Souza, U. P. ; Braga, F. M. S. (2007) Reprodução e alimentação de *Rhamdia quelen* (Quoy & Gaimard, 1824) em rios do Núcleo Santa Virgínia, Parque Estadual da Serra do Mar, São Paulo, SP. Biota Neotropical, v. 7, n. 3, p. 127-133.

[27] Hai, D. Q., Varga, I. S., Matkovics, B., 1995. Effects of an organophosphate on the antioxidant systems of fish tissues. Acta Biol. Hungar. 46, 39-50.

[28] Hasspieler, B. M., Behar, J. V., Carlson, D. B., Watson, D. E., DI Giulio, R. T. (1994) Susceptibility of channel catfish (*Ictalurus punctatus*) and brown bullhead (*Ameriurus nebulosus*) to oxidative stress: a comparative study. Aquatic Toxicological. 28, 53-64.

[29] Hilbig, C. C. ; Bombardelli, R. A. ; Sanches, E. A. ; Oliveira, J. D. S. ; Baggio, D. M. ; Souza, B. E. (2008) Efeito do chumbo sobre a fertilização artificial e incubação de ovos de Jundiá cinza (*Rhamdia quelen*). Acta Sciencie Animal Science, v. 30, n. 2, p. 217-224.

[30] Jesús, T. B., Carvalho, C. E. V. (2008) Utilização de biomarcadores em peixes como ferramenta para avaliação de contaminação ambiental por mercúrio (Hg). Oecologia Brasiliensis, 12:7.

[31] Kang, I. J. ; Yokota, H. ; Oshima, Y. ; Tzuruda, Y. ; Yamaguchi, T. ; Maeda, M. ; Imada, N. ; Tadokoro, H. ; Honjo, T. (2002) Effect of 17β-estradiol on the reproduction of Japanese medaka (*Oryzias latipes*). Chemosphere., 47, 71-80.

[32] Koprucu, K. ; Aydin, R. (2004) The toxic effects of pyrethroiddeltamethrin on the common carp (*Cyprinus carpio L.*) embryos and larvae. Pesticide Biochemistry and Physiology, 80, 47–53.

[33] Kroes, R. ; Kleiner, J. ; Renwick, A. (2005) The Threshold of Toxicological Concern Concept in Risk Assessment. Toxicological Science, 86 (2): 226-230.

[34] Lee, R. F. ; Steinert, S. (2003) Use of the single cell gel electrophoresis/comet assay for detecting DNA damage in aquatic (marine and freshwater) animals. Mutation Research, 544: 43-64.

[35] Lara, L. A. S. ; Duarte, A. A. F. ; Reis, R. M. (2011) Impacto dos disruptores endócrinos na função reprodutiva e sexual de homens e mulheres. Editorial.

[36] Mabilia, R. G. ; Souza, S. M. G. (2006) Efeito do tratamento com diflubenzuron na hematologia de jundiás, *Rhamdia quelen* (Pimelodidae) infestados por *Lernaea cyprinacea* (Copepoda) em banhos de imersão de 24 horas. Acta Sci. Biol. Sci. v. 28, n. 2, p. 159-163.

[37] Mabilia, R. G. ; Souza, S. M. G. ; Oberst, E. R. (2008) Efeito do banho de imersão de diflubenzuron no sêmen do jundiá *Rhandia quelen* parasitado por *Lernaea cyprinacea*. Semina: Ciências Agrárias. v. 29, n. 3, p. 661-666.

[38] Malins, D. C. ; Mccain, B. B. ; Landahl, J. T. ; Myers, M. S. ; Krahn, M. M. ; Brown, D. W. ; Chan; S. L. ; Roubal, W. T. (1988) Neoplastic and other diseases in fish in relation to toxic chemicals: an overview. Aquatic Toxicology,11, 43-67,43.

[39] Mather-Mihaich, E., Di Giulio, R. T. (1991) Oxidant, mixed-function oxidase and peroxisomal responses in channel catfish exposed to a bleached kraft mill effluent. Archives Environmental Contamination Toxicological, 20, 391-397.

[40] Mayon, M. ; Bertrand, A. ; Leroy, D. ; Malbrouck, C. ; Mandiki, S. N. M. ; Silvestre, F. ; Goffart, A. ; Thomé, J-P. ; Kestemont, P. (2006) Multiscale approach of fish responses to different types of environmental contaminations: a case study. Science of the Total Environmental, v. 367, p. 715-731.

[41] Melo, G. C. (2004) Efeitos subletais da ação do organofosforado folidol600 no fígado do peixe de água doce jundiá *Rhamdia quelen* (QUOY & GAIMARD, 1824): uma anál-

ise histopatológica. Thesis. UFPR, Curitiba, Pr, 2004. Dissertação (Mestrado em Biologia Celular e Molecular) - Programa de Pós-Graduação em Biologia Celular e Molecular, Setor de Ciências Biológicas, Universidade Federal do Paraná, Curitiba.

[42] Montanha, F. P. (2010) Efeitos toxicológicos de piretróides (Cipermetrina e Deltametrina) em Jundiá (*Rhamdia quelen*): fase adulta e desenvolvimento embrionário. Curitiba, 2010. Thesis. Pontifícia Universidade Católica do Paraná (Mestrado em Ciência Animal).

[43] Montanha, F. P. ; Pimpão, C. T. (2012) Efeitos toxicológicos de piretróides (Cipermetrina e Deltametrina) em peixes – revisão. Revista Científica Eletrônica de Medicina Veterinária. ISSN: 1679-7353, Ano IX, n. 18.

[44] Moore, A. ; Waring, C. P. (2001) The effects of a synthetic pyrethroid pesticide on some aspects of reproduction in Atlantic Salmon (*Salmo salar* L.). Aquatic Toxicology, v. 52, p. 1-12.

[45] Nacci, D. E. ; Cayula, S. ; Jackim, E. (1996) Detection of DNA damage in individual cells from marine organisms using the single cell gel assay. Aquatic Toxicology, 35: 197-210.

[46] Pamplona, J. H. (2009) Avaliação dos efeitos tóxicos da dipirona sódica em peixe Rhamdia quelen: estudo bioquímico, hematológico e histopatológico. Curitiba, 2009. Thesis. Programa de Pós-graduação em Farmacologia, Setor de Ciências Biológicas, Universidade Federal do Paraná, Curitiba, 2009.

[47] Panter, G. H. ; Thompson, R. S. ; Sumpter, J. P. (1998) Aquatic Toxicology. 1998, 42, 243.

[48] Parra, J. E. G. ; Neto, J. R. ; Veiverbeg, C. A. ; Lazzari, R. ; Bergamin, G. T. ; Pedron, F. A. ; Rossato, S. Sutili, F. J. (2008) Alimentação de fêmeas de Jundiá com fontes lipídicas e sua relação com o desenvolvimento embrionário e larval. Ciência Rural, Santa Maria, v. 38, n. 7, p. 2011-2017.

[49] Parvez, S. ; Raisuddin, S. (2005) Protein carbonyls: novel biomarkers of exposure to oxidative stress-inducing pesticides in freshwater fish *Channa punctata* (Bloch). Environmental Toxicological and Pharmacological., 20:112–117.

[50] Peña-Llopis, S., Ferrando, M. D., Peña, J. (2003) Fish tolerance to organophosphate-induced oxidative stress is dependent on the glutathione metabolism and enhanced by N-acetylcysteine. Aquatic Toxicological . 65, 337–360.

[51] Piancini, L. D. S. (2011) Utilização de biomarcadores genéticos na avaliação aguda do efeito mutagênico dos contaminantes atrazina e cloreto de cobre em *Rhamdia* quelen (Siluriformes, Heptapteridae). Curitiba, 2011. Thesis, Programa de Pós-Graduação em Genética, Setor de Ciências Biológicas, Universidade Federal do Paraná, Curitiba, 2011.

[52] Pimpão, C. T. ; Zampronio, A. R. ; Silva DE Assis, H. C. (2007) Effects os deltamethrin on hematological parameters and enzymatic activity in *Ancistrus multispinis* (Pisces, Teleostei). Pesticide Biochemistry and Physiology, v. 88, p. 122-127.

[53] Ploch, S. A., King, L. C., Kohan, M. J., Di Giulio, R. T. (1998) Comparative in vitro and in vivo benzo[a]pyrene_/DNA adduct formation and its relationship to CYP1A activity in two species of ictalurid catfish. Toxicol. Appl. Pharmacol. 149, 90-98.

[54] Polat, H. ; Erkoç, F. U. ; Viran, R. ; Koçak, O. (2002) Investigation of acute toxicity of beta-cypermethrin on guppies *Poecilia reticulata*. Chemosphere, v. 49, p. 39-44.

[55] Powers, D. A., (1989). Fish as model systems. Science 246, 352/358. Power, M., McCarty, L. S., 1997. Fallacies in ecological risk assessment

[56] Ramsdorf, W. (2011) Avaliação da toxicidade dos compostos fipronil, nitrato de chumbo e naftaleno em peixes. Thesis, Pós-Graduação em Ciências Biológicas da Universidade Federal do Paraná, Curitiba, PR, 2011.

[57] Reys, L. L. (2011) Tóxicos ambientais desreguladores do sistema endócrino. RFML, Grupo de Medicina Preventiva e Ciências Sociais, Faculdade de Medicina de Lisboa, Série III, p. 213-225.

[58] Rice, C. D., Roszell, L. E. (1998) Tributyltin modulates 3,3?,4,4?,5-pentachlorobiphenyl (PCB 126)-induced hepatic CYP1A activity in channel catfish, *Ictalurus punctatus*. Journal Toxicological Environmental Health, 55, 197-212.

[59] Rodrigues, L. C. (2003) Estudo das Glutation S-Transferases hepáticas solúveis do peixe Piaractus mesopotamicus Holmberg, 1887 (Pacu). Thesis. UERJ, RJ, Brazil.

[60] Rodrigues, B. K. (2007) Avaliação dos impactos de agrotóxicos na região do Alto Mogi-Guaçu (MG) por meio de ensaios laboratoriais com *Danio rerio* (Cypriniformes, Cyprinidae). São Carlos, 2007. 138 f. Thesis, Escola de Engenharia de São Carlos, Universidade de São Paulo, 2007.

[61] Routledge, E. J. ; Sheahan, D. ; Desbrow, C. ; Brighty, G. C. ; Waldock, M. ; Sumpter, J. P. (1998) Identification of estrogenic chemicals in STW effluent. 1. Chemical fractionation and *in vitro* biological screening. Environmental Science Technololy. 32 (11): 1549-58.

[62] Salvagni, J. ; Ternus, R. Z. ; Fuentefria, A. M. (2011) Assessment of the genotoxic impact of pesticides on farming communities in the countryside of Santa Catarina State, Brazil. Genetics and Mollecular Biology, 34(1): 122–126.

[63] Sanches, D. C. O. (2006) Desreguladores endócrinos na indução da vitelogenina em peixes nativos. Thesis. UFPR, Curitiba, PR, 2006.

[64] Santos, M., Reyes, F., Areas, M. (2007) Piretróides–uma visão geral. Alimentos e Nutrição Araraquara, v18, n3.

[65] Shih, T. M., Mcdonough Jr., J. H. (1997) Neurochemical mechanisms in soman-induced seizures. J. Appl. Toxicol. 17, 255–264.

[66] Silfvergrip, A. M. C. (1996) A systematic revision of the neotropical catfish genus Rhamdia (Teleostei, Pimelodidae). . 156 f. Thesis (PhD in Zoology)–Stockholm University, Stockholm.

[67] Srivastav, A. K. ; Srivastava, S. K. ; Mishra, D. ; Srivastav, S. ; Srivastav, S. K. (2002) Ultimobranchial Gland of Freshwater Catfish, *Heteropneustes fossilis* in Response to Deltamethrin Treatment. Bulletin Environmental Contamination Toxicology. 68:584–591.

[68] Stegeman, J. J., Lech, J. J., (1991) Cytochrome P450 momooxygenase systems in aquatic species: carcinogen metabolism and biomarkers for carcinogen and pollutant exposure. Environ. Health Perspect. 90, 101/109.

[69] Stumpf, M. ; Ternes, T. A. ; Wilken, R-D; Rodrigues, S. V. ; Baumann,W. (1999) Polar drug residues in sewage and natural waters in the state of Rio de Janeiro, Brazil. The Science Total Environmental, v. 225, p. 135-141.

[70] Sumpter, J. P. (1998) Xenoendorine disrupters--environmental impacts. Toxicology Letters. 102-103, 337.

[71] Tavares-Dias, M., Melo, J. F. B., Moraes, G., Moraes, F. R. (2002). Haematological characteristics of brazilian teleosts. vi. Parameters of jundiá *Rhamdia quelen* (Pimelodidae). Ciência Rural, v. 32, n. 4, 2002.

[72] Ternes, T. A. ; Stumpf, M. ; Mueller, J. ; Haberer, K. ; Wilken, R. D. ; Servos, M. (1999) Behavior and occurrence of estrogens in municipal sewage treatment plants I. Investigation in Germany, Canada and Brazil. Science Total Environmental. 225, 81.

[73] Tramujas, F. F. ; Fávaro, L. F. ; Pauka, L. M. ; Silva De Assis, H. C. (2006) Aspectos reprodutivos do peixe-zebra, *Danio rerio,* exposto a doses subletais de deltametrina. Archives of Veterinary Science, v. 11, n. 1, p. 48-53.

[74] Tsuda, T., Aoki, S., Inoue, T., Kojima, M. (1995) Accumulation and excretion of diazinon, fenthion and fenitrothion by killifish: Comparison of individual and mixed pesticides. Water Res, 29:455-458.

[75] United States Environmental Protection Agency (US EPA). (2011) Fipronil Summary Document Registration Review: Initial Docket June 2011 EPA-HQ-OPP-2011-0448, Washington, D. C. http://www. regulations. gov/> documentDetail;D=EPA-HQ-OPP-2011-0448-0003. 2011. Accessed 30 July 2012.

[76] Van Der Oost, R. et al. (2003) Fish bioaccumulation and biomarkers in environmental risk assessment: a review. Journal Environmental Toxicology and Pharmacology, v. 13, p. 57- 149.

[77] Vieira, V. L. P., Randunz Neto, J., Lopes, P. R. S., Lazari, R., Fonseca, M. B., Menezes, C. (2006). Alterações Metabólicas e Hematológicas em Jundiás (*Rhamdia quelen*) Alimentados com Rações Contendo Aflatoxinas. Ciência Animal Brasileira. Goiânia, v. 7, n. 1, p. 49-55.

[78] Washburn, P. C. ; Di Giulio, R. T. (1989) Biochemical responses in aquatic animals: A review of determinants of oxidative stress. ToxicolChem, 8(12) 1103-1123.

[79] Watson, D. E., Menard, L., Stegeman, J. J., Di Giulio, R. T. (1995) Aminoanthracene is a mechanism-based inactivator of CYP1A in channel catfish hepatic tissue. Toxicological Apply Pharmacological. 135, 208-215.

[80] WHO International Programme on Chemical Safety (IPCS), 1993. Biomarkers and risk assessment: concepts and principles. Environmental Health Criteria 155, World Health Organization, Geneva.

[81] Wilson, C., Tisdell, C. (2001) Why farmers continue to use pesticides despite environmental, health and sustainability costs. Ecolog Econom.,39:449-462.

[82] Wisk, J. D. ; Cooper, K. R. (1990) The stage specific toxicity of 2,3,7,8-tetrachlorodibenzo-p-dioxin in embryos of the japanese medaka (*Oryzias latipes*). Environ Toxicol-Chem, 9 (9), 1159-1169.

[83] Witeck, L. ; Bombardelli, R. A. ; Sanches, E. A. ; Oliveira, J. D. S. ; Baggio, D. M. ; Souza, B. E. (2011) Motilidade espermática, fertilização dos ovócitos e eclosão de ovos de jundiá em água contaminada por cádmo. R. Bras. Zootec. v. 40, n. 3, p. 477-481.

[84] Yang, C. S., Yoo, J. -S. H., Ishizaki, H., Hong, J. (1990). Cytochrome P450IIE1: Roles in nitrosamine metabolism and mechanisms of regulation. Drug Metabol. Rev. 22, 147–159.

[85] Yilmaz, M. ; Gul, A. ; Erbasli, K. (2004) Acute toxicity of alpha-cypermethrin to guppy (*Poecilia reticulate*, Pallas, 1859). Chemosphere, v. 56, p. 381-385.

Chapter 6

The Potential Impacts of Global Climatic Changes and Dams on Amazonian Fish and Their Fisheries

Carlos Edwar de Carvalho Freitas,
Alexandre A. F. Rivas, Caroline Pereira Campos,
Igor Sant'Ana, James Randall Kahn,
Maria Angélica de Almeida Correa and
Michel Fabiano Catarino

Additional information is available at the end of the chapter

1. Introduction

The Amazon River Basin, which encompasses the world's largest remaining tropical rainforest, has the highest diversity of fish species of any region in the world [1]. Some of these species represent highly abundant fish stocks that have supported an important fishery for many decades, or many centuries if the history prior to European colonization is included. The importance of fishing in the Amazon River Basin can easily be observed from the high fish consumption, which is mainly attributed to people who live in rural areas near rivers and lakes (Table 1). Regardless of this importance for food, there is no integrative strategy for fishery management, and the activity in this basin as a whole is highly vulnerable to externalities, including those resulting from environmental changes and man-made interventions.

There is a consensus that fishery production is directly related to biological productivity, which is a function of a set of environmental characteristics in the aquatic system. The majority of the Rio Amazonas and its tributaries are accompanied by large floodplains, which is where most of the biological production occurs. A key factor for biological production in the floodplains of large Amazonian rivers is the flood pulse [9], which generates tremendous variation in the input of nutrients over the course of the year, primarily at river headwaters located in the pre-Andean areas, such as Madeira, Purus, Juruá, and Solimões. The flood pulse is the driving element that structures the landscape of the floodplains adjacent to the river channels, forming a mobile ecotone that is referred to as ATTZ or the aquatic-ter-restrial transition zone [9].

Reference	Sub-basin	Social group	g/per capita.day	Kg/per capita.year
[2]	Rio Negro	urban	53.95	19.69
[3]	Rio Negro	urban	121.70	44.42
[4]	Rio Amazonas	rural	369.00	135.00
[5]	Rio Solimões	rural	510.00 to 600.00	186.00 to 219.00
[6]	Rio Solimões and Rio Japurá	rural	509.00 to 805.00	186.00 to 294.00
[7]	Rio Madeira	rural	243.00	88.00
[8]	Rio Amazonas	rural	511.00 to 643.00	187.00 to 235.00

Table 1. Fish consumption in the Amazon River Basin.

The Amazonian hydrological cycle is annual and quite predictable. The flood intensity and timing is controlled by several factors, including those that act on a global scale. The cyclical phenomenon of warming in the Pacific Ocean near the cost of Peru, termed *El Niño*, is relat-ed to severe drought in the Amazon Basin. Alternatively, *La Niña* is associated with strong floods. The simultaneous occurrence of other climatic phenomena, such as the warming of the Tropical North Atlantic Ocean, has been used to explain extreme climatic events [10].

Despite the lack of models describing the relationship between flood intensity and timing and fishing success, the life strategy of many species of Amazonian fish is synchronized with the hydrological cycle. For example, several species of Characiforms, including *Colosso-ma macropomum, Brycon amazonicus, Prochilodus nigricans, Semaprochilodus insignis, S. taenirus, Piaractus brachypomum* and others, begin their reproductive migration at the beginning of the rainy season when the waters begin flooding [11, 12, 13]. This life strategy was most likely developed to ensure the colonization of the floodplain with newly hatched larvae. The avail-ability of food and places of refuge in the colonized floodplain may determine the strength of the annual recruitment of these species.

There is ample evidence that the Earth's climate is changing more rapidly now than it has in the past [14], with potential effects on the Amazon basin [15, 16]. These effects include phys-ical alterations and changes in nutrient flow [14]. Although uncertainties associated with how local climates change in response to global climate change exist, global circulation mod-els employed by the Intergovernmental Panel on Climate Change (IPCC) have found an in-creased likelihood of a significant increase in the mean global temperature [14]. A rise in sea level is also predicted, with estimates varying between 0.75 m to 1.90 meters by the end of twenty-first century [17]. Other environmental changes in freshwater systems, such as strati-fication, productivity reduction and acidification, have no consistent patterns. These changes are very difficult to generalize based on the available evidence; however, there is a consensus that the impact on fish will be species-specific, that is directly related to the biological characteristics of each species.

Nevertheless, the effects on fisheries should be a result of a series of effects that start at the organism level. At an individual level, all fish have an optimal thermal interval that is limited on the upper and lower boundaries by their critical thermal maxima and minima, respectively [18], thus reducing the analysis of the warming effects. Therefore, fish exposed to temperatures within the sub-lethal interval, excluding the optimal thermal interval, may be affected by warming, and the consequences of this temperature effect should be evident by physiological responses. The high energetic cost necessary to compensate for these unfavorable environmental conditions may affect the growth rates or reproduction success of the fish (Figure 1). Realistically, general effects from water warming can be expected. Because biochemical reaction rates are a function of body temperature, all aspects of an individual fish's physiology, including growth, reproduction and activity, are directly influenced by changes in temperature [19].

When the environment changes, these temperature effects should continue to increase and may be perceptible at the population and community levels. Although different species are affected by environmental change in different ways, the abundance patterns of the entire community, and thus the fishing production, should be influenced (Figure 1).

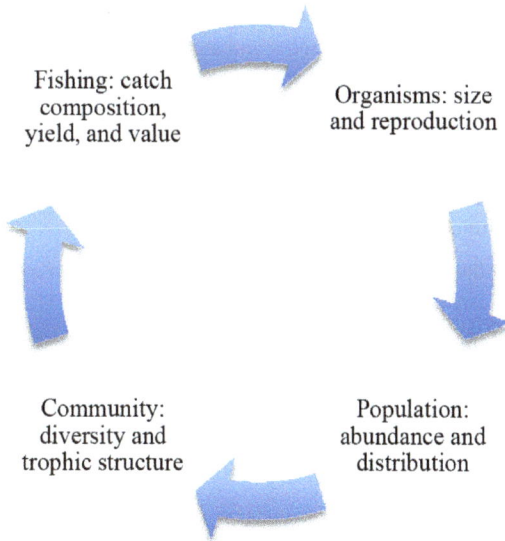

Fishing: catch composition, yield, and value

Organisms: size and reproduction

Community: diversity and trophic structure

Population: abundance and distribution

Figure 1. The effects of global climatic changes on different levels.

Dams cause local changes at the sub-basin level but also have the potential to exhibit regional effects. In essence, the impact of dams on the hydrological river cycle, which primarily involves flood timing, is the most important effect of change in freshwater fisheries because the flood regime is the most important determining force in Neotropical rivers [20]. Dam construction can affect environments and fisheries by changing the timing and quantity of river flows; altering the water temperature, nutrient and sediment transport; reducing adjacent floodplains and other wetlands; and blocking fish migrations.

Currently, there is a large proliferation of hydroelectric dams within the Amazon region. At the western boundary near the Andean and Pre-Andean areas, there are plans for 151 new hydroelectric dams with greater than 2 MW of power over the next 20 years, which is more than a 300% increase [21]. Similarly, in the Brazilian region near the south and southeast boundaries of the Amazon, there are several hydroelectric dam projects that have the potential to completely fragment the river basins with headwaters on the Brazilian Plateau. Similar to climate change, the impact of dams should be associated with the life strategies of different fish species.

The most important species captured by small-scale fisheries in the Amazon basin belong to three groups: Characiforms, which are primarily from the Prochilodontidae, Characidae, and Serrasalmidae families; Siluriforms, which are primarily from the Pimelodidae family and include piramutaba (*Brachyplatystoma vailantii*), dourada (*B. rouseauxii*) and piraíba (*B. filamentosum* and *B. capapretum*); and Perciforms, which are primarily from the genus *Cichla*. Over evolutionary time, members of these groups have developed specific life strategies designed to optimize the survival of Amazonian environmental conditions. Alterations induced by global changes or man-made interventions may directly influence these strategies, with negative effects on both the recruitment and stock abundance of these species, as well as on the socio-economic conditions of the Amazonian people that exploit these fish stocks for food and income.

Therefore, the goals of this chapter are as follows:

1. Review the main scenarios for environmental alterations in the Amazon Basin, which is predicted to be a function of global climatic changes and dams.

2. Identify the potential impacts of different scenarios of environmental alterations in the Amazon Basin on Amazonian freshwater fish populations.

3. Identify the consequences of the predicted impacts on the Amazonian freshwater fish populations, taking into account the main characteristics of the population dynamics.

4. Illustrate the potential social and economic consequences for the local and regional fisheries and the people who depend on these fisheries.

2. Global climatic changes and dams: What we can expect and what are the potential effects

Despite the uncertainties at the local level, it is highly likely that there will be more frequent occurrences of extreme climatic events. Most of the global climate models (GCMs) proposed by

the IPCC [14] project significant Amazonian drying during the 21st century. Pacific sea surface temperature (SST) variation, which is dominated by the El Niño–Southern Oscillation
(ENSO), is the main driving force for wet-season rainfall. However, dry-season rainfall is
strongly influenced by the Tropical Atlantic north-south SST gradient. Therefore, an intensifi-
cation of this gradient from the warming of northern SSTs relative to those of the south would
move the Inter-Tropical Convergence Zone north and strengthen the Hadley Cell circulation.
This change would enhance the duration and intensity of the dry season in much of southern
and eastern Amazonia, which already has occurred in 2005. Studies indicated that the most extreme droughts in Amazonia were a result of the strong events of the El Niño-Southern Oscillation (ENSO), the large temperature increase of the sea surface in the Tropical North Atlantic or
a combination of these events [22]. Changes in precipitation during the dry season are likely
the most critical determinant of the climatic fate of the Amazon [14].

These extreme droughts, even if short in duration, can be catastrophic for aquatic organisms
because of the strong reduction in the area of the aquatic environments. Floodplain lakes are
the most impacted, and the areas of these lakes can be reduced by several orders of magnitude (Figure 2). Although several species of fish are able to relocate to the river channel during the dry season, some lake resident species remain in the lakes and are unable to survive
if the drought is severe. Some studies have observed that fish assemblages seemed to recov-
er rapidly from normal drought seasons [23], but there are indications that extreme
droughts occasionally alter fish assemblages. Some species that are vulnerable to these catastrophic events may disappear at a local level [23].

(a)   (b)

Figure 2. A floodplain area of the Rio Amazonas during the flood season (A) and dry season (B) when an extreme
drought occurred.

Another likely climatic change is global warming [14]. Over the next two decades, a warming of approximately 0.2°C per decade is projected for a range of SRES emission scenarios.
Even if the concentrations of all greenhouse gases and aerosols had been maintained at the

levels present in the year 2000, a further warming of approximately 0.1°C per decade would be expected [14]. This warming represents an increase of 1-7°C in the mean global temperature within the next one hundred years.

Freshwater fish may explore habitats within an optimal thermal interval and thermo-regulate behaviorally and physiologically. Temperature tolerance ranges are species-specific and range from stenothermal species that support only a narrow thermal range to eurythermal species that are able to live in a wide thermal range. Fish populations subjected to changing thermal regimes may increase or decrease in abundance, experience range expansions or contractions or face extinction [19]. There is also an inverse relationship between the temperature and concentration of dissolved oxygen in water. Thus, an increase in the temperature can exacerbate the hypoxia or anoxia conditions naturally observed in some lentic habitats of freshwater fish.

A direct consequence of global warming is a rise in sea level. Despite the uncertainties related to the dynamics of ice sheets and glaciers, there are models that predict a rise in sea level, which were summarized previously [14]. One model proposed a relationship between global sea level variations and the global mean temperature and predicts a rise in sea level ranging from 75 to 190 cm for the period 1990-2100 [17]. However, some scenarios [14] predict a rise of 4.0 meters (Table 2).

What are the potential effects of a rise in sea level for the Amazon Basin? With regard to its physical characteristics, we can anticipate that the sea will be a hydraulic barrier and will flood areas that are not currently flooded but which are primarily within the floodplain adjacent to the river channel. Other environmental consequences of this barrier can also be expected: a reduction in water flow, an increase in the sedimentation rate and an increase of the flooded area. It is possible that the hydrological cycle will also be affected.

These changes to the hydrological cycle can be magnified by the fragmentation of the environment that will occur as a result of the introduction of hydroelectric dams. We can identify at least four phenomena associated with the introduction of dams:

1.   Blockage of the sediment flow in whitewater systems (e.g., Rio Madeira).

2.   Change of the flood pulse.

3.   Blockage of fish migration.

4.   Reduction in oxygen levels both above and below the dams.

Blockage of the sediments might have a large effect on fish communities. Whitewater rivers originate in Pre-Andean areas and are heavily loaded with volcanic soil sediment. The dams act as a barrier and result in a reduction of water speed, thus improving the rate of decantation. The end result is an impoverishment of the river below the dam. Thus, the species that have evolved in the presence of high levels of nutrients will not be able to adapt to the rapid loss in primary productivity that is associated with the reduction of nutrient content. This result will favor a change in the composition of local species and will have serious impacts on fishing activity.

	Temperature change ((C at 2090-2099 relative to 1980-1999)		Sea level rise (m at 2090-2099 relative to 1980-1999)
	Best estimate	Likely range	Model-based range excluding future rapid dynamic changes in ice flow
Constant Year 2000 concentration	0.6	0.3 - 0.9	NA
B1 scenario	1.8	1.1 - 2.9	0.18 - 0.38
A1T scenario	2.4	1.4 - 3.8	0.20 - 0.45
B2 scenario	2.4	1.4 - 3.8	0.20 - 0.43
A1B scenario	2.8	1.7 - 4.4	0.21 - 0.48
A2 scenario	3.4	2.0 - 5.4	0.23 - 0.51
A1F1 scenario	4.0	2.4 - 6.4	0.26 - 0.59

Table 2. Projected global average surface warming and associated sea level rise at the end of the 21st century. Source: [14]

Similarly, Amazonian fish species evolved in a system regulated by an annual and predictable flood pulse, developing life strategies to explore the several habitats available during the hydrological cycle. The elimination or change in the timing or duration of this pulse can destroy signals that trigger reproduction and other life cycle events, which will potentially influence fish recruitment.

The blockage of the fish migration can be critical, with significant impacts for some species that participate in long-distance migrations from the estuary to the headwaters of whitewater rivers to spawn. As a result, some populations may be locally extinct.

The fourth phenomenon concerning the fall in oxygen levels is relatively self-explanatory. The large amount of organic material in the reservoirs will remove a great deal of oxygen from a system that is already low in oxygen content due to the water temperature. The synergy between this phenomenon and global warming is quite evident. The results may include the loss of species with less tolerance for low oxygen conditions.

Clearly, these phenomena can be completely integrated and synergistic, and their effects on fish communities can be magnified and strongly disruptive. As is most often the case with multiple sources of environmental stress, the combined stress resulting from several sources is greater than the sum of the individual stresses. This point is emphasized by [21], who stated that the impact of hydroelectric dams in the Amazon Basin should be considered in a broad perspective, including the planned projects of other Amazonian countries, such as Bolivia, Colombia, Ecuador and Peru. The fragmentation of Amazonian rivers originating in Pre-Andean areas may result in severe nutrient depletion of the rivers because the mountains and associated uplands are the main source of sediments that form the basis for the high primary productivity observed in the Amazonian floodplains.

An unavoidable effect of dams is the shift in species composition and abundance. This shift includes the extreme proliferation of some populations and a reduction, or even elimination, of others [25]. The obstacles in the migratory routes, the loss of natural nursery areas placed upstream of dams and the modification of the hydrological regime downstream of dams, in addition to the rheophilic behavior of the community, are factors directly linked to failures in recruitment and the limited distribution of adults in reservoirs [25], which strongly affect fisheries [26].

3. Life strategies of freshwater amazonian fish

In the Amazon Basin, the life strategies associated with migratory and reproductive processes can be employed to distinguish three fish groups. First, groups can be distinguished by their migration length. The fish species that participate in long-distance migrations are from the family Pimelodidae and belong to a unique genus: *Brachyplatystoma rousseauxii, B. vailantii, B. filamentosum, B. capapretum* and *B. platynemum*[13]. These species migrate up to 3,000 km to complete their life cycle. They migrate from the Amazonian estuary to the border of the Andean mountains in Bolivia, Colombia and Peru [27]. The estuary is the nursery area, and the fish remain there approximately one year prior to beginning their migration. The floodplain areas of the Central Amazon Basin are feeding habitats where the immature fish grow up and store fats prior to their reproductive migration toward the Pre-Andean areas [13, 27, 28, 29]. This process is synchronized with the hydrological cycle. The gonads of *B. rousseauxii* are in an advanced stage of development starting at the beginning of the flood season, while *B. flamentosum, B. platynemum* and *B. vailantii* show the highest reproductive activity at the end of the flood season [28].

The short-distance migratory species belong to several groups, including Siluriforms such as *Pinirampus pirinampu, Calophysus macropterus, Hypophthalmus marginatus, H. edentates, H. fimbriatus, Phractocephalus hemiliopterus, Pseudoplatystoma punticfer, P. fasciatum* and *P. tigrinum* [13, 30, 31]. These species are also called floodplain migratory fish because they participate in short-distance migrations between the main stem of the Amazon River and its tributaries and floodplain lakes. Despite the absence of published studies, evidence from field research indicates that the migrations of this group do not appear to exhibit a pattern associated with reproductive events. Because these fish are predator species, these short-distance migrations have trophic causes and are developed to find prey in general small and medium size characins.

Another short-distance migratory species is a highly diverse group of Characiformes, which are extensively exploited by the small-scale fishing fleet from the Amazon Basin. Species such as *Colossoma macropomum, Prochilodus nigricans, Semaprochilodus insignis, S. taenirus* and several Myleinae and Curimatidae evolved for a life strategy strongly associated with the hydrological cycle of the Amazon Basin [13, 32, 33]. These species build large schools at the beginning of the rainy season and participate short-distance migrations from their feeding habitat, which is generally within black water tributaries, to white water rivers where spawning occurs [13, 32, 33, 34, 35]. The parental schools are very large, containing hundreds of thousands of individuals, and there is no parental care after spawning [13, 36]. Adult fish move toward flooded areas, which are rich in food, aiming to store energy for a new reproductive cycle. There are some differences between the timing of migration for these species; however, the schematic in Figure 3 shows the synchronism with the rainy season when the water starts to rise and the importance of the newly inundated floodplains as a place of refuge and feeding for the young fish [37].

Lastly, there are species that do not need to migrate to complete their life cycle. These fish are a diversified group with species from several orders; however, some cichlids from the genus *Cichla* are highly important for regional fisheries. These cichlids are called peacock bass and are the main target of recreational fisheries that are located primarily in black water rivers. The peacock bass is also a to predator that moves in several environments for trophic reasons [37]. Ornamental fish compose another group of non-migratory species that are exploited by fishing. This group is highly diverse, with species belonging to several orders, including Characiforms, Siluriforms, Perciforms, Osteoglossiforms and Gymnotiforms. In general, these are small sized fish with high levels of endemism.

4. Potential effects of global climatic changes and dams on freshwater amazonian fish

The intensity and direction (positive or negative) of the potential effects of environmental changes will vary among populations and species in the Amazonian fish fauna. Some global scenarios are catastrophic [38], proposing that 75% of global freshwater fish will become extinct before the end of the 21st century due to a reduction in river discharge. Nevertheless, the possible effect is local extinction, which would be a critical event for endemic species. Two species of the small fish *Paracheirodon*, which are exploited as ornamental species, exist in the middle to upper Rio Negro in Brazil and in the upper Rio Orinoco in Colombia and Venezuela. A study conducted at an inter-fluvial palm camp of the Middle Rio Negro found that these two species are rarely observed in the same habitat. The *P. simulans* habitat water temperature ranged from a low of 24.6 to a high of 35.2 ºC, while the *P. axelrodi* habitat temperature varied between 25.1 and 29.9 ºC [39]. The authors propose that because inter-fluvial areas flood as a function of rainfall, a decrease in regional precipitation could alter the hydrologic balance of these wetlands, especially during dry periods, which would lower water levels and increase the water temperature. This scenario would be extremely adverse for *P. simulans*, which exists only in very shallow inter-fluvial areas. A decrease in precipitation could dry out these areas completely, ultimately leading to the local extinction of this species.

At the beginning of rising waters season, the adults move down river from tributaries of black and clear waters to spawning in the turbid and rich environment of white water rivers. After breeding event, these fish move toward the flooded forest for feeding. The larvae are carried by drift toward flooded areas of the floodplain. After six months, when the water starts to recede, large schools of adults and young fish move toward tributaries.

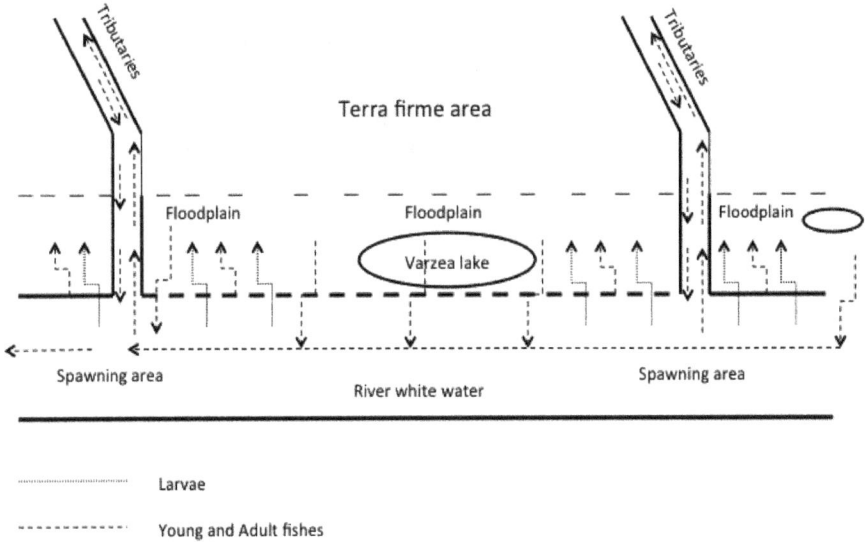

Figure 3. A general description of the Characiforms migrations.

The survivorship or abundance of fish species in a dynamic environment is dependent on three factors: the intensity of change, the velocity at which change will occur and the ability of organisms to adapt in the midst of these changes. Plasticity is a characteristic inherent to each species of fish. However, some common characteristics are useful for classifying the fish into groups and for discussing the most probable effects of environmental changes. For example, the impact of changes in the water temperature should be related to the lethal, sub-lethal and optimal thermal limits of each species. Despite of a scarcity of data on the physiology, life history and behavior of the Amazonian species, some information about population dynamics is available and can be used to hypothesize the effects of the environmental changes resulting from climate change and new dams.

The inverse relationship between temperature and dissolved oxygen in the water may result in an expansion of hypoxia zones. Although the Amazonian fish exhibit a variety of strategies related to oxygen intake in response to hypoxia [40], there are limits to these adaptive strategies that are a result of a long evolutionary time.

Another example of the types of analyses that can be conducted involves the use of the match-mismatch hypothesis (originally proposed by Cushing [41, 42] to describe the relationship between starvation and recruitment), which has clear connections to climate variability. This hypothesis recognizes that early-stage fish need food to survive and grow. It also recognizes that periods of strong food production in the ocean can be variable and are often controlled by climate, which depends on the strength of the wind, the frequency of storms and the amount of heating or fresh water supplied to the surface layers in the ocean. The hypothesis examines the timing match or mismatch between when and where food is available and when and where early-stage fish are able to encounter and consume this food. Assuming that there is a synchrony between the flood pulse and the spawning season of the Amazonian Characiformes, this cycle can be modified by climatic changes with substantial consequences on species that are the most important sources of protein consumption in the Amazon Basin.

In fact, some species or groups of species may be positively affected. A study of the reproduction of fish from the Rio Cuiabá in Upper Pantanal indicated that both the reproductive dynamics and the hydrological regime were closely related. The authors of the study showed that the intense events of floods were positively related with gonadal development of species that participate in long-distance migration and parental care [43].

A potentially useful approach to develop strategies to study the effects of global climate change could be classifying their effects by the type of relationship between the phenomenon and the impact. [44] classified the range of effects that climate change will have on freshwater, estuarine and marine fish into primary, secondary and tertiary categories. These authors found that the primary impacts are climate-related changes that directly affect the behavior, physiology, fitness and survivorship of fish without intermediary causal drivers. Secondary effects are primarily related to changes in the quality or quantity of habitats. Lastly, the tertiary impacts are related to the interactions between several causal factors.

In contrast, impacts from dams are generally related to the fragmentation of the area in which the individual members of a species live, creating obstacles for migration. The decline in the abundance of long-distance migratory species is the most distinct consequence of such filters. The community of these species undergoes seasonal migrations toward spawning habitats located upstream and consequently requires free-flowing stretches of river. Therefore, recruitment success depends on the presence of and accessibility to spawning areas, which are located in the upstream stretches of the main channel and its tributaries, as well as nursery habitats, which are located in flooded areas downstream [25, 45]. The loss of nursery habitat can critically impact several species, including several species of Characins and Perciforms. Dams also alter water temperature and quality [26, 46], which affects the community structure as a whole.

In addition, there are predictable changes in the species composition of the fish assemblages above the dam in the altered environment of the reservoir, and pre-adapted species could become abundant in this location. However, an impoverishment in the fish diversity as a whole would be expected [25, 26]. A study developed to analyze the alterations of the fish communities due to pollution and the damming of highly impacted rivers from Southeast Brazil, which were fragmented and polluted in their upper stretches, and also detected a synergic effect due to these two impact sources [47]. These authors observed a noticeable decrease in species richness in the polluted stretches of the river, with one or two species dominating. However, the artificial control of floods and discharge levels should have direct impacts on recruitment success. An analysis on the influence of the mean annual water level (m), the amplitude (maximum water level of the river in a given year; m) and the flood duration (number of days above 3.5 m; yearly total and for each season; summer and autumn were considered together) on the recruitment of *Prochilodus scrofa* for the fishery conducted at the Itaipu Reservoir and observed that flood duration is more important than flooding amplitude [48].

Table 3 summarizes the effects of global warming, sea level rise and dams on freshwater Amazonian fish, taking into account our level of knowledge. Fish faced with a changing environment must adapt, migrate or perish [19]. In addition to the high level of uncertainty at the species level, some evidence is available to predict that the resulting stress of a temperature increase will affect fauna as a whole, including fish. The effects of the higher energy demand to compensate the stress would start at the physiological level and would include size reduction and reproductive failure. This evolution affects the community structure when the dominant species has more adaptive capacity. Therefore, another possible effect of climate change is the loss of biodiversity through the extinction of specialized or endemic fish species [48]. This pattern of environmental change inducing effects will initiate from a rise in sea level and the introduction of dams.

5. Social and economic effects of climate changes and dams on amazonian fisheries

Fisheries are very important activity worldwide. Gross revenues from marine capture fisheries worldwide are estimated between US$ 80 billion and 85 billion annualy [49]. However, some authors stated that the global marine fisheries are underperforming economically due to overfishing, pollution and habitat degradation [50]. As is the case in many regions of the world, fish are a key source of animal protein, essential amino acids and minerals, mainly for low-income population who live in the Amazon basin [3, 5, 6]. A recent paper examines if marine fisheries and aquaculture can supply fish demand for a growing human population, taking into account climate change [51]. The authors claim that an effective management of fisheries is necessary to assure sustainability for world fish stocks. The authors also called for a reduction in the amount of wild fish employed to produce animal feed.

In general, Brazilian fish production followed the world tendency, with mean rates of growth of 2.48% and 10.82%, for fishing and aquaculture, respectively [52]. Analyzing just Amazonas State, the main producer of fish exclusively from freshwater, we can see that the state follows the same trend of the region, for the last ten years. On average, the Amazonas State contributed 29% of the region's fisheries production.

A closer analysis of the data shed some light on the impacts of environmental changes, as a result of climate changes or dams, on fisheries and its consequences for well being. Figure 5 shows the Amazonas state Gross Domestic Product (GDP) per capita and an index of fish production growth, for the period between 1992 and 2010, taking 1992 as base year.

Impacts	Effects
Global Climate Change – Global warming	Siluriformes–Pimelodidae: long-distance migrations – Alterations in physiological functions to survive in environmental conditions out of an optimal specific interval; – Medium size reduction; – Size of adult stock reduced due to reductions in prey abundance. Characiformes– Short-distance migrations – Alterations in physiological functions to survive in environmental conditions out of an optimal specific interval; – Failure in recruitment due to the mismatch between young fish and food; Perciformes–Cichlidae – Alterations in physiological functions to survive in environmental conditions out of an optimal specific interval; – Failure in recruitment due to impacts on reproductive functions. – Failure in recruitment due to loss of habitat;
– Sea level rise	Siluriformes–Pimelodidae: long-distance migrations – Strong recruitment due to expansion of nursery area at estuary. Characiformes– Short-distance migrations – Reduction of the abundance of less-adapted species for the altered environment; – Changes in the community structure as a response to the alterations in the environment. Perciformes–Cichlidae – Alterations in physiological functions to survive in environmental conditions out of an optimal specific interval; – Failure in recruitment due to impacts on reproductive functions.
Dams	Siluriformes–Pimelodidae: long-distance migrations – Stock abundance reduced due to reductions in the livable area; – Blockage of fish migrations, which create obstacles for freshwater species to complete their life cycles. Characiformes– Short-distance migrations – Change in species abundance because of alterations in the habitat; – Loss of important habitats for young fish (e.g., areas that are seasonally flooded); Perciformes–Cichlidae – Failure in recruitment due to the loss of spawning habitat.

Table 3. The potential impacts of global climate change and dams and their effects on freshwater Amazonian fish

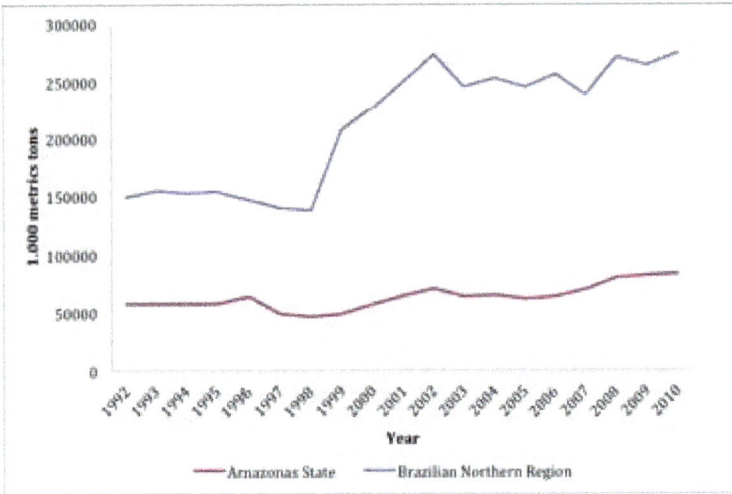

Figure 4. Amazonas State and Brazilian Northern region fisheries production between 1992-2010. (Source: 53, 54).

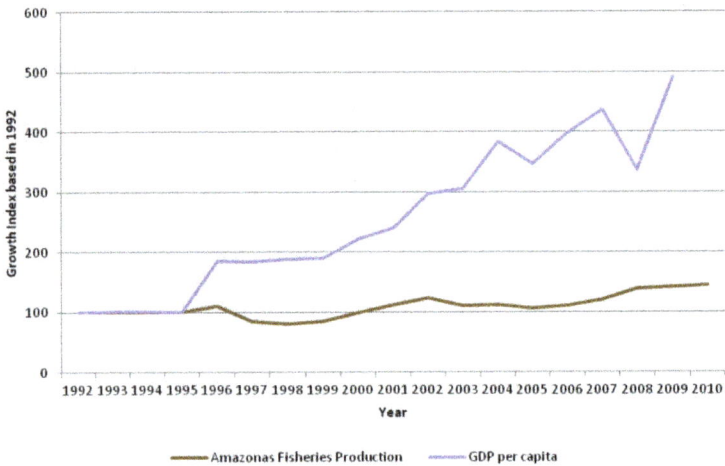

Figure 5. Amazonas State GDP per capita and an index of fish production growth [53, 54].

Taking 1992 as the base year both for a fish production index and GDP, it is possible to see that starting in 1995, GDP per capita grew continuously with fluctuations around the trend due to business cycles. At the same time, the index of fish production also grew but with much less intensity. It is clear from the graph that while GDP per capita presented a strong growth trend, increments in fish production were very slow.

What relationships do these trends have with global warming? It is clear that fish production does not drive GDP growth, so global climate change is not likely to have much impact on the GDP of Amazonas through the impact of global change on fisheries. This is not to imply that there would not be critically important impacts through other sectors of the economy. Of course, impacts on fisheries would have a large impact on the income of those participating in the commercial fishing industry, and it could have a significant impact on the GDP of small cities in Amazonia (5 to 10 thousand inhabitants) that do not have alternative sources of income. This is particular true for small cities that are sufficiently close to Manaus to sell their catch in this large urban market.

It should be noted that GDP is a measure of market output, not economic benefits and there are a number of ways in which fisheries impact social welfare, both within the urban/industrial area and within the rural communities. For example, the output of subsistence fisheries is explicitly excluded from the measure of GDP because the output is not traded in formal markets. Clearly, small communities will suffer immensely if fisheries are highly impacted by climate change. Moreover, Amazonas has a fish culture, as opposed to the beef culture of the rest of Brazil. As Table I indicates, fish consumption is extraordinarily high, even in urban centers. If fish become scarcer and more expensive, the welfare of the urban centers will be diminished as they are forced to substitute meat and poultry for their traditional fish dishes.

The social welfare impacts of global climate change induced impacts on fisheries are difficult to calculate. The reason for this is that the direct impacts on fisheries may change economic behavior, which could then lead to a series of indirect impacts that could compound the impacts of climate change. These reactions could occur between urban and rural communities, within the fishery sector, or within the subsistence communities.

If the impacts of global climate change on fisheries reduce the quality of life of small communities, it could spur additional migration from the small communities to the urban centers. This would increase the urban externalities associated with population increases as a whole, and those associated with immigration of a group of people without training to participate in the service or industrial sectors of the urban center. Moreover, the introduction of more people with a high preference for fish consumption into the urban center will increase the urban demand for fish, putting more pressure on the fisheries near the urban centers, which are already stressed and showing evidence of decline.

People that remain in small communities will continue to be negatively impacted by the decreased fish populations because of the impacts of global climate change. This could lead to several negative impacts. First, they may react to the change by fishing more intensively to try to compensate for the decline in populations. This will further stress the populations that are already stressed by global climate change. Second, they may switch species, trying to capture species that previously were not high priority, but remain more abundant. In general, these species will be smaller, and the trophic cascading associated with the decline in forage fish is difficult to predict. Third, they may turn to more hunting to supply their protein needs, leading to other negative impacts on the ecosystem. In particular, the hunting of caiman could lead to impacts on biodiversity as the controlling predators are eliminated from the ecosystem. This would be in addition to the impacts of reductions in intermediate level aquatic predators such as peacock bass (*Cichla* spp.) which would suffer from the negative impacts of global climate change.

Worse impacts could potentially occur if the rural populations increased their participation in other extractive activities, including agriculture, timbering and non-timber forest products. Although the collection of non-timber forest products, such as fruit and fibers is likely to be have a relatively benign direct impact, areas of the forest that were previously not the subject of economic activity could become the subject of economic activity. The heavy presence of people in these previously unharvested areas could lead to impacts on fisheries and wildlife, interfering with the ability of these areas to serve as a reserve for repopulating depleted areas.

Increased participation in timbering and agriculture will lead to deforestation, which has a negative impact on the biodiversity of the forest. Moreover, it will have a negative impact on the aquatic systems. If communities are successful in developing markets for these extractive products, it could lead to a reverse migration of people from urban areas back to the forest, leading to an increasing cycle of degradation.

Both the direct effects of global climate change and the indirect effects associated with the reaction to global climate change will have negative impacts on the social welfare of both urban and rural populations. It is likely that a feedback cycle could develop where the reaction to degradation is more degradation, dramatically reducing both social welfare and ecosystem function.

6. Conclusions

Climate changes and dams are likely to represent the most important threats to freshwater fish around the world. The effects of climate change on the ecosystem will include alterations of the timing, distribution and form of precipitation, as well as the timing of the flood pulse, and the intensity and frequency of floods and droughts. The impacts of these changes on the fish fauna and fisheries are, at that moment, unpredictable at both the species and ecosystem level. Actually, the degree of uncertainty and the low level of knowledge about the biology of the most of Amazonia fish species, make it hard to determine the current impacts of climate changes for each species and for the ecosystem as a whole, and even harder to predict future impacts. In addition to regional impacts, it will be very difficult to predict the impacts of dams for each fish species from the Amazon basin. As we discussed earlier, dams block fish migration, which could be very critical for many freshwater fish species that need to do migrations to complete their life cycles.

One pertinent question is what we can do to minimize the impacts of global climate changes and dams. Actually, we need to identify clearly the possible strategies to avoid human contribution to the magnitude of both sources of impacts. We realize that any intervention on the crescendo of climate change needs action of a very large scale or a very large package of small scale actions. Both of these necessitate development strategies arising from a coordinated source, such as a global agreement. Unhappily, the global negotiations on this issue have made little progress, and we remain distant from an agreement. On the other hand, the effect associated with potential new dams are in the sphere of national decision-making and these impacts could be avoided if the proposed construction does not take place. Thus, it is not too late to find alternatives to Amazonian hydropower for power supply. A global goal to minimize the impacts of both climate change and dams on freshwater fisheries is needed in order to avoid these severe impacts.

Author details

Carlos Edwar de Carvalho Freitas[1,3*], Alexandre A. F. Rivas[1,3], Caroline Pereira Campos[2], Igor Sant'Ana[1], James Randall Kahn[1,3], Maria Angélica de Almeida Correa[1] and Michel Fabiano Catarino[2]

*Address all correspondence to: cefreitas@ufam.edu.br

1 Federal University of Amazonas, Manaus, Amazonas, Brazil

2 National Institute for Amazonian Research, Manaus, Amazonas, Brazil

3 Washington and Lee University, Manaus, Amazonas, Brazil

References

[1] Freitas CEC, Siqueira-Souza FK, Prado KLL, Yamamoto KC, Hurd LE. Fish diversity in Amazonian floodplain lakes. International Journal of Medical and Biological Frontiers 2010;16:128-142.

[2] Honda EMS, Correa CM, Castelo FP, Zapellini EA. Aspectos gerais do pescado no Amazonas. Acta Amazonica 1975;5(1):87-94.

[3] Shrimpton R, Giugliano R. Consumo de alimentos e alguns nutriente em Manaus, Amazonas. Acta Amazonica 1979;9(1):117-141.

[4] Cerdeira RGP, Ruffino ML, Isaac VJ. Consumo de pescado e outros alimentos pela população ribeirinha do Lago Grande de Monte Alegre. Acta Amazonica 1997;27(3): 213-227.

[5] Batista VS, Inhamuns AJ, Freitas CEC, Freire-Brasil D. Characterization of the fishery in river communities in the low-Solimões/high-Amazon region. Fisheries Management and Ecology 1998;5(5):419-435.

[6]   Fabré N, Alonso J. Recursos ícticos no Alto Amazonas. Sua importância para as populações ribeirinhas. Boletim do Museu Paraense Emilio Goeldi, Serie Zoologia 1998;14(1):19-55.

[7]   Boischio AAP, Henshel D. Fish consumption, fish lore, and mercury pollution - risk contamination for the Madeira River people. Environmental Research 2000;84:108-126.

[8]   Murrieta RSS, Dufour DL. Fish and farinha: protein and energy consumption in Amazonia rural communities on Ituqui Island, Brazil. Ecology of Food and Nutrition 2004;43(3):231-255.

[9]   Junk WJ, Bayley PB, Sparks RE. The flood pulse concept in river floodplain systems. Special publication of the Canadian Journal of Fisheries and Aquatic Sciences 1989;106:110-127.

[10] Oliver LP et al. Drought sensitivity of the Amazon rainforest. Science 2009;323:1344-1347.

[11] Vieira EF, Isaac VJ, Fabré NN. Biologia reprodutiva do tambaqui, Colossoma macropomum Cuvier, 1818 (Teleostei, Serrasalmidae), no baixo Amazonas. Acta Amazonica 1999;29(4):625-638.

[12] Araujo-Lima CARM, Oliveira EC. Transport of larval fish in the Amazon. Journal of Fish Biology 1998;29:1-11.

[13] Araujo-Lima CARM, Ruffino ML. Migratory Fishes of the Brazilian Amazon. In: Carosfeld J, Harvey B, Ross C, Baer A (eds.) Migratory Fishes of South America: Biology, Fisheries and Conservation Status. Ottawa. The World Bank; 2003. p233-302.

[14] IPCC. Summary for Policymakers. Climate Change 2007: The Physical Science Basis. Contribution of Working Group I to the Fourth Assessment Report of the Intergovernmental Panel on Climate Change (Solomon S, Qin D, Manning M, Chen Z, Marquis M, Avryt KB, Tignor M, Miller HL (eds.). Cambridge University Press, Cambridge, UK and New York, USA, 2007, 113p.

[15] Shukla J, Nobre CA, Sellers P. Amazon deforestation and climate change. Science 1990;247:1322-1325.

[16] Malhi Y, Roberts T, Betts RA, Killen TJ, Li W, Nobre CA. Climate Change, Deforestation, and the Fate of the Amazon. Science 2008;319:169-172.

[17] Vermeer M, Rahmstorf S. Global sea level linked to global temperature. PNAS 2009; www.pnas.prg/cgi/doi/10.1073/pnas.0907765106 (acessed 5 July 2012).

[18] Becker CD, Genoway RG. Evaluation of the critical thermal maximum for determining thermal tolerance of freshwater fish. Environmental Biology of Fishes 1979;4:245-256.

[19] Ficke AD, Myrick CA, Hansen LJ. Potential impacts of climate changes on freshwater fisheries. Review in Fish Biology and Fisheries 2007;17:581-613.

[20] Lowe-McConnell RH. Ecological Studies in Tropical Fish Communities. Cambridge University Press, Cambridge, England 1987;382pp.

[21] Finer M, Jenkins CN. Amazon and its implications for Andes-Amazon connectivity. PLoS ONE 2012;7(4): e35126. doi:10.1371/journal.pone.0035126.

[22] Marengo JA, Nobre CA, Tomasella J, Oyama MD, Oliveira GS, Oliveira R, Camargo H, Alves LM, Brown IF. The drought of Amazonia in 2005. Journal of Climatology 2008;21:495-516.

[23] Humphries P, Baldwin DS. Drought and aquatic systems: an introduction. Freshwater Biology 2003;48:1141-1146.

[24] Matthews WJ, Marsh-Matthews E. Effects of drought on fish across axes of space, time and ecological complexity. Freshwater Biology 2003;48:1232-1253.

[25] Agostinho AA, Pelicice FM, Gomes LC. Dams and the fish fauna of the Neotropical region: impacts and management related to diversity and fisheries. Brazilian Journal of Biology 2008;68(4,suppl.):1119-1132.

[26] Agostinho AA, Gomes LC, Veríssimo S, Okada EK. Flood regime, dam regulation and fish in the Upper Paraná River: effects on assemblages attributes, reproduction and recruitment. Reviews in Fish Biology and Fisheries 2004;14(1):11-19.

[27] Barthem RB, Goulding M. The catfish connection: ecology, migration, and conservation of Amazon predators. Biology and Resource Management in the Tropics Series. Columbia Press, New York 1997;144 p.

[28] Agudelo E, Salinas Y, Sánches CL, Muñoz-Sosa DL, Alonso JC, Arteaga ME, Rodrígues OJ, Anzola NR, Acosta LE, Núñez M, Valdés H. Bagres da la Amazonia Colombiana: Uno recurso sin fronteras. Fabré NN, Donato JC, Alonso JC (Eds). Instituto Amazónico de Investigaciones Científicas SINCHI. Programa de Ecosistemas Acuáticos. Editorial Scripto, Bogotá 2000;252p.

[29] Barthem RB, Ribeiro MCLB, Petrere M. Life strategies of some long-distance migratory catfish in relation to hydroelectric dams in the Amazon Basin. Biological Conservation, 1991;55:339–345.

[30] Ruffino ML, Isaac VJ. Dinamica populacional do Surubim-Tigre, Pseudoplatystoma tigrinum (Valenciennes, 1840) no médio Amazonas (Siluriformes, Pimelodidae). Acta Amazonica 1999;29(3):463-476.

[31] Pérez A, Fabré NN. Seasonal growth and life history of the catfish Calophysus macropterus (Lichtenstein, 1819) (Siluriformes: Pimelodidae) from the Amazon floodplain. Journal of Applied Ichthyology 2009;25:343–349.

[32] Loubens G, Panfili J. Biologie de Colossoma macropomum (Teleostei: Serrasalmidae) dans le bassin du Mamoré (Amazonie bolivienne). Ichthyological Exploration of Freshwaters 1997;8:1–22.

[33] Ribeiro MCLB, Petrere Jr. M. Fisheries ecology and management of the Jaraqui (Sem-aprochilodus taeniurus, S. insignis) in Central Amazonia. Regulated Rivers: Research and Management 1990;5:195–215.

[34] Fernandes CC. Lateral migrations of fishes in Amazon floodplain. Ecology of Freshwater Fish 1997;6:36-44.

[35] Mota SQ, Ruffino ML. Biologia e pesca do curimatá (Prochilodus nigricans Agassiz, 1829) (Prochilodontidae) no Médio Amazonas. Revista UNIMAR 1997;19:493-508.

[36] Ruffino ML, Isaac VJ. Life cycle and biological parameters of several Brazilian Amazon fish species. The ICLARM Quaterly 1995;18:41-45.

[37] Jepsen DB, Winemiller KO, Taphorn DC. Age structure and growth of peacock cichlids from rivers and reservoirs of Venezuela. Journal of Fish Biology 1999;55:433-450.

[38] Xenopoulos MA, Lodge DM, Alcamo J, Märker M, Shulze K, Van Vuuren DP. Scenarios of freshwater fish extinctions from climate change and water withdrawl. Global Change Biology 2005;11:1557-1564.

[39] Marshall BG, Forsberg BR, Hess LL, Freitas CEC. Water temperature differences in interfluvial palm swamp habitats of Paracheirodon axelroldi and P. simulans (Osteichthyes: Characidae) in the middle Rio Negro, Brazil 2011;22(4):377-383.

[40] Almeida-Val V, Val AL, Duncan WP, Souza FCA, Paula-Silva MN, Land S. Scaling effects on hypoxia tolerance in the Amazon fish Astronotus ocellatus (Perciformes: Cichlidae): contribution of tissue enzyme levels. Comparative Biochemistry and Physiology;125B:219-226.

[41] Cushing DH. Climate and fisheries. Academic Press, London 1982;373p.

[42] Cushing DH. Plankton production and year-class strength in fish populations: an update of the match/mismatch hypothesis. Advances in Marine Biology 1990;26:249-293.

[43] Bailly D, Agostinho AA, Suzuki HI. Influence of the flood regime on the reproduction of fish species with different reproductive strategies in the Cuiabá River, Upper Pantanal, Brazil. River Research and Applications 2008;24:1218-1229.

[44] Koehn JD, Hobday AJ, Pratchett MS, Gillanders BM. Climate change and Australian marine and freshwater environments, fishes and fisheries: synthesis and options for adaptation. Marine An Freshwater Research 2011;62:1148-1164.

[45] Agostinho AA, Vazzoler AEAM, Gomes LC, Okada EK Estratificacíon especial y comportamiento de Prochilodus scrofa en distintas fases del ciclo de vida, en la planicie de inundacíon del alto río Paraná y embalse de Itaipu, Paraná, Brazil. Revue due Hydrobiologie Tropicale 1993;26:79-90.

[46] Barrela W, Petrere Jr M. Fish community alterations due to pollution and damming in Tietê and Paranapanema Rivers (Brazil). River Research and Applications 2003;19:59-76.

[47] Porto L, McLaughlin R, Noakes D. Low-head barrier dams restrict the movements of fishes in two lake Ontario streams. North American Journal of Fisheries and Management 1999;4:1028-1036.

[48] Gomes LC, Agostinho AA. Influence of the flooding regime on the nutritional state and juvenile recruitment of the curimba, Prochilodus scrofa, Steindachner, in Upper Paraná river, Brazil. Fisheries Management and Ecology 1997;4:263-274.

[49] Angermeier PL. Ecological attributes of extinction-prone species: loss of freshwater fishes of Virginia. Conservation Biology 1995;9:143-158.

[50] Food and Agriculture Organization. The State of World Fisheries and Aquaculture 2010. Rome 2011.

[51] Sumaila UR, Cheung WWL, Lam VWY, Pauly D, Herrick S. Climate change impacts on the biophysics and economics of world fisheries. Nature Climate Change 2011; doi: 10.1038/NCLIMATE1301.

[52] Merino G, Barange M, Blanchard JL, Harle J, Holmes R, Allen I, Allison EH, Badjeck MC, Dulvy NK, Holt J, Jennings S, Mullon C, Rodwell LD. Can marine fisheries and aquaculture meet fish demand from a growing human population in a changing climate? Global Environment Change 2012;22:795-806. Doi: 10.1016/j.gloenvcha.2012.03.003.

[53] Food and Agriculture Organization. Statistical databases. Available in: <http://www.fao.org>. Acessed on September 2012.

[54] Instituto Brasileiro do Meio Ambiente e dos Recursos Naturais Renováveis – IBAMA. Estatística da Pesca 2007. Brasil – Grandes Regiões e Unidades da Federação 2007.

[55] Ministério da Pesca e Aquicultura. Produção Pesqueira e Aquícola – Estatística 2008 e 2009 – 2010.

Chapter 7

Functional and Structural Differences in Atria Versus Ventricles in Teleost Hearts

Christine Genge, Leif Hove-Madsen and
Glen F. Tibbits

Additional information is available at the end of the chapter

1. Introduction

Across vertebrates, the fish heart is structurally relatively simple. The heart of teleosts is unique in structure, composed of four chambers in series: venous sinus, atrium, ventricle and bulbus arteriosus. The two chambers acting as pumps are the atrium and ventricle, a simplified version of that seen in tetrapods. These chambers develop from a simple linear tube [1] and differ not only morphologically but also physiologically with different characteristic rates of contractility [2]. Fish heart design is reflective of the needs for oxygen delivery to working skeletal muscle in an often oxygen poor environment. The heart is the first definitive organ to develop and become functional, as embryological survival depends on its proper function. The vertebrate-specific development of multiple chambers with enough muscle to generate higher systemic pressures allowed for perfusion of larger and more complex tissues.

Fish show great variability in the development of the heart and modification of heart performance capacities to meet tissue perfusion demands imposed by differences in life history. Research has tended to focus on inotropic or chronotropic responses of the heart to stress such as temperature in larger fish such as trout, or on development through the use of zebrafish mutants. Between these sources much has been revealed about the contractile properties of the two chambers. This research will be reviewed in the context of roles of the atrium and ventricle in fish in achieving variability in myocardial contractility.

2. Gene expression

In mammals, 23% of cardiac-specific genes show differences in expression between the atria and the ventricle [3]. Key morphogenetic events during heart ontogenesis are conserved across vertebrates [4]. Gene expression programs specific to each chamber are likely regulating key aspects of chamber formation. The focus on fish heart developmental patterns has primarily been due to the use of zebrafish as a model organism of vertebrate development. The morphological differences created during chamber development are guided by changes in gene expression patterns [1], a process well conserved in vertebrates, regardless of whether there are one or two atria or ventricles. Genetic screens in zebrafish have revealed genes and pathways underlying development in the heart. Since many stages of development are conserved across vertebrates, this information has generated considerable insight into congenital heart defects in humans. In the context of this chapter, timing and localization of gene expression can also contribute to our understanding of the differentiation of chambers both in morphology as well as through variation in contractility.

The T-box genes (Tbx) are responsible, in large part, for patterning that distinguishes chamber myocardium from non-chamber myocardium [5]. Subregionalization is important for atrial or ventricular identity, either via selective transcription or repression of genes. The physical interaction of Nkx2.5, Gata4 and Tbx5 in particular activates the expression of chamber-specific genes to stimulate the differentiation of primary myocardial cells into chamber myocardium [1]. By the time cells are restricted to cardiac lineage, they are already destined for either the atrium or the ventricle. During early gastrulation, myocardial progenitors already have axial coordinates that lead to chamber assignment [6]. There are two sources of these myocardial progenitors, or two heart fields. In the first heart field, cardiomyocyte differentiation initiates in the ventricle and allows continuous addition of new cardiomyocytes to the venous pole of the heart tube progressing towards the atrium [7]. Differentiation at the arterial pole in the second heart field occurs only in the later stages of cardiac development. By 72 hours post-fertilization (hpf) individual heart chambers are morphologically differentiated with the ventricular wall being thicker and ventricular cardiomyocytes being larger than those of the atrium [8].

The chambers express different subsets (or paralogs) of sarcomeric proteins such as myosin heavy chains (MHC) and myosin light chains (MLC) [9]. Variation in paralog expression (frequently referred to in the literature as isoforms) may confer differing contractile properties to the atrium and the ventricle, such as with altered MHC expression being correlated with altered myofibrillar ATPase activity (crucian carp - [10]). MHC expression is chamber-specific by 22 hpf, preceding any obvious chamber formation but corresponding to the initiation of the heart beat [6]. These isoforms are not only divided spatially but also by differences in timing of expression. Atrial-specific myosin gene expression is induced after ventricle-specific myosin gene expression in zebrafish [11]. This time-dependent expression may be due to repression of the promoters of these genes [1].

The timing of these developmental pathways is critical, as the development of one chamber can influence the other. Ventricular morphology can change with developmental atrial dysfunction seen with zebrafish mutations in atrial-specific MHC [11]. For example, in the wea mutant, a lack of a functional atrial MHC results in defects in atrial myofilament organization and contractility. Despite the mutation being seen only in a chamber-specific protein, the ventricle changes as well in response to these atrial defects. Effective embryonic circulation requires appropriate relative sizes of the atrium and ventricle. Bone morphogenetic protein (BMP) signaling has been shown to be involved with the regulation of chamber size in zebrafish [12]. However, mutations affecting BMP signaling only reduce atrium size but not the ventricle. During this stage of development, the size of chambers may be relatively independent of one another.

3. Morphology

In fish the heart is arranged in series with the ventricle primarily filled by contraction of the atrium [13], rather than the mammalian system in which the thin-walled atrium contributes only a small amount to ventricular filling under resting conditions. While the atrium and ventricle compose the main muscular components of the fish heart, there are other chambers with important roles in guiding appropriate blood flow through the heart. Generally in fish blood flows through the sinus venosus to the atrium to the ventricle and then out to the bulbus anteriosus. More recently, the atrioventricular region and conus arteriosus have been identified as morphologically discrete segments present throughout modern teleosts as well [14, 15].

3.1 General morphology

As with most vertebrates, the cardiac tube is S-shaped with a dorsally positioned atrium and ventrally positioned ventricle to allow for appropriate momentum of the bloodstream towards the arterial pole. The morphology of the heart sets up an alternating arrangement of slow conducting segments (e.g. inflow tract, atrioventricular region and outflow tract) with fast conducting contractile segments (atrium and ventricle). Blood first enters the thin-walled sinus venosus. This is an independent, but non-muscular chamber in fish that acts as a drainage pool for the venous system. The composition of the sinus venosus varies greatly between species, from primarily connective tissue in *Danio rerio* (zebrafish) to mainly myocardium in *Anguilla anguilla* (European eel) to even smooth muscle cells in *Cyprinus carpio* (carp) [16, 17, 18]. The presence of the sinus venosus has thus been suspected to have allowed the development of the atrium as the principal driving force for ventricular filling [19]. Blood then moves into the atrium, an active contractile chamber in fish that again shows considerable variability in size and shape between species [18]. A general trend throughout teleost species is the similarity in volume of the two main muscular chambers is indicative of the importance of the role of the atrium in filling the ventricle. While atrial volume is closer in size to that of the ventricle in fish, the atrium is still less developed muscularly and possibly representative of a more primitive heart [20-22].

Separating the atrium and the ventricle is the atrioventricular region, a ring of cardiac tissue supporting the AV valves. The degree of isolation of this region from the surrounding musculature varies between species [23] but in all species it plays a critical role in regulating the pattern of electrical conduction. The ventricle has a mass and wall thickness up to five times that of the atrium [18]. The thickly muscular ventricle varies in terms of structural organization, histology, coronary distribution and relative mass between species as well. This varia-tion reflects the fact that different modes of heart performance are attributed to different ventricle types since this chamber is responsible for the active cardiac output [18]. This clas-sification will be discussed further in this section. From the ventricle, blood moves into the outflow tract leading into the beginning of the dorsal aorta. In more basal fish there are two segments: the proximal muscular conus arteriosus and the distal arterial-like bulbus arterio-sus [24]. While later teleosts were thought to have lost the conus arteriosus, more recent evi-dence shows it is a distinct segment in many families [14, 15]. The bulbus arteriosus is more prominent and serves as the main component of the outflow tract. This elastic chamber ex-pands to store large portions of the stroke volume, and gradually releases a steady non-damaging flow to gill vascular [18]. The role of these segments is variable between species and has been reviewed in more detail [25].

The AV-region, bulbus arteriosus and conus arteriosus are unique to fish though homolo-gous regions may be seen in early embryological stages of higher vertebrates. While, the at-rium and ventricle are present across vertebrates, their diversity in morphology and function and how they contribute to contractility will be the focus of the remainder of this Chapter.

3.2. Myocardial arrangement

The teleost myocardium is composed of two distinct layers: compact and spongy. The at-rium tends to be primarily spongy, with more open space between fibers and greater vari-ability in orientation [26]. The external rim is myocardium surrounding a complex network of pectinate muscles. Collagen provides support, being predominant both in sur-rounding the myocardium and the internal trabecular network [15]. Meanwhile, while the ventricle composition varies greatly between species, the periphery of the ventricle tends to be primarily compact tissue with the luminal being primarily spongy [15, 27]. The com-pact layer is composed of typically well-organized fibers composing well-arranged bun-dles covering the inner ventricle. The spongy layer is composed of long finger-like branching cells in close proximity to luminal fluids. Not all fish species have distinct lay-ers in the ventricle to the same degree as the salmonids and scombroids in which most of the experiments characterizing the morphology of the chambers has been done. For exam-ple, the relatively sedentary flatfish has a heart with virtually no lumen and is composed almost entirely of spongy tissue [28]. These phenotypes have been subdivided into four subclasses (Types I-IV). A Type I ventricle consists of an entirely spongy ventricle, while Type II has both compact and spongy myocardium but with coronary vessels restricted to the compact layer. Type III and Type IV have vascularized trabeculae, with coronary ves-sels penetrating spongy and compact myocardium. Types III and IV are distinguished from one another by Type III having less than 30% of the heart mass composed of com-pact myocardium (exemplified by many elasmobranchs) and Type IV representing ventri-cles with greater amounts of compact myocardium [29].

3.3. Cardiomyocytes

On the cellular level, cardiomyocytes are more cuboidal in the ventricle compared to the more squamous shape seen in the atrium [30]. Ventricular myocytes from ecothermic vertebrates more closely resemble myocytes from neonatal mammals rather than adults in terms of structure and function. They are long (>100 μm), thin (<10 μm) and generally lacking in t-tubules (trout - [18]; trout - [31]; neonatal rabbit - [32]). These myocytes appear in electron microscopy images to have less extensive sarcoplasmic reticulum (SR) than mammals [33] but some peripheral couplings with sarcolemmal membrane are observed [34]. Atrial muscle cells contain well-developed sarcomeric reticulum, peripheral couplings but no t-tubules [35]. The narrow extended shape of teleost cardiomyocytes from both chambers decreases the diffusion distance from the sarcolemma to the center of the cell [36-38]. This high surface-to-volume ratio of the cells may be related to increased efficacy of E-C coupling. These morphological distinctions contribute to the chamber-specific contractile properties which will be discussed further in this Chapter.

3.4. Lifestyle demands

These morphological differences between species are reflective of a diversity of lifestyle demands. Active fish have a larger relative heart mass than that of less active species; the relative heart mass in the skipjack tuna is 10 times larger than in the sluggish flounder [39]. However, a similar atrium to ventricle mass ratio exists in species with very different lifestyles (0.2 in active Atlantic salmon, Atlantic cod and inactive winter flounder - [28]). More active fish also show a difference in whole heart shape with active fish (e.g. salmonid and scombrid families) exhibiting pyramid-shaped hearts whereas inactive fish hearts are more rounded and sac-like in shape [40]. The pyramidal shape normally correlates with a thicker myocardial wall and high cardiac output. This allows for high levels of stroke work and the maintenance of relatively high blood pressures [18].

The amount of compact myocardium is generally increased in fish with greater cardiac output demands. With pyramidal ventricles, the myocardial wall is normally formed primarily by the compact layer, which is complex and thicker in these higher performance fish [18] although this phenotype is also found in the relatively sedentary Antarctic teleosts [26]. This suggests that shape and degree of compact myocardium do not always correlate with lifestyle and in fact the shape of the chamber does not always correlate with the inner architecture [41]. This active phenotype may also display plasticity, as trout that are rendered inactive will display an abnormal rounded shape [42]. This may also be the case in aquaculture conditions, where fish become less active and may be overfed [43]. The proportion of compact myocardium has also been shown to vary with seasons [18] and growth [44].

Active fish rely on coronary circulation to allow for increased cardiac output. More compact tissue in the ventricle generally requires greater O_2 perfusion. In general, sluggish species lack this coronary circulation and can rely completely on O_2 contained in the venous circulation [45]. Hence thicker ventricles seen in more active fish have a more developed coronary distribution to the compact myocardium, and a greater percentage of compact myocardium. Very active fish like scombroids and elasmobranchs also have a coronary supply to the spongy myocardium and atrium [46]. However, this increased demand for O_2 supply for cardiac function means that these species are at a disadvantage in hypoxic water. The more primitive hagfish heart composed almost entirely of spongy myocardium can meet cardiac ATP requirements for resting conditions even in anoxic water [47].

4. Filling patterns

Atrial filling occupies a substantial proportion (up to 75%) of the cardiac cycle and as discussed above, the atrial chamber volume in fish may be comparable to that of the ventricle. Substantial debate exists in the literature as to the mechanism of both atrial and ventricular filling. Though noninvasive echocardiography has greatly contributed to research on flow patterns, it does require anaesthetization and restraint of the fish. Because the heart is acutely sensitive to venous pressure, and venous pressure is acutely sensitive to gravitational forces and extramural pressure, considerable variation may be seen from study to study. Certain studies have shown a more monophasic filling of both atrium and ventricle via a vis-a-tergo mechanism (e.g. trout - [48]). In this model, ventricular pressure is derived from arterial flow and venous capacitance is the primary determinant of cardiac output [13]. Other studies have demonstrated a more biphasic filling pattern via a vis-a-front mechanism in which cardiac suction plays a prominent role in ventricular pressure (e.g. elasmobranchs - [49]). Even triphasic patterns have been observed, with a third wave of filling possibly involving sinus contraction (e.g. leopard shark - [49]).

These filling patterns of the chambers may be both a reflection of phylogeny and external conditions. Less active fish such as the sea raven require above-ambient filling pressures that may require vis-a-tergo filling [50]. The venous pressure control mechanisms are not seen to be as developed in elasmobranchs as in teleosts resulting in a greater necessity of vis-a-front filling [51]. Meanwhile with increased stroke volumes during exercise resulting in more positive filling pressures, teleosts may further shift to a vis-a-tergo cardiac filling mode for increased activity [52]. However, the relative proportion of early to late phase filling (or passive vs. atrial active) suggests that the late phase is greater than the early phase in biphasic [49] and the third phase in triphasic does not change this pattern. With late phase (active atrial contraction) playing such an important role regardless of filling mode, the trend still holds that fish hearts are still more dependent on atrial contraction for ventricular filling than mammals.

5. Physiological differences between chambers

It has been proposed that a key regulator of cardiac performance is the end diastolic volume (EDV) in both mammals and fish. Many of the morphological differences between chambers may have evolved in order to meet the high physiological demands of variation in metabolic rate and increased tolerance to a wide range of environmental conditions. Several of these will be highlighted in this section.

5.1. Force generation

Cardiac output may be modulated through variations in stroke volume and heart rate, but the relative contributions vary substantially among fish species. In general fish respond to greater hemodynamic loads by increasing cardiac output mainly through increased stroke volume rather than heart rate. Stroke volume can vary up to 2-3 fold with exercise (salmonids - [53]) but not all species exhibit the same ability [54-55]. End systolic volume (ESV) of the fish heart relative to stroke volume is very small [56-57] unlike that of the mammalian heart. However, end diastolic volume in fish is greatly determined by atrial contraction rather than central venous pressure seen in mammals [13]. The modulation of atrial contractility and Frank-Starling mechanisms are thus particularly important in determining ventricle filling.

The whole heart displays a greater sensitivity to filling pressure than in mammals, in part attributed to greater myocardial extensibility coupled with an increase in myofilament calcium sensitivity over a large range of sarcomere lengths [58]. The Starling principle dictates that the strength of contraction is, in part, a function of the length of the muscle fiber (and hence sarcomere length) prior to contraction (i.e. preload). Stroke volume increases as a function of end diastolic volume since muscle fibers and sarcomeres within the fibers are stretched towards their optimum. Between various species of fish there is considerable variation in length-dependent sensitivity. For example, flounder heart is very sensitive to filling pressures, needing smaller increases in preload to achieve higher values of stroke volume. This more pronounced Starling curve is further impacted by the high distensibility of its chambers based on pressure-volume curves [28]. This length-dependent strength of contraction is also directly affected by temperature in certain teleost species (e.g. trout- [58]). With fish cardiac myocytes showing enhanced length-dependent activation, and calcium sensitivity increasing with sarcomere length [59], calcium handling is an important consideration in the basis of the length dependence of the strength of cardiac contraction in fish. The unique properties of E-C coupling in the chambers of the fish heart will be discussed in greater detail further in this Chapter. The reduced relative duration of diastole seen with active atrial contraction in fish may have evolved in response to the relatively slow rate of the excitation-contraction coupling processes.

The importance of strong atrial contraction is reflected in the fact that atrial muscle generates greater maximal force (F_{max}) than ventricular muscle in a variety of teleost species (e.g. rainbow trout- [60]; yellowfin tuna - [61]; brown trout - [62]). Although systolic blood pressure measurements show the atrium to be a weaker pump, the higher force generated in excised strips suggests the relative weakness of the atrium may simply be reflective of the anatomical difference in wall thickness between the chambers. The greater F_{max} of the atrial cardiomyocytes individually does not account for the difference in overall force generation by the greater quantity of cardiomyocytes in the thicker ventricle wall.

Cardiac performance is directly related to the contractile properties of cardiomyocytes. As mentioned previously, ventricular and atrial muscles express different paralogs (or isoforms) of myosin heavy chain (MHC) [63]. The maximum rate of force produced by muscle fibers is determined to a large degree by MHC paralog content. Alterations in paralog composition therefore are associated with changes in myofibrillar function as well as myosin

ATPase activity [64]. Because of this, the myosin ATPase reaction is higher in atrial than in ventricular cardiomyocytes [65]. Across vertebrates, from the primitive dogfish right up to mammals, the atrium has a higher concentration of MHC protein content than the ventricle [4]. The increased levels of MHC may be contributing to the increased F_{max} of individual atrial cardiomyocytes relative to ventricular. However, MHC expression pattern differences between chambers may also be species-specific. In goldfish, both chambers have electrophoretically similar MHC while in other fish species such as pike, ventricular myocardium exhibited electrophoretically faster MHC [65]. Linking chamber-specific isoform usage with functional differences across a broader phylogenetic range as well as temperatures may help to clarify the importance of this heterogeneity.

5.2. Action potential

As in other vertebrates, the teleost action potential can be divided into a rapid upstroke of the action potential (phase 0), a rapid initial repolarization of the AP (phase 1) followed by a plateau phase (phase 2). The plateau phase is followed by repolarization of the action potential (phase 3) returning it to the resting membrane potential (phase 4). Teleost atrial and ventricular myocytes also have distinct AP morphologies with the atrial action potential having a shorter plateau phase and a slower repolarization of the AP. Among the ionic currents contributing to the differences in fish atrial and ventricular AP morphologies, ventricular myocytes have prominent inwardly rectifying potassium current I_{K1} and a small I_{Kr} while the opposite is true for atrial myocytes [66-67]. Other currents expected to affect the AP plateau such as the L-type calcium current (I_{Ca}) or the Na^+/Ca^{2+} exchanger (NCX) current (I_{NCX}) have not been compared systematically in the teleost heart.

Action potentials have been recorded in a number of teleost species (carp - [68]; trout - [34]; zebrafish - [69]) with a considerable variability in AP morphology with the atrial AP duration ranging from 197 ms in carp atrium at 27 °C to 281 ms in trout atrium at 15 °C, 601 ms in carp at 11 °C and 1284 ms at 4 °C and ventricular AP duration ranging from 407 ms at 10 degrees to 143 ms at 20 °C in trout ventricle [71-72] and from 1042 ms at 11 °C to 737 ms at 18 °C in Burbot ventricle [70,72], likely reflecting species-dependent differences as well as variations in physiological-experimental conditions such as temperature or salinity. Although these variations in AP-duration are larger than 10-fold, they allow the teleost heart to adapt to a wide temperature range [18] while the mammalian heart becomes refractory when temperature is lowered. The greater adaptability of the teleost heart to temperature is also reflected in the properties of ionic channels and ion transporters the determine excitability and the action potential shape. For example the change in trout NCX activity is only weakly modified by temperature changes (Q_{10} = 1.2 - 1.5) as compared to mammals (Q_{10} = 2.5) [73-74] and calcium uptake and leak from the sarcoplasmic reticulum is slowed in parallel, resulting in preservation of calcium release from the SR in trout [75-76]

In recent years, action potentials have also been recorded in myocytes isolated from teleost atrial and ventricular myocytes with several reports showing that the resting membrane potential in atrial myocytes is significantly more depolarized than in ventricular myocytes [66,

77]. It was proposed that this could be due to the low I_{K1} density in trout atrial myocytes [66]. However, membrane potentials recorded in atrial tissue preparations [71] are comparable to values recoded in trout ventricular preparations [70, 78] under comparable experimental conditions. Studies have showed that the resting membrane potential recorded with current-clamp technique in trout atrial myocytes was highly sensitive to the holding current applied [70]. Moreover, recordings of potassium conductance in these myocytes suggested that the loss of the resting membrane potential in isolated trout atrial myocytes is likely due to the lack of activation of the acetyl choline-activated potassium current (I_{KACh}).

5.3. Excitation-contraction coupling

Calcium handling is an important consideration as the primary basis for regulation of cardiac contraction in all species, including fish, since it is determined by delivery and removal of calcium to and from cardiac troponin C (cTnC), the thin filament protein which senses changes in cytosolic Ca^{2+} concentrations and initiates the contractile process. Therefore, the amplitude and kinetics of the cytosolic calcium transient will modulate the kinetics and force of contraction, and this in turn will set the limit for the contraction frequency. Mammalian cardiac muscle has intracellular calcium stores in the SR in close proximity to T-tubules that provide the majority of the calcium that binds to cTnC. In ectotherms, the activation of calcium release from the SR and its contribution to contractile activation is controversial with some reports showing that the majority of calcium is derived from extracellular space [79, 36, 34, 58] while others report a considerable contribution of the SR both to the activation of contraction [38, 80] and to calcium removal from the cytosol during relaxation [58, 80-81]. Interestingly, calcium handling by the SR has also been reported to differ between fish atrial and ventricular myocytes and these differences as well as species dependent differences are likely to account for part of this controversy on the role of the SR in the teleost heart. Additionally, a significant number of reports examining the role of the SR in the teleost heart relied heavily on pharmacological inhibition of the SR calcium release channel (also termed ryanodine receptor or RyR2) in multicellular tissue preparations. As discussed below, this approach may not be optimal to evaluate the contribution of the SR to the activation of contraction if other calcium delivering mechanisms are capable of compensating for inhibition of the RyR2. Instead, quantification of the contribution of each mechanism involved in the delivery and removal of calcium to and from cTnC is necessary to estimate its capacity and contribution.

5.4. Activation of contraction

In mammalian cardiomyocytes the key contributors to the activation of contraction are the L-type calcium channel that mediates calcium entry across the sarcolemma and the SR, which releases the majority of the calcium from the SR. During the upstroke of the action potential, reverse mode NCX may also contribute [82]. Since the initial reports on I_{Ca} measurements in fish cardiomyocytes [73, 83-84] there is now substantial information on I_{Ca} density in atrial and ventricular myocytes from a number of fish species under different experimental conditions [77, 85]. Again, differences in experimental conditions and species are abundant and there are substantial differences in the measurements of calcium entry though I_{Ca} among species at different temperatures; Burbot : I_{Ca}, 0.81±0.13 pA pF^{-1} at 4°C, and 1.35±0.18 pA pF^{-1} at 11°C [55]; trout: 1, 2 and 4 pA pF^{-1} at 7, 14 and 21 °C, respectively [77]. Among the experimental conditions that have may have the largest impact on I_{Ca} measurements, ruptured-patch technique with a high (5 mM) EGTA-buffered patch pipette solution has been shown to result in overestimation of I_{Ca} when compared to measurements performed with low EGTA (25 μM) or perforated patch-clamp technique [75]. Therefore, estimations of the contribution of the calcium carried by I_{Ca} to the activation of contraction also varies from as little as 25-30% in trout atrial myocytes using perforated patch technique [75,86], to more than 50% in zebrafish ventricular myocytes subjected to perforated patch-clamp technique [87] and close to 100% using ruptured patch-clamp technique with EGTA buffered patch pipettes in carp ventricular myocytes [37] and in trout atrial myocytes subjected to β-adrenergic stimulation [87]. The latter does not, however, necessarily imply that I_{Ca} is the sole source of calcium delivery in these preparations but merely indicates that it is sufficiently large to activate contraction if no other calcium sources are available. As mentioned above, chamber specific differences in the I_{Ca} density have not been systematically investigated in the teleost heart.

Theoretically, reverse mode NCX could also contribute to the activation of contraction at depolarized membrane potentials and/or when sodium channels produce a large local increase in the subsarcolemmal Na concentration. Several studies have shown that fish cardiomyocytes have a high NCX activity than mammalian cardiomyocytes, and the direct contribution of the NCX to the activation of contraction has been estimated to range from 25-45% under physiological conditions [86-87, 90]. However, the relative contribution of I_{NCX} to total sarcolemmal calcium entry depends critically on the membrane potential and the intracellular Na$^+$ concentration [74, 85-86]. Moreover, there is evidence that reverse mode NCX can trigger calcium release from the SR in trout atrial myocytes [90].

Finally, calcium release from the SR has also been reported to contribute to the activation of contraction in fish cardiac myocytes with a contribution that varies between 0 and more than 50% depending on species, tissue and experimental temperature. Some of the first publications to examine the effects of ryanodine, a compound that modulates the opening of the SR calcium release channel, on contraction in multicellular tissue preparations were [91-92] but the reported effects on contraction were only marginal at physiological temperatures and stimulation frequencies. Ryanodine was, however, reported to inhibit post rest potentiation, a phenomenon caused by calcium release from the SR [78], and in 1992 it was reported to have a strong effect on contraction in trout [93] and skipjack tuna [94] at 25 °C, suggesting a temperature- and/or species-dependent component of the effect of ryanodine. In 1998, it was shown that steady-state and maximal SR calcium content in trout ventricular myocytes was about 10 times higher than values reported in human cardiomyocytes and that uptake rates were sufficiently fast for the SR to participate in the regulation of calcium handling on a beat-to-beat basis [75] and in 1999 it was shown that brief membrane depolarizations were capable of eliciting calcium release from the SR in trout atrial myocytes corresponding to a contribution of up to 50% of the total calcium required for the activation of contraction [80].

Subsequently it was shown that this contribution of the SR to the activation of contraction was equally important at 21 and 7°C in trout atrial myocytes [95]. These findings were confirmed in trout using simultaneous recordings of membrane currents and calcium transients [89] and with confocal calcium imaging, which also demonstrated that local spontaneous calcium release from the SR (known as calcium sparks in mammalian cardiomyocytes) are also present in trout atrial and ventricular myocytes as well as in zebrafish ventricular myocytes [96]. This study also showed that the spark frequency in trout atrial myocytes was comparable to that recorded in human atrial myocytes under similar experimental conditions [97], but twice as fast as in trout ventricular myocytes. In line with this, the frequency of spontaneous membrane depolarizations, induced by SR calcium release, were also about 6-fold more frequent in trout atrial than ventricular myocytes, suggesting that SR calcium is more labile in atrial than ventricular myocytes. A higher contribution of the SR Ca^{2+} release has also been reported for the tuna heart [38, 67].

Experiments attempting to identify the triggers of SR calcium release showed that both ICa and reverse mode NCX are capable of triggering calcium release from the SR [90] and, unexpectedly, this study showed that the gain of I_{Ca} and I_{NCX}-induced calcium release was similar in trout atrial myocytes. Moreover, it was shown that the relative contribution of the two mechanisms depended strongly on the membrane potential and the intracellular Na^+-concentration. Considering that an intracellular $[Na^+]$ between 13 and 14 mM has been reported in trout cardiomyocytes [98] reverse mode NCX could may be expected to contribute significantly to the triggering of calcium release from the SR in trout. While in mammals $[Na^+]_i$ is higher in atrial than ventricular cells [99], no difference has been seen between chambers in rainbow trout [98]. However, with loss of function of NCX in mutant zebrafish (tre mutant) more cardiac fibrillation is seen in the atrium than the ventricle [100]. This suggests that the atrium is more susceptible to spontaneous Ca^{2+} release from the SR as a result of disruptions of normal calcium transients.

5.5. Relaxation of contraction

Under steady-state conditions, the calcium delivered to the myofilaments during the activation of contraction has to be removed from the cytosol during relaxation. Moreover, if equilibrium is to be maintained, the calcium entering across the sarcolemma must be extruded from the cell and calcium released from intracellular organelles must be re-sequestered by the same organelle. Based on the estimates of sarcolemmal calcium entry in the fish heart, sarcolemmal calcium extrusion would therefore be expected to vary between 50 and 100% in fish cardiomyocytes depending on the species and tissue. In the mammalian heart, forward mode NCX generally accounts for the vast majority of the sarcolemmal calcium extrusion [101] and experiments using rapid caffeine application in 0Na/0Ca-solution showed that this is also true for trout ventricular myocytes [84]. Assuming that the tail current elicited upon repolarization of the membrane potential is primarily carried by forward mode NCX, the time integral of this current can be used to estimate calcium extrusion by the NCX [75]. With this approach calcium extrusion by the NCX was estimated to account for about 55% of the total increase in calcium during activation of contraction suggesting that intracellular recycling of calcium during contraction must account for about 45% of the total calcium transient [86]. On the other hand, measurements of the NCX activity showed a forward NCX-rate under physiological conditions of 36 amol pF^{-1} s^{-1}, which is sufficient to remove the total calcium required for the activation of contraction in a few hundred ms [76].

In agreement with the notion that calcium cycling by the SR takes place on a beat-to-beat basis and could contribute to the activation of contraction, measurements of SR calcium load with rapid caffeine application and specific protocols used to measure steady-state and maximal SR calcium loading, showed that the SR calcium content in trout ventricular and atrial myocytes can be up to 10 times higher than in mammalian myocytes [58,61,77,84], and that the rate of SR calcium uptake under physiological conditions is also sufficient to remove the total calcium required for the activation of contraction between beats at physiological beating rates and temperatures in trout atrial and ventricular myocytes [75,95]. In the presence of β-adrenergic stimulation, the relationship between SR calcium uptake and the cytosolic calcium level was shifted towards lower calcium concentrations, but when normalized to the total calcium transport SR calcium uptake was only slightly increased from 35 to 41% by β-adrenergic stimulation [102].

It is interesting to note that the SR calcium load in fish cardiomyocytes is more than 10-fold higher than the amount of calcium released on each beat [84, 95, 103]. The reason for such an excessive SR calcium load compared to the calcium requirements on a beat-to-beat basis remains elusive, but it definitely provides fish species with an intracellular calcium reservoir that allows the heart to function under adverse conditions where SR calcium refilling is impaired for extended periods of time. This also explains why the inhibition of SR calcium uptake (with cyclopazonic acid) or increasing SR calcium leak (with ryanodine) may have a much slower effect on contraction in fish than in mammalian hearts.

Atrial cardiomyocytes have both greater SR Ca^{2+} loads [105-106] and make greater use of SR Ca^{2+} than ventricular cardiomyocytes [60, 105]. The degree of atrio-ventricular difference in SR Ca^{2+} load and usage is also species specific, being larger in rainbow trout than crucian carp or burbot [105].

The reliance on intracellular Ca^{2+} leads to more rapid contraction kinetics of the fish atrium due to faster calcium release and reuptake rates via the SR relative to ventricular tissue in a variety of teleost species (e.g. rainbow trout - [60]; burbot - [105]). Lowering of cytosolic Ca^{2+} concentration in this manner is via sarco-endoplasmic reticulum Ca^{2+}-ATPase (SERCA), and thus the higher SERCA protein expression found in atrial contributes to this shorter duration of isometric contraction relative to ventricular myocardium [107]. In addition, the strength of contraction in teleost atrial muscle has greater ryanodine sensitivity than that in ventricles [60, 62] and is similar in magnitude to what is typically seen in many mammals [82].

Interestingly, the maximal capacity and steady-state SR Ca^{2+} load does not necessarily correlate with SERCA activity or ryanodine sensitivity in all ectotherms. Inverse relationships between ryanodine sensitivity and SR Ca^{2+} load have been found in rainbow trout, burbot and crucian carp [105]. While atrial contraction is more strongly inhibited by ryanodine than ventricular contraction [60], the increased activity of SR Ca^{2+} ATPase may indicate that atria are able to load the SR with Ca^{2+} in the time between contractions. These factors may contribute to a faster recovery of myocardial contractility or increased mechanical restitution rates in atrial muscle. This may mean increased importance of SR Ca^{2+} for contraction during periods of environmental stress. More studies are necessary to clarify the role of SR Ca^{2+} handling in fish cardiomyocytes, especially with the species-specific and chamber-specific vagaries.

Differences in chamber-specific expression of the contractile element responsible for calcium sensitivity may also guide differences between contractility of the atrium and ventricle. Myofilament Ca^{2+} sensitivity is a critical factor in regulating cardiac contractility. The cardiac TnT subunit of the zebrafish troponin complex has been shown to be required for the contraction of both the atrium and the ventricle [106] as it is in mammals. However, zebrafish possess two copies of the gene for cardiac troponin C, with one demonstrated to be predominantly expressed in the heart of the zebrafish embryo. Loss of function of this gene results in abnormal atrial and ventricular chamber morphology, but the second cTnC gene enables function to be regained in the atrium [107]. This suggests a possible difference in expression of cTnC between chambers that could potentially lead to variation in calcium sensitivity. Gene duplication has led to multiple copies of sarcomeric proteins, and variation between these fish-specific paralogs may be a method of coping with varying environmental conditions.

6. Environmental influences

The interspecific variation of fish cardiac anatomy and physiology is guided by evolutionary adaptation to different habitats, modes of life and activity levels. Because ectothermic fish are exposed to different temperatures either seasonally or via the vertical thermocline, temperature-dependent regulation of cardiac contractility is crucial. Many fish face unpredictable temperature changes, meaning that fish hearts must have intrinsic mechanisms or rely on post-translational modifications to protect against acute temperature changes when there is not enough time for transcriptional regulation to take effect [108, 34].

Cardiac performance may change through independent changes in stroke volume and heart rate, with relative contributions of each varying substantially among fish. During exercise, more athletic fish modulate cardiac output preferentially by increasing SV, while less active species preferentially modulate HR in different species [54-55]. With environmental hypoxia, reflex bradycardia occurs and increased stroke volume is necessary to maintain cardiac output [109-110]. Higher temperatures have been shown to increase heart rate while colder temperatures decrease stroke volume in crucian carp [10]. For this review, the impact of temperature on cardiac contractility will be focused on. Mechanisms of control of both stroke volume and heart rate must be considered when looking at temperature sensitivity of the fish heart.

There appears to be seasonal variation with the activity of pacemaker cells (marine plaice - [111]). Heart rate is maintained despite the lower temperatures without the need for extrinsic modification. Changes in heart rate seen with temperature are not solely dependent on pacemaker cells. Therefore, changes that are seen may be due to ion channel temperature sensitivity. Chronic temperature stress can also modify K^+ channel conductances and reduce action potential duration to enable higher heart rates (trout - [112]). Thermal acclimation modifies functional properties and subunit composition of K_{IR2} channels [113]. Cold temperatures induce decreased I_{K1}, possibly increasing the excitability of cardiomyocytes but increased I_{Kr} leading to limited AP duration and decreased refractoriness of the heart [114]. These changes in I_K appear when it is the predominant current, namely in ventricular cardiomyocytes [66] while the changes in I_{Kr} appear in both atrial and ventricular cardiomyocytes [114]. The density in particular of ventricular I_{K1} increases in warm-acclimated vs. cold acclimated fish (trout - [113]).

The plasticity of E-C coupling in fish has been suggested to be based, at least in part, on these temperature-induced changes in I_K [114]. These modifications lead to changes in Ca^{2+} influx and hence variation in E-C coupling. Ca^{2+} pumps and ion channels themselves are also known to be sensitive to acute temperature change [103]. The large SR calcium stores and low calcium sensitivity of ryanodine receptors have been proposed to play a critical role in maintaining contraction at lower temperatures than mammals [81]. Higher SR Ca^{2+} content may be necessary to initiate and maintain SR Ca^{2+} release events at lower temperatures [80,115].

Different species may have variations in response to temperature depending on their evolutionary history. Eurythermal fish such as trout must be able to cope with seasonal temperature fluctuations as well as more acute temperatures through the thermal gradient. Other stenothermal species are only able to cope with a more limited temperature range, though this range may be considered on the extreme end of the range for a eurythermal fish (e.g. Antarctic fish). Interspecies variation in excitation-contraction coupling seen with temperature acclimation may be indicative of the capacity for thermal niche expansion. This substantial variation in SR Ca^{2+} release relative to SL Ca^{2+} influx ratio exists between species as well as being dependent on acute and chronic temperature changes. Cold-tolerant active fish have a larger capacity to store Ca^{2+} in the SR than mammals (e.g. active teleosts - [115]; scombroids - [116]). The atrio-ventricular differences in Ca^{2+} loading are also more pronounced in cold-acclimated than warm acclimated fish (trout - [112].

However, the contribution of the SR to cytosolic Ca^{2+} management varies with temperature in fish hearts. With cold acclimation, SR function is also enhanced more strongly in atrial than ventricular myocytes (trout - [60]). More cold tolerant species may also possess greater SR Ca^{2+} content to begin with (bluefin tuna and mackerel - [112]). The degree of modification of I_{Ca} to offset the negative effects of colder temperatures appears to be species specific, with many teleosts such as trout demonstrating relatively temperature-insensitive Ca^{2+} flux [103] whereas certain scombroids show significant reductions in I_{Ca} with decreasing temperatures [112]. This temperature sensitivity may differ between atrial and ventricular cardiomyocytes, with ventricular maximum SR Ca^{2+} load varying only in the ventricle of most scombrids [112].

As previously mentioned, isoform compositional changes can lead can lead to variation in contractility. Modulation of MHC isoform expression patterns with temperature also leads to changes in myofibrillar ATPase activity (crucian carp - [10]). This may also be a species-related difference, since not all species display the same changes. The degree to which each chamber varies is still unclear. Differing paralog usage between the chambers may be linked to temperature-induced Ca^{2+} sensitivity differences. Examples of this may exist in important of the contractile element. Troponin I, the inhibitory subunit of troponin, has multiple paralogs expressed in cardiac tissue. The relative expression of these paralogs in the ventricle shifts with temperature, possibly leading to changes in the calcium affinity of the entire troponin complex [117]. However, while paralog profiles vary between tissues (heart, slow skeletal) no information is known about the atrium. TnC isoform differences between chambers [107] may also lend insight into Ca^{2+} sensitivity changes with temperature as the affinity of cTnC in trout is temperature dependent [118]. As previously mentioned, the very presence of multiple fish-specific paralogs of important components of the contractile element may be indicative of how fish have evolved mechanisms to thrive in variable environments. Incorporating the chamber-specific variation with temperature response reveals the complexity of these mechanisms in achieving survival.

7. Summary

The unique structure of the fish heart accommodates the variability in function that have allowed species to exploit many different environments. The increased importance of the atrial contraction in guiding changes in cardiac output indicates that some of this physiological versatility may be due to atrio-ventricular differences.These two main chambers of the fish heart differ in terms of contractile properties from the molecular level of E-C coupling and AP morphology, and the whole heart morphology and function reflect this. The fish heart is representative of early embryological stages of higher vertebrates, these studies on both the development and functioning of chamber-specific properties lends insight into proper functioning of the heart not only in fish. However, the degree of atrio-ventricular differences vary between species making it very difficult to generalize species-specific results across phylogeny. The species-specific differences in these two chambers lend insight into how evolutionary history guides responses to environmental factors such as ambient temperature. Fish-specific genome duplications may have allowed for multiple chamber-specific genes with variable functions that allow for flexibility in function. Further work is required to determine how genome duplication has shaped the structure and function of the fish heart and to clarify how this wide-range of species-specific phenotypes has been maintained.

Author details

Christine Genge[1], Leif Hove-Madsen[2] and Glen F. Tibbits[1,3*]

*Address all correspondence to: tibbits@sfu.ca

1 Molecular Cardiac Physiology Group, Simon Fraser University, Burnaby, Canada

2 Cardiovascular Research Centre CSIC-ICCC, Hospital de Sant Pau, Barcelona, Spain

3 Cardiovascular Sciences, Child and Family Research Institute, Vancouver, Canada

References

[1] Moorman, A.F. and Christoffels, V.M. (2003) Cardiac chamber formation: develop‐ ment, genes and evolution. Physiological Reviews 83: 1223-1267

[2] Satin, J., Fujii, S. and DeHaan, R. (1988) Development of cardiac beat rate in early chick embryos is regulated by regional cues. Developmental Biology 129: 103-113.

[3] Tabibiazar, R., Wagner, R., Liao, A. and Quertermous, T. (2003) Transcriptional profiling of the heart reveals chamber-specific gene expression patterns. Circulation Research 93: 1193-1201.

[4] Franco, D., Gallego, A., Habets, P., Sans-Coma, V. and Moorman, A. (2002) Species‐ specific differences of myosin content in the developing cardiac chambers of fish, birds and mammals. Anatomical Record 268: 27-37.

[5] Evans, S., Yelon, D., Conlon, F. and Kirby, M. (2010) Myocardial lineage develop‐ ment. Circulation Research 107: 1428-1444.

[6] Stanier,D. and Fishman, M. (1992) Patterning the zebrafish heart tube: acquisition of anteroposterior polarity. Developmental Biology 153: 91-101.

[7] de Pater, E., Clijsters, L., Marques, S., Lin, Y., Garavito-Aquilar, Z., Yelon, D. and Bakkers, J. (2009) Distinct phases of cardiomyocyte differentiation regulate growth of the zebrafish heart. Development 136, 1633-1641.

[8] Warren, K., Wu, J., Pinet, F. and Fishman, M. (2000) The genetic basis of cardiac func‐ tion: dissection by zebrafish (Danio rerio) screens. Phil. Trans. Royal Society of Lon‐ don B 355: 939-944.

[9] Yelon, D., Horne, S. and Stainier, D. (1999) Restricted expression of cardiac myosin genes reveals regulated aspects of heart tube assembly in zebrafish. Developmental Biology 214: 23-37.

[10] Vornanen, M. (1994) Seasonal adaptation of crucian carp (Carassius carassius L.) heart: glycogen stores and lactate dehydrogenase activity. Canadian Journal of Zoology, 1994, 72(3): 433-442.

[11] Berdougo, E., Coleman, H., Lee, D.H., Stanier, D.Y. and Yelon, D. (2003) Mutation of weak atrium/atrial myosin heavy chain disrupts atrial function and influences ventricular morphogenesis in zebrafish. Development 130: 6121-6129.

[12] Marques, S. and Yelon, D. (2009) Differential requirement for BMP signaling in atrial and ventricular lineages establishes cardiac chamber proportionality. Developmental Biology 328: 432-482.

[13] Cotter, P.A. Han, A.J., Everson, J.J. and Rodnick, K.J. (2008) Cardiac hemodynamics of the rainbow trout (Oncorhynchus mykiss) using simultaneous Doppler echocardiography and electrocardiography. Journal of Experimental Zoology Part A: Ecological Genetics and Physiology 309: 243-254.

[14] Schib, J.L., Icardo, J.M., Durán, A.C., Guerrero, A., López, D., Colvee, E., de Andrés, A.V. and Sans-Coma, V. (2002) The conus arterious of the adult gilthead seabream (Sparus auratus). Journal of Anatomy 201:395–404.

[15] Icardo, J.M. (2006) Conus arteriosus of the teleost heart: dismissed, but not missed. Anatomical Record A 288: 900-908.

[16] Santer, R. and Cobb, J. (1972) The fine structure of the heart of the teleost, Pleuronectes platessa L. Zeitschrift für Zellforschung und mikroskopische Anatomie 131: 1-14.

[17] Yamauchi, A. (1980) Fine structure of the fish heart. In: Bourne G (ed) Heart and Heart-like Organs, vol 1. Academic, New York, pp 119–148.

[18] Farrell A.P. and Jones, D.R. (1992) The heart. In: Hoar WS, Randall DJ, Farrell AP (eds) Fish physiology, Vol XII, The cardiovascular system, Part A. Academic, San Diego, pp 1–87.

[19] Johansen, K. (1965) Cardiovascular dynamics in fishes, amphibians, and reptiles. Annals of the New York Academy of Sciences 127 Comparative Cardiology: 414-442.

[20] Forster, M. and Farrell, A. (1994) The volumes of the chambers of the trout heart. Comparative Biochemisty and Physiology A.109: 127-132.

[21] Sudak, F.N. (1965) Intrapericardial and intercardiac pressures and the events of the cardiac cycle in Mustelus canis (Mitchill). Comparative Biochemical Physiology 14, 689-705.

[22] Hu, N., Yost, H.J. and Clark, E. (2001) Cardiac morphology and blood pressure in the adult zebrafish. Anatomical Record 264: 1-12.

[23] Icardo, J.M. and Colvee, E. (2011) The atrioventricular region of the teleost heart. A distinct heart segment. Anatomical Record 294: 236-242.

[24] Icardo, J.M., Colvee, E., Cerra, M. and Tota, B. (2002) The structure of the conus arteriosus of the sturgeon (Acipenser naccarii) heart II. The myocardium, the subepicardium, and the conus-aorta transition. Anatomical Record 268: 388-398.

[25] Icardo, J.M. (2012) The teleost heart: A morphological approach. In Ontogeny and Phylogeny of the Vertebrate Heart ed. Wang, T. and Sedmera D. Springer-Verlag New York Inc.

[26] Santer, R.M., Walker, M.G., Emerson, L. and Witthames, P.R. (1983) On the morphology of the heart ventricle in marine teleost fish (Teleostei). Comparative Physiology and Biochemistry 76A:453–457.

[27] Icardo, J., Imbrogno, S., Gattuso, A., Colvee, E. and Tota, B. (2005) The heart of Sparus auratus: a reappraisal of cardiac functional morphology in teleosts. Journal of Experimental Zoology 303A: 665-675.

[28] Mendonca, P., Genge, A., Deitch, E. and Gamperl, A. (2007) Mechanisms responsible for the enhance pumping capacity of the in situ winter flounder heart (Pseudopleuronectes americanus). American Journal of Physiology Regulatory and Integrative Comparative Physiology 293:2112-2119.

[29] Tota, B. (1989) Myoarchitecture and vascularization of the elasmobranch heart ventricle. Journal of Experimental Zoology:122–135.

[30] Rohr, S., Otten, C. and Abdelilah-Seyfried, S. (2008) Asymmetric involution of the myocardial field drives heart tube formation in zebrafish. Circulation Research 102: 12-19.

[31] Vornanen, M. (1998) L-type Ca^{2+} current in fish cardiac myocytes: effects of thermal acclimation and β-adrenergic stimulation. Journal of Experimental Biology 201, 533–547.

[32] Huang, J., Hove-Madsen, L. and Tibbits, G. (2004) Na^+/Ca^{2+} exchange activity in neonatal rabbit ventricular myocytes. American Journal of Physiology: Cell Physiology 288: C195-C203.

[33] Santer, R.M. (1974) The organization of the sarcoplasmic reticulum in teleost ventricular myocardial cells. Cell and Tissue Research 151: 395-402.

[34] Vornanen, M., Shiels, H. and Farrell, A. (2002) Plasticity of excitation-contraction coupling in fish cardiac myocytes. Comparative Biochemical Physiology 132A, 827-846.

[35] Saetersdal, T.S., Justesen, N.P. and Krohnstad, A.W. (2003) Ultrastructure and innervation of the teleostean atrium. Journal of Molecular and Cellular Cardiology 6: 415-437.

[36] Tibbits G.F., Moyes C.D. and Hove-Madsen L. (1992) Excitation–contraction coupling in the teleost heart. In: Hoar W.S., Randall D.J., Farrell A.P. (eds) Fish physiology. Academic Press, New York, pp 267–296.

[37] Vornanen, M. (1997). Sarcolemmal Ca^{2+} influx through L-type Ca channels in ventricular myocytes of a teleost fish. American Journal of Physiology 272, R1432–R1440.

[38] Di Maio, A. and Block, B.A. (2008) Ultrastructure of the sarcoplasmic reticulum in cardiac myocytes from Pacific bluefin tuna. Cell and Tissue Research 334:121-34.

[39] Brill, R., and Bushnell, P. 1991 Effects of open and closed system temperature changes on blood oxygen dissociation curves of yellowfin tuna (Thunnus albacares) and skipjack tuna (Katsuwonus pelamis). Canadian Journal of Zoology 69:1814-1821.

[40] Santer, R.M. (1985) Morphology and innervation of the fish heart. Advances in Anatomy, Embryology and Cell Biology 89, 1–102.

[41] Simoes, K., Vicentini, C.A., Orsi, A.M. and Cruz, C. (2002) Myoarchitecture and vasculature of the heart ventricle in some freshwater teleosts. Journal of Anatomy 200:467–475.

[42] Claireaux, G., McKenzie, D., Genge, A., Chatelier, A., Aubin, J. and Farrell, A. (2005) Linking swimming performance, cardiac pumping ability and cardiac anatomy in rainbow trout. Journal of Experimental Biology 208: 1775-1784.

[43] Gamperl, A. and Farrell, A. (2004) Cardiac plasticity in fishes: environmental influences and intraspecific differences. Journal of Experimental Biology 207: 2539-2550.

[44] Cerra, M.C., Imbrogno, S., Amelio, D., Garofalo, F., Colvee, E., Tota, B., and Icardo, J.M. (2004) Cardiac morphodynamic remodelling in the growing eel (Anguilla Anguilla L.). Journal of Experimental Biology 207:2867–2875.

[45] Graham, M.S. and Farrell, A.P. (1990) Myocardial oxygen consumption in trout acclimated to 5°C and 15°C. Physiological Zoology 63:536-554.

[46] Davie, P. and Farrell, A. (1991) The coronary and luminal circulations of the myocardium of fishes. Canadian Journal of Zoology 69: 1993-2001.

[47] Forster, M., Axelsson, M., Farrell, A. and Nilsson, S. (1991) Cardiac function and circulation in hagfishes. Canadian Journal of Zoology 69: 1985-1992.

[48] Minerick, A., Chang, H., Hoagland, T. and Olsen, K. (2003) Dynamic synchronization analysis of venous pressure-driven cardiac output in rainbow trout. American Journal of Physiology Regulatory and Integrative Comparative Physiology 285: 889-896.

[49] Lai, N., Shebatai, R., Graham, J., Hoit, B., Sunndrhagen, K. and Bhargava, V. (1990) Cardiac function of the leopard shark, Triakis, semifasciatus. Journal of Comparative Physiology: Biochemical and System Environmental Physiology 160: 259-268.

[50] Farrell, A.P. (1984). A review of cardiac performance in the teleost heart: intrinsic and humoral regulation. Canadian Journal of Zoology 62, 523-536.

[51] Sandblom, E., Cox, G., Perry, S. and Farrell, A. (2009) The role of venous capacitance, circulating catecholamines and heart rate in the hemodynamic response to increased

temperature and hypoxia in the dogfish. American Journal of Physiology: Regulatory and Integral Comparative Physiology 296: 1547-1556.

[52] Sandblom, E. and Axelsson, M. (2006) Adrenergic control of venous capacitance during moderate hypoxia in the rainbow trout (Oncorhynchus mykiss): role of neural and circulating catecholamines. American Journal of Physiology: Regulatory and Integral Comparative Physiology 291: 711-718.

[53] Kiceniuk, J.W. and Jones, D.R. (1977) The oxygen transport system in trout (Salmo gairdneri) during exercise. Journal of Experimental Biology 69, 247-260.

[54] Cooke, S., Grant, E., Schreer, J. Philipp, D. and DeVries, A. (2003) Low temperature cardiac response to exhaustive exercise in fish with different levels of winter quiescence. Comp. Biochem. Physiol. Mol. Integr. Physiol. 134: 157-165.

[55] Clark, T. and Seymour, R. (2006) Cardiorespiratory physiology and swimming energetics of a high-energy demand teleost, the yellowtail kingfish (Seriola lalandi). Journal of Experimental Biology 209: 3940-2951.

[56] Franklin, C. E. and Davie, P. S. (1992). Dimensional analysis of the ventricle of an in situ perfused trout heart using echocardiography. Journal of Experimental Biology 166, 47-60.

[57] Coucelo, J., Joaquim, N. and Couvelo J. (2000) Calculation of volumes and systolic indices of heart ventricle from Halobatrachus didactylus: echocardiographic noninvasive method. Journal of Exp. Zool. 286: 585-595.

[58] Shiels, H.A. and White, E. (2005) Temporal and spatial properties of cellular Ca2+ flux in trout ventricular myocytes. American Journal of Physiology: Regulatory and Integral Comparative Physiology 288: R1756-R1766.

[59] Patrick, S., White, E. and Shiels, H. (2010) Rainbow trout myocardium does not exhibit a slow inotropic response to stretch. Journal of Experimental Biology 214: 1118-1122.

[60] Aho, E. and Vornanen, M. (1999) Contractile properties of atrial and ventricular myocardium of the heat of the rainbow trout Oncorhynchus mykiss: effects of thermal acclimation. Journal of Experimental Biology 202: 2663-267.

[61] Shiels H.A., Freund E.V., Farrell A.P. and Block B.A. (1999) The sarcoplasmic reticulum plays a major role in isometric contraction in atrial muscle of yellowfin tuna. Journal of Experimental Biology 202:881–890.

[62] Mercier, C., Axelsson, M., Imbert, N., Claireaux, G., Lefrancois, C., Altimiras, J. and Farrell A.P. (2002) In vitro cardiac performance in triploid brown trout at two acclimation temperatures. Journal of Fish Biology 60: 117-133.

[63] Karasinski, J. (1993) Diversity of native myosin and myosin heavy chain in fish skeletal muscles. Comparative Biochemisty and Physiology Part B: Comparative Biochemistry 106: 1041-1047.

[64] Bottinelli, R., Canepari, M., Pellegrino, M. and Reggiani, C. (1995) Force-velocity properties of human skeletal muscle fibres: myosin heavy chain isoform and temperature dependence. Journal of Physiology 495: 573-586.

[65] Karasinski, J., Sokalski, A. and Kilarski, W. (2001) Correlation of myofibrillar ATPas activity and myosin heavy chain content in ventricular and atrial myocardium of fish heart. Folia Histochem Cytobiol 39: 23-28.

[66] Vornanen, M., Ryokkynen, A. and Nurmi, A. (2002) Temperature-dependent expression of sarcolemmal K^+ currents in rainbow trout atrial and ventricular myocytes. American Journal of Physiology Regulatory and Integrative Comparative Physiology 282: R1191-R1199.

[67] Galli, G., Lipnick, M. and Block, B. (2009) Effect of thermal acclimation on action potentials and sarcolemmal K^+ channels from Pacific bluefin tuna cardiomyocytes. American Journal of Physiology: Regulatory and Integrative Comparative Physiology 297:502-509.

[68] Vornanen, M. and Tuomennoro, J. (1999) Effects of acute anoxia on heart function in crucian carp: effects of cholinergic and purinergic control. American Journal of Physiology 277: R465-R475.

[69] Nemtsas, P., Wettwer, E., Christ, T., Weidinger, G. and Ravens, U. (2009) Adult zebrafish heart as a model for human heart? An electrophysiological study. Journal of Molecular and Cellular Cardiology 48:161-71.

[70] Moller-Nielsen, T. and Gesser, H. (1992) Sarcoplasmic reticulum and excitation contraction coupling at 20 and 10 degrees in rainbow trout myocardium. Journal of Comparative Physiology B 162: 526-534.

[71] Molina, C., Gesser, H., Llach, A. and Hove-Madsen, L. (2007) Modulation of membrane potential by an acetylcholine-activated potassium current in trout atrial myocytes. American Journal of Physiology: Regulatory and Integrative Comparative Physiology 292: R388-R395.

[72] Haverinen, J. and Vornanen, M. (2009) Responses of action potential and K^+ currents to temperature acclimation in fish hearts: phylogeny or thermal preferences? Physiological Biochemical Zoology 82: 468-482.

[73] Tibbits G. F., Philipson K. D. and Kashihara H. (1992) Characterization of myocardial Na^+-Ca^{2+} exchange in rainbow trout. American Journal of Physiology: Cell Physiology 262: C411–C417.

[74] Xue, X., Hryshko, L., Nicoll, D., Philipson, K.D. and Tibbits, G.F. (1999) Cloning, expression and characterization of the trout cardiac Na^+/Ca^{2+} exchanger

[75] Hove-Madsen, L. and Tort, L. (1998). L-type Ca2+ current and excitation- contraction coupling in single atrial myocytes from rainbow trout. American Journal of Physiology: Regulatory and Integrative Comparative Physiology 275, R2061-R2069.

[76] Hove-Madsen, L. and Tort, L. (2001). Characterization of the relationship between Na^+-Ca^{2+} exchange rate and cytosolic calcium in trout cardiac myocytes. Pflugers Arch. 441, 701-708.

[77] Shiels, H., Vornanen, M. and Farrell, T. (2002) Effects of temperature on intracellular $[Ca^{2+}]$ in trout atrial myocytes. Journal of Experimental Biology 205: 3641-3650.

[78] Hove-Madsen, L. and Gesser, H. (1989) Force frequency relation in the myocardium of rainbow trout. Effects of K^+ and adrenaline. Journal of Comparative Physiology B 159: 61-69.

[79] Fabiato, A. (1983) Calcium-induced release of calcium from the cardiac sarcoplasmic reticulum. American Journal of Physiology: Cell Physiology 245, C1-C14.

[80] Hove-Madsen, L., Llach, A. and Tort, L. (1999) Quantification of calcium release from the sarcoplasmic reticulum in rainbow trout atrial myocytes. Pflügers Arch 438: 545–552.

[81] Shiels, H. A., Thompson, S. and Block, B. A. (2011) Warm fish with cold hearts: thermal plasticity and excitation-contraction coupling in bluefin tuna. Proceedings of the Royal Society of London B: Biological Sciences 278, 18-27.

[82] Bers, D. (2002) Cardiac excitation-contraction coupling. Nature 415, 198-205.

[83] Vornanen, M. (1997). Sarcolemmal Ca^{2+} influx through L-type Ca channels in ventricular myocytes of a teleost fish. American Journal of Physiology: Regulatory and Integrative Comparative Physiology 272, R1432–R1440.

[84] Hove-Madsen, L., Llach, A. and Tort, L. (1998) Quantification of Ca^{2+} uptake in the sarcoplasmic reticulum of trout ventricular myocytes. American Journal of Physiology: Regulatory and Integrative Comparative Physiology 275: R2070-R2080.

[85] Haverinen, J. and Vornanen, M. (2005) Significance of Na^+ current in the excitability of atrial and ventricular myocardium of the fish heart. Journal of Experimental Biology 209: 549-557.

[86] Hove-Madsen, L., Llach, A. and Tort, L. (2000) Na^+/Ca^{2+} exchange activity regulates contraction and SR Ca^{2+} content in rainbow trout atrial myocytes. American Journal of Physiology: Regulatory and Integrative Comparative Physiology 279:R1856-R1864.

[87] Zhang, P.C., Llach, A., Sheng, X.Y., Hove-Madsen, L. and Tibbits, G.F. (2011) Calcium handling in zebrafish ventricular myocytes. American Journal of Physiology: Regulatory and Integrative Comparative Physiology 300: R56-R66.

[88] Shiels, H.A., Vornanen, M. and Farrell, A.P. (2003) Acute temperature change modulates the response of I_{Ca} to adrenergic stimulation in fish cardiomyocytes. Physiological and Biochemical Zoology 76(6):816–824.

[89] Vornanen M (1999) Na^+/Ca^{2+} exchange current in ventricular myocytes of fish heart: contribution to sarcolemmal Ca^{2+} influx. Journal of Experimental Biology 202:1763–1775.

[90] Hove-Madsen, L., Llach, A., Tibbits, G.F. and Tort, L. (2003) Triggering of sarcoplasmic reticulum Ca^{2+} release and contraction by reverse mode Na^+/Ca^{2+} exchange in trout atrial myocytes. American Journal of Physiology: Regulatory and Integrative Comparative Physiology 284: R1330–R1339.

[91] Driedzic, W. and Gesser, H. (1988) Differences in force-frequency relationships and calcium dependency between elasmobranch and teleost hearts. Journal of Experimental Biology 140, 227-241.

[92] el-Sayed, M.F and Gesser, H. (1989) Sarcoplasmic reticulum, potassium, and cardiac force in rainbow trout and plaice. American Journal of Physiology: Regulatory and Integrative Comparative Physiology 257, R599–R604, 1989.

[93] Hove-Madsen, L. (1992) The influence of temperature on ryanodine sensitivity and the force-frequency relationship in the myocardium of rainbow trout. Journal of Experimental Biology 167: 47-60.

[94] Keen, J.E., Farrell, A.P. Tibbits, G.F. and Brill, R.W. (1992) Cardiac physiology in tunas II: Effect of ryanodine, calcium, and adrenaline on force–frequency relationships in atrial strips from skipjack tuna, Katsuwonus pelamis. Canadian Journal of Zoology 70(6): 1211-1217.

[95] Hove-Madsen, L., Llach, A. and Tort, L. (2001) The function of the sarcoplasmic reticulum is not inhibited by low temperatures in trout atrial myocytes. American Journal of Physiology: Regulatory and Integrative Comparative Physiology 281: R1902-R1906.

[96] Llach, A., Molina, C., Alvarez-Lacalle, E., Tort, L., Benitez, R. and Hove-Madsen, L. (2011) Detection, properties, and frequency of local calcium release from the sarcoplasmic reticulum in teleost cardiomyocytes. PLoS ONE 6(8): e23708. doi:10.1371/journal.pone.0023708.

[97] Christ T., Boknik P., Wohrl S. (2004) L-type Ca^{2+} current down-regulation in chronic human atrial fibrillation is associated with increased activity of protein phosphatases. Circulation. 110:2651–2657.

[98] Birkedal, R. and Shiels, H. (2007) High $[Na^+]_i$ in cardiomyocytes from rainbow trout. American Journal of Physiology Regulatory and Integrative Comparative Physiology 293: 861-866.

[99] Wang, G., Schmied, R., Ebner, F. and Korth, M. (1993) Intracellular sodium activity and its regulation in guinea-pig atrial myocardium. Journal of Physiology 465: 73-84.

[100] Langenbacher, A., Dong, Y., Shu, X., Choi, J., Nicoll, D., Goldhaber, J., Philipson, K. and Chen J. (2005) Mutation in sodium-calcium exchanger 1 (NCX1) causes cardiac fibrillation in zebrafish. Proceedings of the National Academy of Sciences of the United States of America 102: 17699-17704.

[101] Bassani J., Bassani R. and Bers D.M. (1994) Relaxation in rabbit and rat cardiac cells: species-dependent differences in cellular mechanisms. Journal of Physiology 476: 279-293.

[102] Llach, A., Huang, J., Sederat, F., Tort, L., Tibbits, G.F. and Hove-Madsen, L. (2004) Effect of beta-adrenergic stimulation on the relationship between membrane potential intracellular [Ca^{2+}] and sarcoplasmic reticulum Ca2+ uptake in rainbow trout atrial myocytes. Journal of Experimental Biology 207: 1369-1377.

[103] Shiels H.A., Vornanen M. and Farrell A.P. (2000) Temperature-dependence of L-type Ca^{2+} channel current in atrial myocytes from rainbow trout. Journal of Experimental Biology 203:2771–2780.

[104] Rantin, F., Guerra, C., Kalinin, A. and Glass, M. (1998) The influence of aquatic surface respiration (ASR) on the cardio-respiratory function of the serrasalmid fish Piaractus mesopotamicus. Comparative Biochemistry and Physiology 119A: 991-997.

[105] Haverinen, J. and Vornanen, M. (2009) Comparison of sarcoplasmic reticulum calcium content in atrial and ventricular myocytes of three fish species. American Journal of Physiology: Regulatory and Integral Comparative Physiology 297, R1180-R1187.

[106] Sehnert, A.J., Huq, A., Weinstein, B.M., Walker, C., Fishman, M. and Stanier, D.Y. (2002) Cardiac troponin T is essential in sarcomere assembly and cardiac contractility. Nature Genetics 31: 106-110.

[107] Sogah, VM, Serluca, FC, Fisherman, MC, Yelon, DL, Macrae, CA and Mably, JD (2010) Distinct troponin C isoform requirements in cardiac and skeletal muscle. Developmental Dynamics 239 (11), 3115-3123.

[108] Tiitu, V. and Vornanen, M. (2003) Ryanodine and dihydropyridine receptor binding in ventricular cardiac muscle of fish with different temperature preferences. Journal of Comparative Physiology B 173, 285-291.

[109] Gamperl, A.K. and Driedzic, W.R. (2009) The cardiovascular systems. In Fish physiology vol 27 edited by J.G. Richards, A.P. Farrell and C.J. Brauner. Academic Press, San Diego, Calif. pp.487-503.

[110] Perry, S. and Desforges, P. (2006) Does bradycardia or hypertension enhance gas transfer in rainbow trout (Oncorhynchus mykiss)? Comparative Biochemistry and Physiology 144A: 163-172.

[111] Harper, A., Newton I. and Watt, P. (1995) The effect of temperature on spontaneous action potential discharge of the isolate sinus venosus from winter and summer plaice (Pleuronectes platessa). Journal of Experimental Biology 198: 137-140.

[112] Haverinen, J. and Vornanen, M. (2009) Responses of action potential and K+ currents to temperature acclimation in fish hearts: phylogeny or thermal preferences? Physiological Biochemical Zoology 82: 468-482.

[113] Hassinen, M., Paajanen and Vornanen, M. (2008) A novel inwardly rectifying K+ channel Kir2.5, is upregulated under chronic cold stress in fish cardiac myocytes. Journal of Experimental Biology 211: 2162-2171.

[114] Aho, E. and Vornanen, M. (2002) Cold acclimation increases basal heart rate but decreases its thermal tolerance in rainbow trout (Oncorhynchus mykiss). Journal of Comparative Physiology B 171, 173–179.

[115] Hassinen, M., Laulaja, S., Paajanen, V., Haverinen, J. and Vornanen, M. (2011) Thermal adaptation of the crucian carp (Carassius carassius) cardiac delayed rectifier current, I_{Ks}, by homomeric assembly of K v7.1 subunits without MinK. American Journal of Physiology Regulatory and Integrative Comparative Physiology 301: R255–R265.

[116] Galli, G., Lipnick, M., Shiels, H. and Block, B. (2011) Temperature effects on Ca2+ cycling in scombrid cardiomyocytes: a phylogenetic comparison. Journal of Experimental Biology 214, 1068-1076.

[117] Alderman, S., Klaiman, J., Deck, C. and Gillis, T. (2012) Effect of cold acclimation on troponin I isoform expression in striated muscle of rainbow trout. American Journal of Physiology Regulatory and Integrative Comparative Physiology 303: R168-R176.

[118] Gillis, T.E., Marshall, C.R., Xue, X.H., Borgford, T.J. and Tibbits, G.F. (2000) Ca^{2+} binding to cardiac troponin C: effects of temperature and pH on mammalian and salmonid isoforms. American Journal of Physiology Regulatory and Integrative Comparative Physiology 279:R1707–R1715.

Chapter 8

Ontogenetic Dietary Shifts in a Predatory Freshwater Fish Species: The Brown Trout as an Example of a Dynamic Fish Species

Javier Sánchez-Hernández, María J. Servia,
Rufino Vieira-Lanero and Fernando Cobo

Additional information is available at the end of the chapter

1. Introduction

The brown trout (*Salmo trutta*) is a species of Eurasian origin, but at has become naturalized in many other parts of the world. It has an outstanding socio-economic importance, both in commercial and sport fisheries, and it is frequently used as tourist attraction [1,2]. The case of brown trout is a clear example of a 'dynamic' fish species, as its diet and feeding behaviour can vary greatly among individuals, age classes, seasons and rivers. The composition of brown trout diet is strongly influenced by environmental and biotic factors. For example, water temperature plays an important role, as it influences food intake and the activity of fishes [3], but also the emergence and activity of aquatic insects or other potential prey items. Also water flow rate can be extremely important for drifting feeders such as brown trout, as they regulate food availability. The are many abiotic factors that influence feeding behaviour, but in general, biotic factors such as the locomotor ability of fishes, accessibility, abundance and antipredator behaviour of prey are thought to be the most important factors in the determination of the diet and feeding strategies in fishes. Usually not all the available prey is consumed by the predator, a feature that allows biologists to distinguish between trophic base and trophic niche (Figure 1). The trophic base consists of all potential prey items that the brown trout is able to consume and it is determined by the feeding habits of the fish, the size of the mouth and the anatomical characteristics of its digestive tract. However, the trophic niche is the variety of organisms that are really consumed by the predator, which depends on different factors that play an important role when choosing criteria prey items as, for example, prey abundance, including site-specific prey accessibility, prey size, energetic selection criteria and prey preference. In this context, the trophic niche for brown trout is very flexible and is usually broader in adults than juveniles.

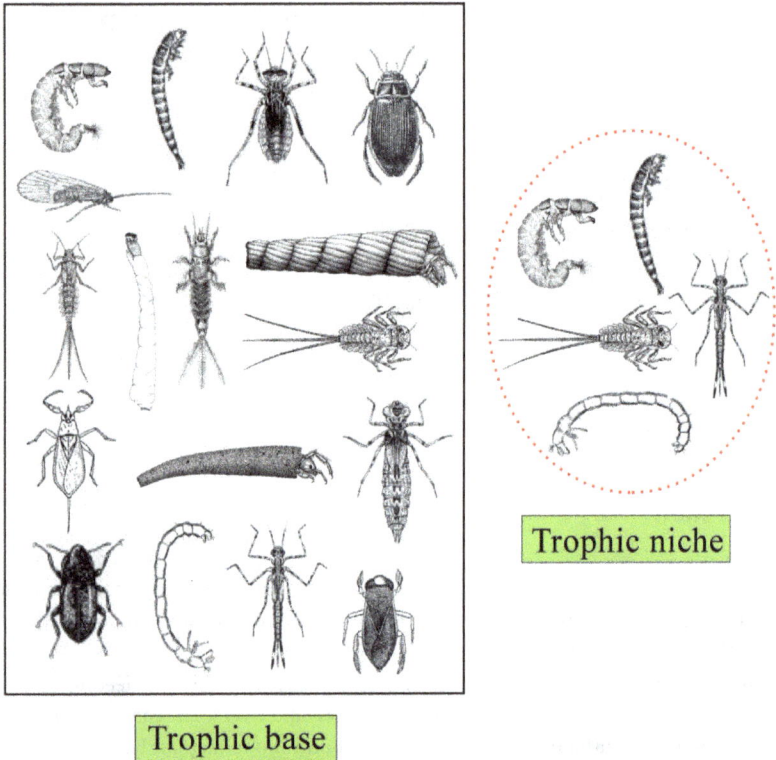

Trophic niche

Trophic base

Figure 1. Graphical example of the trophic base and niche for brown trout.

A knowledge of the foraging ecology of fishes is fundamental to understanding the processes that function at the individual, population and community levels since the factors that influence the acquisition and assimilation of food can have significant consequences for the condition, growth, survival and recruitment of fishes [4]. In this context, the development of effective conservation programmes requires a clear understanding of fish ecological require-ments, so the knowledge of its feeding habits is essential to achieve this objective. For example, the knowledge on how food is shared among individuals of the same population is critical for understanding its functioning. Hence, conclusions of field studies on feeding could help wildlife managers to take measures to preserve fish populations, especially for threatened and exploited species. In this chapter we will briefly discuss the variables that are involved in the feeding behaviour of brown trout as an example of a predatory freshwater fish species.

2. Methodology and types of analysis employed in feeding and ontogenetic dietary shifts studies

The majority of researchers have conducted feeding studies in feral fish populations based on diet descriptions of the stomach contents, using occurrence, numerical, gravimetric and volumetric methods. The main disadvantage of feeding studies is that fish are systematically killed in order to study their stomach contents. However, due to the decline of many natural fish populations, the studies that use non-lethal methods are now more frequent. Different techniques have been used to collect stomach contents without harming the fish such as gastric lavage, emetics or forceps [5-7]. The effectiveness of the gastric lavage is not related to the size of the trouts, but rather to the prey's own morphological characteristics, the degree of repletion of the stomach and the extent of digestion of the food [8]. The effectiveness of this method is inversely related to the degree of repletion [8].

2.1. Prey selection analysis

Prey selection is an important part of fish feeding ecology. In order to study prey selection of fishes, several indices have been employed, such as the Savage index [9] and Ivlev's selectivity index [10]. The Savage index varies from zero (maximum negative selection) to infinity (maximum positive selection), whereas possible values of Ivlev's selectivity index range from –1 to +1, with negative values indicating rejection or inaccessibility of the prey, zero indicating random feeding, and positive values indicating active selection. Moreover, several researchers have demonstrated that studies based on food selection provide insight into factors involved in prey choice of brown trout [e.g. 11-13].

2.2. Stomach content analysis

In the early 80s Hyslop reviewed the methods used to study the feeding behaviour of fishes and their application to stomach content analyses [14]. Hyslop pointed out the difficulties in the application of these methods and, where appropriate, proposed alternative approaches. Food overlap between age classes can be assessed with Schoener's overlap index [15]. The overlap index has a minimum of 0 (no prey overlap), and a maximum of 1 (all prey items in equal proportions), and diet overlap is usually considered significant when value of the index exceeds 60% [16]. A chi-square ($\chi 2$) test can be used to test for significant differences in the diet composition between age classes [e.g. 17].

2.3. Graphical methods

Graphical methods proposed by Costello [18] and Tokeshi [19] were used to illustrate the relative importance of prey species and to assess the feeding strategy of fish species. Amundsen and collaborators designed an alternative method of Costello graphical method, by plotting prey abundance (Ai) (y - axis) against the frequency of occurrence in diet (Fi) (x - axis) for each prey species. Information on prey importance, feeding strategy and niche breadth can be obtained by examining the distribution of points along the diagonals and axes of the graph [20] (Figure 5 and Section 4.1).

2.4. Niche breadth indexes

Marshall and Elliott compared univariate and multivariate numerical and graphical techniques for determining inter- and intraspecific feeding relationships in estuarine fish [21] and on the basis of this study, different indices have been employed by ichthyologists to study niche breadth and diet specialisation. Generally, the Shannon diversity index was combined with the Levin's index to assess niche breadth [21] and the evenness index was used to evaluate diet specialisation, these being indices employed to study feeding habits in brown trout populations [22,23]. However, stable isotope analysis is a potentially powerful method of measuring trophic niche width, particularly when combined with conventional approaches [24]. For this reason, over the past two decades this methodology has been employed to study the trophic interactions and dietary niche in different fish species, and it has been recently used to study ontogeny and dietary specialization in brown trout [25,26].

2.5. Multivariate approaches

Recently prey trait analysis has been proposed as a functional approach to understand mechanisms involved in predator–prey relationships [27,28]. Despite the disadvantages of this methodology [29 and references therein], it has been used in order to get a deeper insight into the mechanisms that regulate diet composition and feeding habits of fishes, providing extremely valuable ecological information and complementing traditional diet analysis [23,29,30]. For the application of prey trait analysis, researchers have to use the same trait database and trait analyses as de Crespin de Billy [27]. To evaluate the potential vulnerability of invertebrates to fish predation, de Crespin de Billy and Usseglio-Polatera created a total of 71 different categories for 17 invertebrate traits [(1) macrohabitat, (2) current velocity, (3) substratum, (4) flow exposure, (5) mobility/attachment to substratum, (6) tendency to drift in the water column, (7) tendency to drift at the water surface, (8) trajectory on the bottom substratum or in the drift, (9) movement frequency, (10) diel drift behaviour, (11) agility, (12) aggregation tendency, (13) potential size, (14) concealment, (15) body shape (including cases/tubes), (16) body flexibility (including cases/tubes) and (17) morphological defences] [28]. The information of this trait database is structured using a 'fuzzy coding' procedure; thus, a score is assigned to each taxon describing its affinity for each category of each trait, with '0' indicating 'no affinity' to '5' indicating 'high affinity'. The taxonomic resolution (order, family and genus) use in the classification process corresponds to the lowest possible level of determination of taxa in fish gut contents. When identification to genus is not possible or in the case of missing information for a certain genus, the value assign for a trait is that of the family level, using the average profile of all other genus of the same family. Additionally, all the taxa and their assigned scores for each category can be found in previous works [27,28]. Prey trait analysis should be carried out with the software R (version 2.11.1), its ADE4 library for the analysis in R is free and downloadable at http://cran.es.r-projet.org/. Finally, the analysis of prey traits has provided ichthyologists with important clues for understanding the ontogenetic dietary shifts in freshwater fish species. As shown in sections 4 and 5, it is an important tool to disentangle the food resource partitioning among both sympatric age classes and fish species.

3. Diet composition of newly emerged brown trout fry

In brown trout populations there is strong evidence for a critical period with high mortality in the first few weeks after fry emergence [3]. Furthermore, the most critical stage for population regulation in the whole life cycle is the density-dependent mortality of young trout in the first few weeks of the life cycle soon after the young fish start to feed [3 and references therein]. Thus, first feeding of newly emerged fry is very important for brown trout survival in this phase of the life cycle, and in newborns of brown trout first feeding can occur even prior to emergence [13,31,32]. In this sense, the feeding behaviour of newly emerged brown trout fry has been studied in both laboratory conditions and in natural spawning areas. Results of those studies show that feeding in recently emerged fry can be initiated before complete yolk exhaustion [13,31,32]. Zimmerman and Mosegaard observed that alevins of brown trout began feeding in experimental conditions when yolk constituted approximately 40% of the total alevin dry weight [31]. Other researchers have indicated that brown trout fry under natural conditions start feeding when having almost 30% of yolk sac remaining compared to the presumed original size of the yolk sac at hatching [32], while in a recent study no food particles have been found in the stomachs of fry having >10% of the yolk sac remaining [13].

The optimal foraging theory (OFT) explains adaptation via natural selection through quantitative models, which led to a better understanding of foraging behaviour. Hence, OFT predicts that predators should select prey that maximise the energetic gains available in relation to the energetic costs of capturing, ingesting and digesting the prey [33,34]. In this context, Many researchers have found that chironomid larvae and baetid nymphs seem to be the most important food items for newborns in different geographical areas [e.g. 13,32,35]. These are probably the most accessible invertebrates living in the gravel interstices on nesting grounds at the moment of emergence, providing over 80% of the energetic input [13]. However, although chironomid larvae and baetid nymphs seem to be the most important food items for newborns, newly emerged brown trout fry can show differences in the selection of these prey items. Although Baetidae is abundant in the benthos, this taxon is negatively selected according to Ivlev's selectivity index, whereas Chironomidae remains positively selected (Figure 2), demonstrating that abundance of prey items in the benthos is not the only factor explaining the complex mechanism that operates in the food selection during this phase of the ontogeny. Thus, prey size may affect the prey ingestion in early fish larvae, and much literature focuses on the relationship between prey size and mouth size as the primary factor of prey selection [e.g. 36]; but in general, other factors apart from size, such as locomotor skills of fish or accessibility and antipredator behaviour of prey items play an important role in feeding behaviour. These hypotheses that could explain the absence of some items in the stomachs in spite of their abundance in the benthos [13].

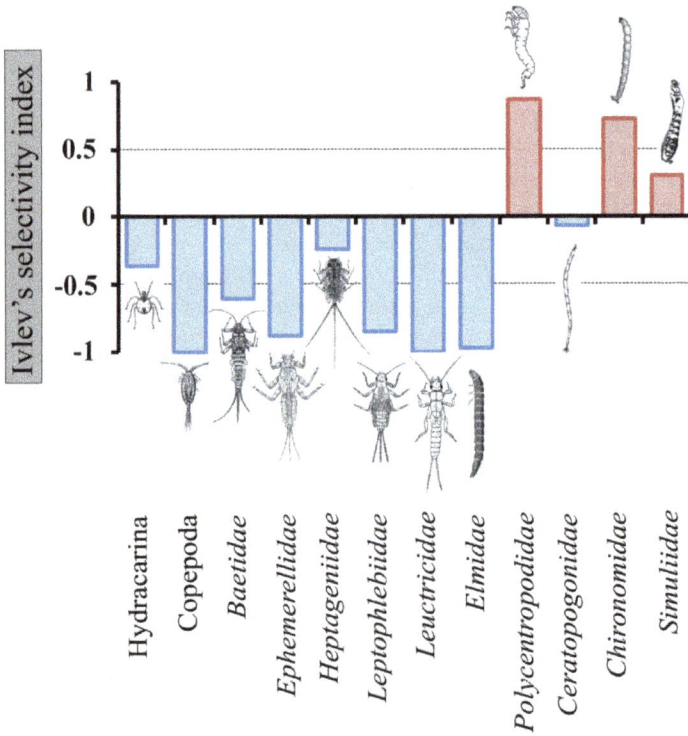

Figure 2. Prey selectivity according to Ivlev's selectivity index of newly emerged brown trout fry in the River Iso (NW Spain) (modified from [13]).

The feeding diversity of juvenile fishes is generally greater than during the larval period, and there is often an increase in the importance of species-specific dietary traits [4]. However, recent studies have demonstrated that at the moment of complete yolk absorption, the fry shows a dramatic shift in niche breadth, which might be related to the improvement of swimming and handling ability of fry for capturing and ingesting both aquatic invertebrates and aerial imagoes [13].

A common practice in many countries associated with river restoration is the rehabilitation of spawning sites with different techniques [37], but recently different authors have emphasized the importance of the complete recolonization of spawning grounds by benthic macroinvertebrates, including first instars, in order to assure the presence of the required amount of prey for the feeding of young fry after restoration works [13]. Hence, at the moment of hatching, a certain density of small prey should be present in the gravel, as searching for food is limited to the nest area and fry forage on available prey [13].

4. Diet changes with age: Food resource partitioning and change in piscivorous behaviour

In Salmonid populations, dominant fishes may exclude less aggressive individuals, limiting their access to resources within patches. For example, dominant Atlantic salmon *Salmo salar* Linnaeus, 1758 may exclude subordinates from high-quality patches by intimidation or direct aggression [38,39]. However, subordinate fish may gain access to food by using high-quality patches when dominants are absent [40] or may be constrained to foraging in marginal areas [41]. When brown trout and Atlantic salmon co-occurred, trout has been observed to be dominant over salmon, holding feeding stations by swimming actively in the central regions of food patches, whereas salmon occupied the margins, generally remaining stationary on the stream bed [41]. Moreover, in habitats in which food is patchily distributed in time and space, fishes can benefit by moving between patches [42,43], with subordinate animals moving little in comparison with dominant fishes [39].

During their life history brown trout undergo ontogenetic habitat shifts [44 and references therein] due to changes in habitat selection operating at multiple spatial scales [44]. These shifts during fish life stage transitions may be accompanied by a marked reduction in intra-specific competition in the fish population, facilitating the partitioning of resources [e.g. 45,46]. Moreover, dietary analyses usually show high values of diet overlap among age classes, but the differences in the use of feeding habitat and behavioural feeding habits are important adaptive features that may reduce the intra-specific competition in the population [23]. Thus, although the diet comparison among age classes can show a remarkable similarity in their prey utilization patterns, sometimes the high overlap values may not indicate competition, since fishes can adopt different strategies to overcome competence, i.e. resource partitioning among age classes can occur at five different levels: (1) diet composition; (2) prey selection; (3) prey size; (4) habitat utilization for feeding; and (5) niche breadth. Also stomach fullness can vary among age classes as shown in section 4.6.

4.1. Changes in diet composition with age

In brown trout, as in many other fish species, there is normally a change in the diet composition during the life of the fish. Thus, juveniles mainly consume prey items linked to the bottom of the river, many of them interstitial, i.e. living among grains of sand or gravel. Opposite, terrestrial invertebrates and fishes are important resources for large trouts. The contribution of these food items to fish diets increases with predator size or age because larger fish can feed on a wider range of preys as shown in Figures 3 and 4. Within a population, the percentage of the most important prey items change with age. In one study of a river in Italy the percentage of plecopteran nymphs in the diet tended to increase with the individual's age [17]. In another study, *Baetis* spp. dominated in all age classes in different proportions, whereas the percentage of caddisflies with cases, (*Allogamus* sp.) and mayflies (*Ecdyonurus* spp.) tended to increase with age [23]. Thus, as shown in Table 1, each age class consumes significantly different prey items, Chironomidae being the most frequently consumed prey item in age 0+ and age 3+ (44.41% and 29.84% respectively) and aerial imagoes of Ephemeroptera in age 1+ and age 2+ (77.49% and 49.08% respectively).

Regarding the changes in the proportion of terrestrial invertebrates consumed by trouts during the ontogeny, previous studies have demonstrated that aquatic invertebrates dominated the diet in all age classes [23,47], but it seems clear that terrestrial invertebrates were more frequently consumed by older trout [17,23,48], terrestrial organisms being important prey during warmer seasons in salmonids [49,50].

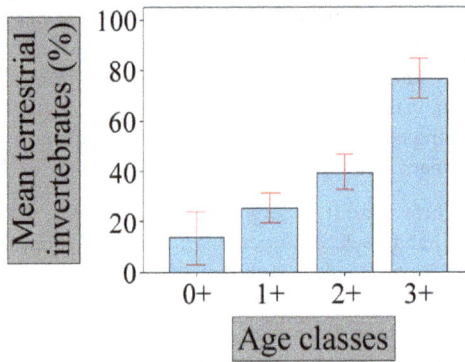

Figure 3. Percentage of terrestrial invertebrates (in terms of relative abundance) consumed by each age class of *Salmo trutta* in the River Anllóns (NW Spain) during summer. Error bars represent 95% confidence intervals.

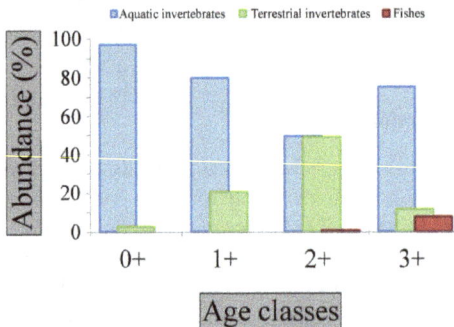

Figure 4. Diet composition consumed by each age class of *Salmo trutta* in the River Lengüelle (NW Spain) during summer.

| | Age 0+ | | | Age 1+ | | | Age 2+ | | | Age 3+ | | |
| | Diet | Ivlev | | Diet | Ivlev | | Diet | Ivlev | | Diet | Ivlev | |
	(%)	Benthos	Drift	(%)	Benthos	Drift	(%)	Benthos	Drift	(%)	Benthos	Drift
Aquatic invertebrates												
Oligochaeta gen. sp.	0.14	-0.68	1	0	-1	—	0	-1	—	0	-1	—
Ancylidae	2.07	0.04	1	0.15	-0.85	1	0	-1	—	0	-1	—
Hydrobiidae	2.48	0.90	0.82	2.11	0.88	0.79	0.92	0.74	0.58	4.84	0.95	0.90
Lymnaeidae	1.66	-0.18	-0.01	1.96	-0.10	0.07	20.64	0.79	0.85	29.03	0.85	0.89
Sphaeriidae	0	-1	—	0.53	0.59	1	1.38	0.82	1	0	-1	—
Hydracarina gen. sp.	0.41	0.64	-0.61	0.23	0.43	-0.76	0	-1	-1	0	-1	-1
Ostracoda gen. sp.	0.55	1	1	0.23	1	1	0	—	—	0	—	—
Baetidae	1.93	0.49	-0.06	0.15	-0.63	-0.87	8.72	0.86	0.60	0	-1	-1
Caenidae	5.10	0.94	0.50	0.60	0.59	-0.48	0.46	0.49	-0.57	0	-1	-1
Ephemerellidae	0	-1	-1	0.23	-0.77	-0.86	2.29	0.14	-0.12	0	-1	-1
Ephemeridae	0	—	—	0.08	1	1	0	—	—	0	—	—
Leuctridae	0.55	0.85	1	0.08	0.25	1	0.46	0.82	1	0	-1	—
Aeshnidae	0	—	—	0.08	1	1	0	—	—	0	—	—
Calopterygidae	6.07	1	1	3.78	0.45	0.17	8.72	0.72	0.53	0	-1	-1
Coenagrionidae	0.28	0.61	1	0	-1	—	0	-1	—	0	-1	—
Gomphidae	0.14	-0.13	1	0	-1	—	0	-1	—	0	-1	—
Aphelocheiridae	0.41	-0.27	-0.28	0.08	-0.81	-0.81	0	-1	-1	0.81	0.05	0.05
Gerridae	0.14	1	-0.28	0	—	-1	0	—	-1	0.81	1	0.54
Sialidae	0.14	1	1	0.23	1	1	0	—	—	0	—	—
Dytiscidae	0	—	—	0	—	—	0.46	1	1	0	—	—
Elmidae	0.41	-0.58	-0.56	0.15	-0.83	-0.81	0.46	-0.55	-0.52	0	-1	-1
Brachycentridae	0	-1	—	0.08	-0.64	1	0	-1	—	0	-1	—
Hydropsychidae	2.76	-0.61	0.17	0.45	-0.92	-0.62	0.92	-0.85	-0.36	0.81	-0.87	-0.41
Leptoceridae	0.28	1	1	0	—	—	0	—	—	0	—	—
Limnephilidae	0.55	0.92	1	0.45	0.91	1	0.46	0.91	1	0.81	0.95	1
Philopotamidae	0.14	0.51	1	0	-1	—	0	-1	—	0	-1	—
Polycentropodidae	0	-1	—	0.30	0.25	1	0	-1	—	0	-1	—
Rhyacophilidae	0	-1	—	0	-1	—	0.46	0.49	1	0	-1	—
Sericostomatidae	0.14	1	1	0.15	1	1	0.46	1	1	0	—	—
Chironomidae	44.41	0.31	0.18	4.08	-0.71	-0.76	1.83	-0.86	-0.89	29.84	0.12	-0.01
Simuliidae	15.17	-0.51	-0.18	1.89	-0.92	-0.84	0.92	-0.96	-0.92	12.90	-0.56	-0.26
Tipulidae	0	-1	—	0.08	0.54	1	0	-1	—	0	-1	—
Terrestrial invertebrates												
Ephemeroptera gen. sp.	12.14	—	1	77.49	—	1	49.08	—	1	2.42	—	1
Trichoptera gen. sp.	0.14	—	-0.28	0.45	—	0.30	0	—	-1	0	—	-1
Chironomidae	0.69	—	-0.93	2.04	—	-0.82	0	—	-1	13.71	—	-0.20
Simuliidae	0.69	—	1	0.68	—	1	0	—	—	2.42	—	1

	Age 0+			Age 1+			Age 2+			Age 3+		
	Diet (%)	Ivlev		Diet (%)	Ivlev		Diet (%)	Ivlev		Diet (%)	Ivlev	
		Benthos	Drift		Benthos	Drift		Benthos	Drift		Benthos	Drift
Arachnida gen. sp.	0	—	—	0.08	—	—	0	—	—	0	—	—
Acanthosomatidae	0	—	—	0.08	—	1	0.46	—	1	0	—	—
Psyllidae	0	—	—	0.15	—	1	0	—	—	0	—	—
Chloropidae	0.14	—	-0.28	0	—	-1	0	—	-1	0	—	-1
Diptera gen. sp.	0	—	-1	0.08	—	-1	0	—	-1	0	—	-1
Syrphidae	0	—	-1	0	—	-1	0	—	-1	0	—	-1
Xylomyidae	0	—	—	0.15	—	1	0	—	—	0	—	—
Cynipidae	0	—	-1	0	—	-1	0	—	-1	0.81	—	0.25
Formicidae	0	—	-1	0	—	-1	0	—	-1	0.81	—	0.54
Carabidae	0	—	—	0.08	—	1	0	—	—	0	—	—
Chrysomelidae	0.28	—	-0.28	0	—	-1	0	—	-1	0	—	-1
Coleoptera gen. sp.	0	—	—	0	—	—	0.46	—	1	0	—	—
Other prey items												
Pseudochondrostoma duriense	0	—	—	0.45	—	—	0.46	—	—	0	—	—
Eggs	0	—	—	0.15	—	1	0	—	—	0	—	—

Table 1. Diet composition and prey selection according to Ivlev's selectivity index in each age class of *Salmo trutta* in the River Furelos (NW Spain) during summer.

The maximum and mean prey size eaten generally increases with size in predatory fish species [51]. Piscivorous behaviour is most frequent in large brown trout, and studies show that it occurs in older individuals with a size of 20–30 cm. Trout in smaller size classes have rarely/ never been recorded eating other fish [23,26,52,53]. In contrast, Sánchez-Hernández and collaborators have recorded piscivorous behaviour in an age-0 trout (the individual found was 8.5 cm in fork length) [29]. This behaviour could be related with the hypothesis of Mittelbach and Persson, who stated that fish species that had larger mouth gapes became piscivorous at younger ages and at smaller sizes [51], and demonstrating that mouth gape is not a limitation to use fishes by small trouts [29]. One possible advantage for small trout in diversifying into eating fish is a reduction in competition with other individuals in the same size classes.

The feeding strategy among age classes may be illustrated with graphical methods, such as the modified Costello graphical method [20] based on the relative importance of prey species. Figure 5, shows plots of prey abundance (A_i) against frequency of occurrence of prey in the diet (F_i) for three different age classes. The plots show feeding strategy differs between age classes, varying in degrees of specialization and generalization on different prey types. For all age classes, *Baetis* spp. were the most important prey, always being eaten by more than eighty-five percent of the individuals (F_i = 87.5% in age-1+ to 100% in age-2+) and represented a high contribution in specific abundance (A_i = 23.93% in age-1+ to 68.95% in age-0+). However, the majority of the prey items presented low values for both F_i and A_i (lower left quadrant) for all age classes (Figure 5), displaying evidence of a generalist strategy for these prey (e.g. *Chimarra marginata*, Simuliidae (adult), Chironomidae (adult) and *Ophiogomphus* sp.).

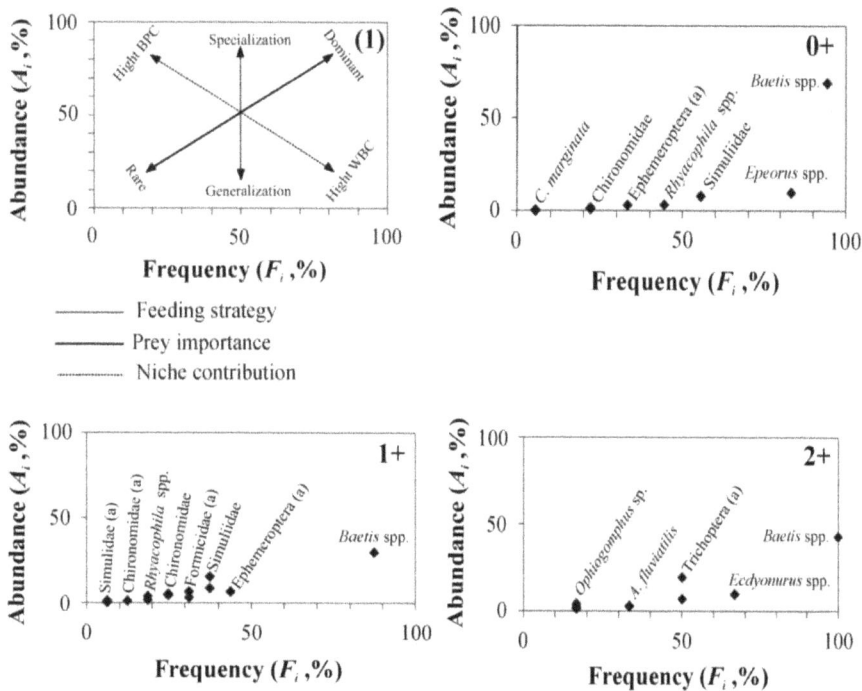

Figure 5. Feeding strategy diagram. (1) Explanatory diagram of the modified Costello method according to Amundsen and collaborators [20]. Data are presented for each age class.

Finally, although in some occasions the diet comparison among age classes can show a remarkable similarity in their prey utilization patterns, sometimes the high overlap values may not indicate competition, since the differences in the behavioural feeding habits are important adaptive features that may reduce the intra-specific competition in the population [23]. Previously, the differences in the behavioural feeding habits among age classes had been studied using prey trait analysis [23], obtaining important advances to disentangle food resource partitioning among cohorts. Details and information needed for the elaboration of prey trait analysis can be found in the introduction section and bibliography [27-29]. Thus, in the reference [23] 'diel drift behaviour' trait showed that age-2+ is clearly separated from the other age classes. Age-2+ tended to feed on prey with no tendency in diel drift behaviour as shown by the presence of *Ancylus fluviatilis*, Gerridae, Coleoptera and Formicidae in their stomachs. On the contrary, age-0+ preferred to feed on prey with nocturnal diel drift behaviour tendency due to the presence of Philopotamidae (*C. marginata*). Age-0+ showed a wider distribution of values in the fuzzy principal component analysis (FPCA) of the 'trajectory' trait; age-0+ tended to feed on prey with oscillatory and by random trajectory (Lumbriculidae, Simuliidae and Chironomidae), whereas age-2+ tended to feed on prey with lineal trajectory due to the presence of Gerridae [23]. On the other hand, no clear differences have been found in the traits 'tendency to drift in the water column' and 'tendency to drift at the water surface' for prey among age classes [23], however as shown in Figure 6 (fuzzy principal component analysis calculated with the values reported in Table 1), age-3+ is clearly separated from the other age classes, showing that the differences ability to feed at different depths of the water column is possible among cohorts.

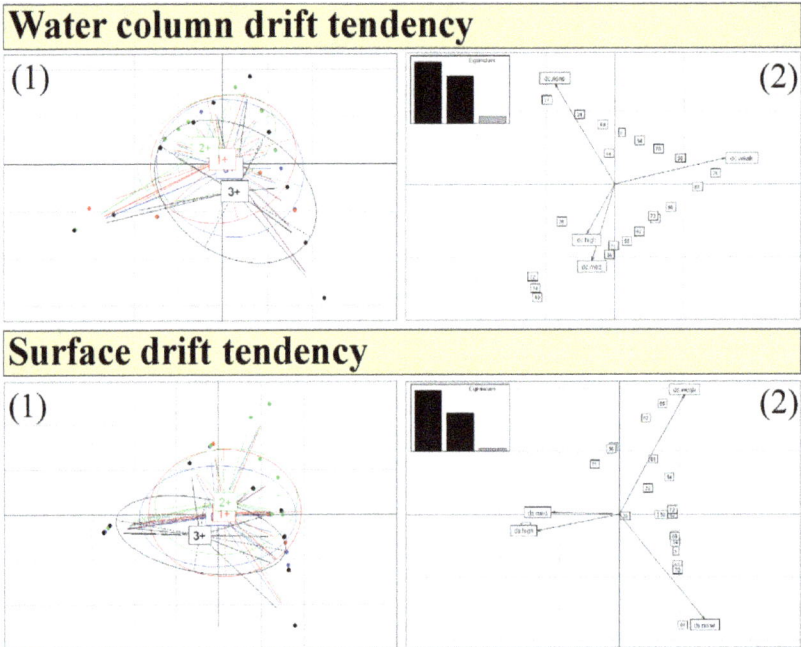

Figure 6. Biplot of gut contents obtained from a fuzzy principal component analysis (FPCA) based on behavioural feeding habits of the four age classes. (1) Similarity results among age classes according to the gut contents of Table 1. Data are presented for each age class. 0+: age-0+, 1+: age-1+, 2+: age-2+ and 3+: age-3+. Ellipses envelop weighted average of prey taxa positions consumed by age classes: Labels (0+, 1+, 2+ and 3+) indicate the gravity centre of the ellipses (2) Factorial correspondence analysis between both traits and prey items and their spatial distribution with histogram of eigenvalues. Details needed for the elaboration of these graphics can be found in the introduction section and bibliography [27-29].

4.2. Changes in prey selection with age

Prey abundance should be an important factor involved in prey choice. However, fishes do not always consume the most abundant taxa available in the environment [13,29,54]. The total abundance and biomass of invertebrates drifting during the day describe the potential prey available to juvenile brown trout better than abundance and biomass of benthic invertebrates do [55]. Different authors have shown that active choice guided by energetic optimization criteria appeared to be of limited importance in determining the size composition of prey eaten by trouts [11]. These authors also stated that the operating mechanisms of prey-size selection are probably not independent of the characteristics of the size-frequency distribution of the available prey. Moreover, in brown trout spatial and temporal variations in prey selection are possible, due to the preference for the different prey items related to the site-specific prey accessibility [12]. As shown in section 3, the size-frequency distribution of potential prey items in the benthos was different to that of prey in the stomachs of newly emerged brown trout [13], a result that agrees with observations made in other fish species [e.g. 54] and observations made in age-0+ individuals of *S. trutta* during the autumn [29]. In spite of the high abundance of Elmidae in the benthos, this prey item was negatively selected [29]. The rejection of the elmid beetle may be due to their low energetic value, as they have an intense sclerotisation, but it may also be due to their bad taste [56-58]. Hence, other factors, besides prey abundance, including site-specific prey accessibility, prey size, energetic selection criteria and prey preference of fishes, play an important role in the feeding behaviour of freshwater fishes [54].

Fochetti and collaborators found a high preference for species of Trichoptera by trout younger than 3+, a preference for plecopteran species by those older than three years, and a general negative avoidance for species of Ephemeroptera by all age classes [17]. As shown in Table 1, prey selection is clearly different among age classes and important prey items such as Chironomidae in age 0+ and age 3+ and aerial imagoes of Ephemeroptera in age-1+ and age-2+ are positively selected in the benthos and drift respectively. Overall, as different researchers have demonstrated, the mechanisms involved in prey selection among age classes in the same population are complex and may be related at different levels: (1) prey selection in fishes is related to prey characteristics, such as size, locomotor skills, accessibility or anti-predator behaviour, (2) prey selection in fishes is related to fish characteristics, like prior experience, locomotor skills, stomach fullness, mouth gape, sensory capabilities and fish size and (3) prey selection in fishes is related to physical habitat characteristics, as flow patterns and structural complexity of habitat [12,13,34,54].

4.3. Changes in prey size with age

Ontogenetic dietary shifts may also occur at the level of prey size [23,54]. Steingrímsson and Gíslason showed a consistent, but moderate, shift towards larger prey with increased body size in brown trout [59]. Several researchers have found that mean prey size increases as predator size increases [e.g. 23,48,54,60]. The size-frequency distributions of the available terrestrial prey were always greatly dominated (75–90%) by the two smallest size classes (1–2 and 2–3 mm long), prey over 4 mm long being extremely scarce, while size distributions of aquatic prey were less skewed [11]. Thus, in general it is correct to state that trout fed mainly on prey within 1–4 mm size range [11,48], with 2–3 mm prey being the most commonly consumed [48]. On the contrary, Rincón and Lobón-Cerviá showed that organisms 1–2 mm long were generally the most numerous [11]. By age classes the average prey size consumption is different with higher values in age-2+ (8.4 mm ± 1.62) than age-0+ (4.2 mm ± 0.25) and age-1+ (5.9 mm ± 0.51), but there are no significant differences between age-0+ and age-1+ [23]. Figure 7 illustrates the age-related variation in prey size, showing that mean prey size tends to increase with age.

In conclusion, prey-size selection is probably dependent on the characteristics of the size-frequency distribution of the available prey [11], and the size-related differences in the diet of trout can be related to gape-limitations, increasing mean prey size and maximum prey-size with trout size [48].

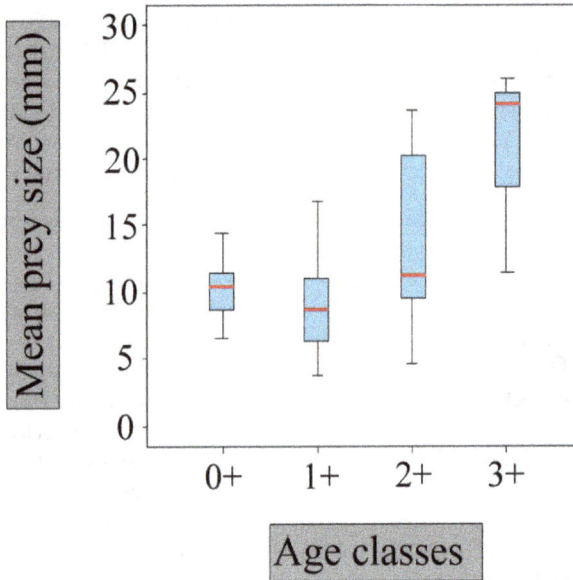

Figure 7. Box plots of the age-related variation in prey size of *Salmo trutta* in the River Furelos (NW Spain) during summer. The solid line within each box represents the median, the bottom and top borders indicate the 25th and 75th percentiles, the notches represent the 95% confidence intervals.

4.4. Changes in the habitat used for feeding with age

In fishes, patches used for feeding and refuges are normally different, as shown by several researchers [61-63], due to brown trout being a habitat generalist. Also, patterns in habitat selection have been shown to be driven by physical and environmental factors operating at multiple spatial scales [44].

Many organisms exhibit ontogenetic shifts in their diet and habitat use, which often exert a large influence on the structure and expected dynamics of food webs and ecological communities [64]. Special attention has been given to ontogenetic shift in habitat preference in brown trout populations [e.g. 44,65-67]. It is well-known that in brown trout populations habitat use changes during ontogeny, preferring deeper and slower flowing water as they increased in size [e.g. 44,65].

Habitat patches used by brown trout can be monitored by radio telemetry [e.g. 68] and although these studies have shown that brown trout feed on young white suckers *Catostomus commersonii* (Lacepède, 1803) at night in shallow habitats, little information was obtained about the habitat used for feeding. Also, microhabitat use of freshwater fishes has been studied by snorkel observations in previous studies [e.g. 69]. This methodology could be used to study feeding habitat requirements of fish species. However, there are still gaps to be filled before snorkel surveys can be fully adopted in fish diet studies. In fact, one of the main disadvantages of this approach is the need for good visibility. This has been one of the main handicaps because, although brown trout normally yields a bimodal (crepuscular) pattern of activity with a major peak at dawn and a lower one around dusk [70], brown trout can also feed at night [71]. Another limiting factor for the application of snorkel surveys to study feeding habitat require-ments is related to the physical characteristic of the river such as current, depth or turbidity. It is well-known that different macroinvertebrates have different preferences for habitats [72] and so prey trait analysis has been proposed as a functional approach to understand mecha-nisms involved in predator-prey relationships [27-29]. Consequently it may be useful for understanding inter-species interactions and the mechanisms that determine food partitioning between them [29,30]. Nowadays, they have been recently used to provide interesting results about differences in feeding habitat requirements among age classes in brown trout [23], for example, young of the year (0+) tend to capture prey living in moderate current velocities, whereas other age classes (1+ and 2+) tend to feed on prey living in fast current velocities [23].

A previous study on habitat choice in a littoral zone of Lake Tesse (Norway) showed that small trout had a strong association with the bottom and larger trout occurred more frequently higher up in the water column, and suggested that this difference in vertical distribution was also reflected in food choice [73]. In streams, competition among fish species may also be reduced by vertical segregation [e.g. 41,69,74]. In spite that terrestrial invertebrates, as component of the diet, are more important in adults than juveniles of brown trout [23], these same authors found that the ability to feed at different depths of the water is similar among age classes. Nevertheless, it is possible that different age classes of fish may become vertically segregated by concentrating on different prey types living in different parts of the water column as shown in section 4.1. Hence, additional studies are needed in order to clarify whether vertical segregation among cohorts is related to the ability to feed at different depths of the water column.

Finally differences in the use of feeding habitat are important adaptive features that may reduce the intra-specific competition in the population. In this context, fuzzy principal component analysis (FPCA) has shown that age-0+ tended to feed on prey living in moderate current velocities, although overlap was higher between age-1+ and age-2+, preferring to feed on prey living in fast current velocities [23]. Moreover, these researchers have found that age-0+ showed a higher spectrum of prey, which revealed a greater ability to prey on different macrohabitats, whereas age-1+ and age-2+ preferred to feed on epibenthic prey living in erosional macrohabitats. In our case with the values reported in Table 1 and as shown in Figure 8, 'current velocity' trait shows no clear differences for prey of the four age classes. On the contrary, 'macrohabitat' trait shows that age-1+ has the most ample spectrum for this trait, tending to feed on epibenthic prey living in erosional macrohabitats, whereas age-3+ tends to feed on prey items available in the water column (Figure 8).

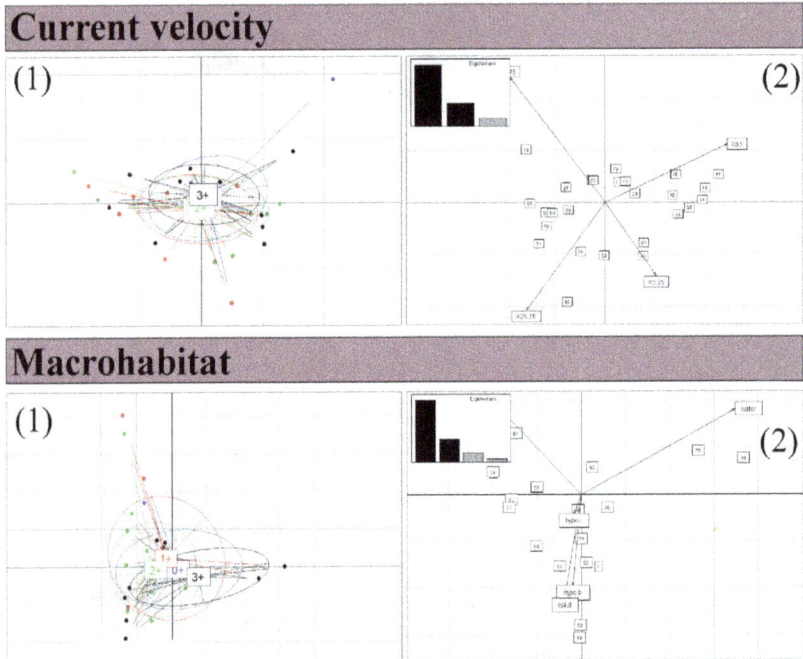

Figure 8. Biplot of gut contents obtained from a fuzzy principal component analysis (FPCA) based on preferential habitat utilization for feeding of the four age classes. (1) Similarity results among age classes according to the gut contents of Table 1. Data are presented for each age class. 0+: age-0+, 1+: age-1+, 2+: age-2+ and 3+: age-3+. Ellipses envelop weighted average of prey taxa positions consumed by age classes: Labels (0+, 1+, 2+ and 3+) indicate the gravity centre of the ellipses. (2) Factorial correspondence analysis between both traits and prey items and their spatial distribution with histogram of eigenvalues. Details needed for the elaboration of these graphics can be found in the introduction section and bibliography [27-29].

4.5. Changes in the niche breadth with age

Deady and Fives showed that niche breadth decreases with fish length in corkwing wrasse, *Symphodus (Crenilabrus) melops* (Linnaeus, 1758), indicating an increase in dietary specialization with increasing length [75]. Magalhães found dietary shifts throughout the ontogeny in an endemic cyprinid of the Iberian Peninsula (*Squalius pyrenaicus* (Günther, 1868)), including shifts from soft-bodied to hard-shelled prey and decreased animal prey breadth [76]. In contrast, several other researchers have found that niche breadth increases with body size [e.g. 23,77]. Oscoz and collaborators found that larger fish have a higher number of potential prey items available and a wider niche breadth, as indicated by their higher trophic diversity index values [77]. As mentioned in section 3, analysis of diet changes on newly emerged brown trout fry suggests a dramatic shift in niche breadth at the moment of complete yolk absorption, which might be related to the improvement the fry's swimming and handling ability in capturing and ingesting prey [13]. In a recent study on brown trout populations, it has been demonstrated that niche breadth, measured as Levin's index, increases with fish length in *Salmo trutta*. However, no differences were found in the Shannon diversity and evenness indices of prey eaten among age classes [23].

As can be seen in Figure 9; the number of different prey types consumed by S. trutta increases during ontogeny. Age-0+ shows the smallest prey spectrum, whereas in age-1+ a significant increase in the prey types consumed by juveniles is observed. However, no significant differences in the dietary niche among ages-1+, 2+ and 3+ are observed in Figure 9. Hence, although no clear results have been observed in the variation of the diversity indices of prey among age classes [23], it could be argued that there is a tendency to increase dietary niche with increasing length or age, at least when the niche breadth of juveniles (0+), subadults (1+) and adults (≥2+) is compared.

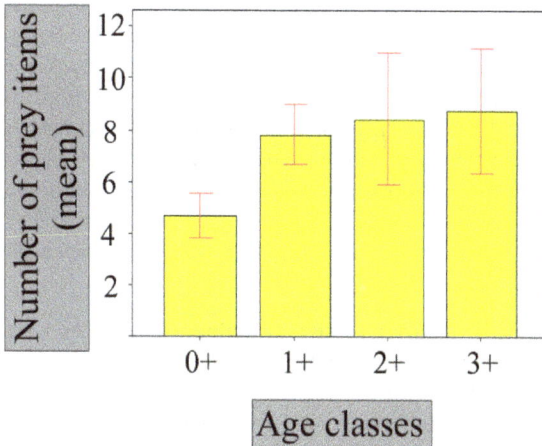

Figure 9. Age-related variation in the number of different prey types consumed by *Salmo trutta* in the River Lengüelle (NW Spain) during summer. Error bars represent the 95% confidence intervals.

4.6. Changes in the fullness index with age

Several researchers have found in different fish species that the stomach fullness index (defined as the weight of the stomach contents (in grams) divided by the weight of the predator (in grams) and multiplied by 100) varies during ontogeny [78,79]. In salmonids the results are contradictory: in brook charr *Salvelinus fontinalis* (Mitchill, 1814) no differences have been found in the stomach fullness [80], whilst other researchers have demonstrated that the stomach fullness of brown trout varies among size classes [47]. Brown trout between the size of 40 mm and 320 mm fed more intensively, whilst the intensity declined above 320 mm length [47]. In the River Furelos (NW Spain), we have found that stomach fullness during the summer is different among age classes (Kruskal-Wallis test; $p < 0.001$), being higher in age-0+ ($9\% \pm 0.64$) than age-1+ ($1.1\% \pm 0.14$), age-2+ ($1\% \pm 0.25$) and age-3+ ($1.1\% \pm 0.24$) (all Mann–Whitney U test, $p < 0.001$) but no differences have been found between ages-1+, 2+ and 3+ (all Mann–Whitney U test, $p > 0.05$). Moreover, stomach fullness decreases with fish size ($r = -0.72$; $p < 0.001$) (unpublished data). Hence, stomach fullness can vary among age classes; however additional studies are needed in order to clarify whether stomach fullness varies during ontogeny in brown trout.

5. Competion for food between brown trout and other sympatric fish species

Trophic interactions between species are important factors structuring animal communities. Brown trout are top-consumers in freshwater habitats and play an important role as carriers of energy from lower to higher trophic levels (i.e. predators). Many freshwater fish species tend to occupy a specific type of habitat but there are lots of exceptions. For example, spatial niche overlap is considerable where Atlantic salmon and brown trout co-occur, although young Atlantic salmon tend to occupy faster flowing and shallower habitats [e.g. 74]. Moreover, when both fish species co-occur, the habitat used by Atlantic salmon is restricted through interspecific competition by the more aggressive brown trout, indicating an interactive segregation between fish species [e.g. 41,74]. Indeed, it is well known that brown trout is a territorial drift feeder [3,81], and several authors have reported the behavioural dominance of trout over cyprinids in streams [82-84].

The competitive coexistence between species occupying similar niches may be facilitated by a generalisation of niche width as predicted by the optimal foraging theory (OFT), rather than the specialised niche width predicted by the classic niche theory as a response to interspecific competition [85]. However, studies on food partitioning in fish communities have obtained contradictory results. Whereas several authors have found differences in diet composition among sympatric fish species [e.g. 86,87], other researchers concluded that the same food resource can be shared by several species [29,30,85]. In these cases, the differences in behavioural feeding habits, handling efficiency and feeding habitat utilization are important adaptive features that may reduce the inter-specific competition in the fish community and permit the partitioning of food that allows coexistence [29,30]. Thus, sympatric fish species can adopt different strategies to overcome competion and food resource partitioning can occur at different levels.

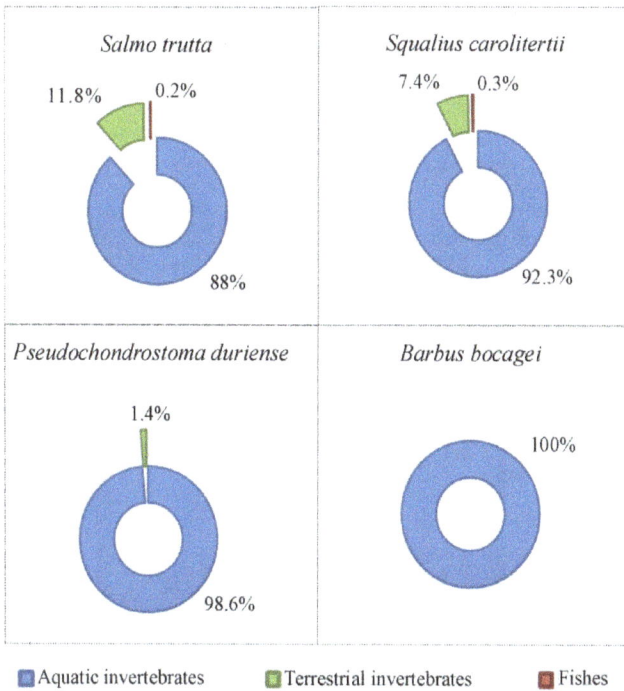

Figure 10. Diet composition consumed by each fish species in the Tormes River (Central Spain) during summer.

Firstly, the use of microhabitats is often different between species, due to segregation of microhabitats, an important factor in reducing the effects of competition for food [69,88,89]. For example, *Barbus bocagei* Steindachner, 1865 occupied deeper habitats and selected lower positions in the water column than *Pseudochondrostoma polylepis* (Steindachner, 1865), and *Squalius pyrenaicus* (Günther, 1868), *P. polylepis* occupied microhabitats with greater velocities than the other two species and *S. pyrenaicus* selected shallower habitats than the other two species [69]. In another study, *S. trutta* showed wider diversity in the habitat used for feeding than *Squalius carolitertii* (Doadrio, 1988), *Pseudochondrostoma duriense* (Coelho, 1985) and *B. bocagei* [30]. Hence, differences were found among species in their ability to feed at different depths of the water column [29,30] as shown in snorkelling studies into microhabitat use in fish [69].

Secondly, different species may specialise in different resources. For example, many cyprinid fish sympatric with trout feed on a significant amount of detritus and plant material not used by trout, leading to reduced inter-specific competition [29,30]. Moreover, resource partitioning may also occur at the level of prey size [29,30,90], although it is not clear whether this size selective strategy is adopted to reduce interspecific competition or it is the result of foraging behaviour and/or morphological constraints such as gape size [29,91]. Also, terrestrial prey are present primarily on the stream surface and although tend to be absent from the diets of benthic feeders such as *B. bocagei* (Figure 10), terrestrial inputs may constitute an important food resource for freshwater fish species and especially for brown trout. Thus, the utilization of allochthonous food resources such as terrestrial invertebrates by fishes may reduce competition facilitating the partitioning of resources [30].

Thirdly, diel segregation is possible among fish species, and this may also lead to reduced interspecific competition between fish [29,92,93]. According to macroinvertebrate trait analyses, sticklebacks (*Gasterosteus aculeatus* Linnaeus, 1758) and *P. duriense* show a slight preference for prey that drift during the day, whilst age-0 *S. trutta* seem to prefer to feed at dusk, whereas *Achondrostoma arcasii* (Steindachner, 1866) differs from the other three species due to its preference to feed on prey on organisms with weak or no tendency to drift [29]. However, the "diel drift behaviour" of macroinvertebrate prey of brown trout and three sympatric cyprinids is similar [30]. Hence, the differences in the diel feeding behaviour among sympatric fish species might only be adopted in highly competitive communities, where food is a more limiting resource.

6. Conclusion

To summarize, the present study supports the hypothesis differences in the feeding habits and habitat utilization of different age classes of trout could reduce competition for food, by allowing food resource partitioning. Hence, age-related diet shifts occur at five different levels: (1) diet composition changes with fish age; (2) prey selection varies with fish age, probably due to prey-size selection which is in turn dependent on the size-frequency distribution of the available prey; (3) mean prey size increases with fish size and age; (4) habitat utilization for feeding may be different among age classes; (5) niche breadth tends to increase with age and fish size. Finally, also the stomach fullness can vary among age classes. However, additional studies are needed in order to clarify whether stomach fullness varies during the ontogeny in brown trout.

Acknowledgements

Dr. Adrian Seymour and Josué Sánchez are acknowledged for valuable comments and grammar corrections on the manuscript.

Author details

Javier Sánchez-Hernández[1,2*], María J. Servia[3], Rufino Vieira-Lanero[2] and Fernando Cobo[1,2]

*Address all correspondence to: javier.sanchez@usc.es

1 Department of Zoology and Physical Anthropology, Faculty of Biology, University of Santiago de Compostela, Spain

2 Station of Hydrobiology "Encoro do Con", Castroagudín s/n, Vilagarcía de Arousa, Pontevedra, Spain

3 Department of Animal Biology, Vegetal Biology and Ecology, Faculty of Science, University of A Coruña, Spain

References

[1] Aas O, Haider W, Hunt L. Angler responses to potential harvest regulations in a Norwegian sport fishery: a conjoint-based choice modeling approach. North American Journal of Fisheries Management 2000;20(4) 940-950. DOI: 10.1577/1548-8675(2000)020<0940:ARTPHR>2.0.CO;2

[2] Butler JRA, Radford A, Riddington G, Laughton R. Evaluating an ecosystem service provided by Atlantic salmon, sea trout and other fish species in the River Spey, Scotland: the economic impact of recreational rod fisheries. Fisheries Research 2009;96(2-3) 259-266. DOI: 10.1016/j.fishres.2008.12.006

[3] Elliott JM. Quantitative Ecology and the Brown Trout. Oxford: Oxford University Press; 1994.

[4] Nunn AD, Tewson LH, Cowx IG. The foraging ecology of larval and juvenile fishes. Reviews in Fish Biology and Fisheries 2012;22(2) 377-408. DOI: 10.1007/s11160-011-9240-8

[5] Seaburg KG. A stomach sampler for live fish. The Progressive Fish-Culturist. 1957;19(3) 137-139. DOI: 10.1577/1548-8659(1957)19[137:ASSFLF]2.0.CO;2

[6] Wales JH. Forceps for removal of trout stomach content. The Progressive Fish-Culturist 1962;24(4) 171. DOI: 10.1577/1548-8659(1962)24[171:FFROTS]2.0.CO;2

[7] Jernejcic F. Use of emetics to collect stomach contents of Walleye and Large mouth Bass. Transactions of the American Fisheries Society 1969;98(4) 698-702. DOI: 10.1577/1548-8659(1969)98[698:UOETCS]2.0.CO;2

[8] Sánchez-Hernández J, Servia MJ, Vieira-Lanero R, Cobo F. Evaluación del lavado gástrico como herramienta para el análisis de la dieta en trucha común. Limnetica 2010;29(2) 369-378.

[9] Savage RE. The relation between the feeding of the herring off the cast coast of England and the plankton of the surrounding waters. Fishery Investigation, Ministry of Agriculture, Food and Fisheries, Series 2, 1931;12 1-88.

[10] Ivlev VS. Experimental ecology of the feeding of fishes. Translated from the Russian by Douglas Scott. New Haven: Yale University Press; 1961.

[11] Rincón PA, Lobón-Cerviá J. Prey-size selection by brown trout (Salmo trutta L.) in a stream in northern Spain. Canadian Journal of Zoology 1999;77(5) 755-765. DOI: 10.1139/z99-031

[12] Johnson RL, Coghlan SM, Harmon T. Spatial and temporal variation in prey selection of brown trout in a cold Arkansas tailwater. Ecology of Freshwater Fish 2007;16(3) 373-384. DOI: 10.1111/j.1600-0633.2007.00230.x

[13] Sánchez-Hernández J, Vieira-Lanero R, Servia MJ, Cobo F. First feeding diet of young brown trout fry in a temperate area: disentangling constraints and food selection. Hydrobiologia 2011a;663(1) 109-119. DOI: 10.1007/s10750-010-0582-3

[14] Hyslop EJ. Stomach contents analysis—a review of methods and their application. Journal of Fish Biology 1980;17(4) 411-429. DOI: 10.1111/j.1095-8649.1980.tb02775.x

[15] Schoener TW. Nonsynchronous spatial overlap of lizards in patchy habitats. Ecology 1970;51(3) 408-418. DOI: 10.2307/1935376

[16] Wallace RK Jr. An assessment of diet overlap indexes. Transactions of the American Fisheries Society 1981;110(1) 72-76. DOI: 10.1577/1548-8659(1981)110<72:AAODI>2.0.CO;2

[17] Fochetti R, Argano R, Tierno de Figueroa JM. Feeding ecology of various age-classes of brown trout in River Nera, Central Italy. Belgian Journal of Zoology 2008;138(2) 128-131.

[18] Costello MJ. Predator feeding strategy and prey importance: a new graphical analysis. Journal of Fish Biology 1990;36(2) 261-263. DOI: 10.1111/j.1095-8649.1990.tb05601.x

[19] Tokeshi M. Graphical analysis of predator feeding strategy and prey importance. Freshwater Forum 1991;1 179-183.

[20] Amundsen P-A, Gabler HM, Staldvik FJ. A new approach to graphical analysis of feeding strategy from stomach contents data – modification of the Costello (1990) method. Journal of Fish Biology 1996;48(4) 607-614. DOI: 10.1111/j.1095-8649.1996.tb01455.x

[21] Marshall S, Elliott M. A comparison of univariate and multivariate numerical and graphical techniques for determining inter- and intraspecific feeding relationships in estuarine fish. Journal of Fish Biology 1997;51(3) 526-545. DOI: 10.1111/j.1095-8649.1997.tb01510.x

[22] Oscoz J, Leunda PM, Campos F, Escala MC, Miranda R. Diet of 0+ brown trout (Salmo trutta L., 1758) from the river Erro (Navarra, North of Spain). Limnetica 2005;24(3-4) 319-326.

[23] Sánchez-Hernández J, Cobo F. Summer differences in behavioural feeding habits and use of feeding habitat among brown trout (Pisces) age classes in a temperate area. Italian Journal of Zoology 2012a;79(3) 468-478. DOI: 10.1080/11250003.2012.670274

[24] Bearhop S, Adams CE, Waldron S, Fuller RA, MacLeod H. Determining trophic niche width: a novel approach using stable isotope analysis. Journal of Animal Ecology 2004;73(5) 1007-1012. DOI:10.1111/j.0021-8790.2004.00861.x

[25] Grey J. Ontogeny and dietary specialization in brown trout (Salmo trutta L.) from Loch Ness, Scotland, examined using stable isotopes of carbon and nitrogen. Ecology of Freshwater Fish 2001;10(3) 168-176. DOI: 10.1034/j.1600-0633.2001.100306.x

[26] Jensen H, Kiljunen M, Amundsen, P-A. Dietary ontogeny and niche shift to piscivory in lacustrine brown trout Salmo trutta revealed by stomach content and stable isotope analyses. Journal of Fish Biology 2012;80(7) 2448-2462. DOI: 10.1111/j. 1095-8649.2012.03294.x

[27] de Crespin de Billy V. Régime alimentaire de la truite (Salmo trutta L.) en eaux courantes: rôles de l 'habitat physique des traits des macroinvertébrés. PhD thesis. Université Claude Bernard, Lyon; 2001

[28] de Crespin de Billy V, Usseglio-Polatera P. Traits of brown trout prey in relation to habitat characteristics and benthic invertebrate communities. Journal of Fish Biology 2002;60(3) 687-714. DOI: 10.1111/j.1095-8649.2002.tb01694.x

[29] Sánchez-Hernández J, Vieira-Lanero R, Servia MJ, Cobo F. Feeding habits of four sympatric fish species in the Iberian Peninsula: keys to understanding coexistence using prey trais. Hydrobiologia 2011b;667(1) 119-132. DOI: 10.1007/s10750-011-0643-2

[30] Sánchez-Hernández J, Cobo F. Summer food resource partitioning between four sympatric fish species in Central Spain (River Tormes). Folia Zoologica 2011;60(3) 189-202.

[31] Zimmerman CE, Mosegaard H. Initial feeding in migratory brown trout (Salmo trutta L.) alevins. Journal of Fish Biology 1992;40(4) 647-650. DOI: 10.1111/j. 1095-8649.1992.tb02612.x

[32] Skoglund H, Barlaup BT. Feeding pattern and diet of first feeding brown trout fry under natural conditions. Journal of Fish Biology 2006;68(2) 507-521. DOI: 10.1111/j. 0022-1112.2006.00938.x

[33] Pyke GH, Pulliam HR, Charnov EL. Optimal foraging: a selective review of theory and tests. Quarterly Review of Biology 1977;52(2) 137-154.

[34] Gerking SD. Feeding ecology of fish. San Diego: Academic Press; 1994.

[35] McCormack JC. The food young trout (Salmo trutta) in two different necks. Journal of Animal Ecology 1962;31(2) 305-316.

[36] Cunha I, Planas M. Optimal prey size for early turbot larvae (*Scophthalmus maximus* L.) based on mouth and ingested prey size. Aquaculture 1999;175(1-2) 103-110. DOI: 10.1016/S0044-8486(99)00040-X

[37] Hunter CJ. Better trout habitat: a guide to stream restoration and management. Washington DC: Island Press; 1991.

[38] Gotceitas G, Godin J-GJ. Foraging under the risk of predation in juvenile Atlantic salmon (*Salmo salar* L.): effects of social status and hunger. Behavioral Ecology and Sociobiology 1991;29(4) 255-261. DOI: 10.1007/BF00163982

[39] Griffiths SW, Armstrong JD. Kin-biased territory overlap and food sharing among Atlantic salmon juveniles. Journal of Animal Ecology 2002;71(3) 480-486. DOI: 10.1046/j.1365-2656.2002.00614.x

[40] Alanärä A, Burns MD, Metcalfe NB. Intraspecific resource partitioning in brown trout: the temporal distribution of foraging is determined by social rank. Journal of Animal Ecology 2001;70(6) 980-986. DOI: 10.1046/j.0021-8790.2001.00550.x

[41] Höjesjö J, Armstrong J, Griffiths S. Sneaky feeding by salmon in sympatry with dominant brown trout. Animal Behaviour 2005;69(5) 1037-104. DOI: 10.1016/j.anbehav.2004.09.007

[42] Armstrong JD, Braithwaite VA, Huntingford FA. Spatial strategies of wild Atlantic salmon parr: exploration and settlement in unfamiliar areas. Journal of Animal Ecology 1997;66(2) 203-211.

[43] Ruxton GD, Armstrong JA, Humphries S. Modelling territorial behaviour of animals in variable environments. Animal Behaviour 1999;58(1) 113-120. DOI: 10.1006/anbe.1999.1114

[44] Ayllón D, Almodóvar A, Nicola GG, Elvira B. Ontogenetic and spatial variations in brown trout habitat selection. Ecology of Freshwater Fish 2010;19(3) 420-432. DOI: 10.1111/j.1600-0633.2010.00426.x

[45] Elliott JM. The food of trout (*Salmo trutta*) in a Dartmoor stream. Journal of Applied Ecology 1967;4(1) 59-71.

[46] Amundsen P-A, Bøhn T, Popova OA, Staldvik FJ, Reshetnikov YS, Kashulin N, Lukin A. Ontogenetic niche shifts and resource partitioning in a subarctic piscivore fish guild. Hydrobiologia 2003;497(1-3) 109-119. DOI: 10.1023/A:1025465705717

[47] Kara C, Alp A. Feeding habits and diet composition of brown trout (*Salmo trutta*) in the upper streams of river Ceyhan and river Euphrates in Turkey. Turkish Journal of Veterinary and Animal Sciences 2005;29 417-428.

[48] Montori A, Tierno de Figueroa JM, Santos X. The diet of the brown trout *Salmo trutta* (L.) during the reproductive period: size-related and sexual effects. International Review of Hydrobiology 2006;91(5) 438-450. DOI: 10.1002/iroh.200510899

[49] Nakano S, Kawaguchi Y, Taniguchi Y, Miyasaka H, Shibata Y, Urabe H, Buhara N. Selective foraging on terrestrial invertebrates by rainbow trout in a forested headwater stream in northern Japan. Ecological Research 1999;14(4) 351-360. DOI: 10.1046/j. 1440-1703.1999.00315.x

[50] Utz RM, Hartman KJ. Identification of critical prey items to Appalachian brook trout (*Salvelinus fontinalis*) with emphasis on terrestrial organisms. Hydrobiologia 2007;575(1) 259-270. DOI: 10.1007/s10750-006-0372-0

[51] Mittelbach GG, Persson L. The ontogeny of piscivory and its ecological consequences. Canadian Journal of Fisheries and Aquatic Sciences 1998;55(6) 1454-1465. DOI: 10.1139/ f98-041

[52] Kahilainen K, Lehtonen H. Brown trout (*Salmo trutta* L.) and Arctic charr (*Salvelinus alpinus* (L.)) as predators on three sympatric whitefish (*Coregonus lavaretus* (L.)) forms in the subarctic Lake Muddusjärvi. Ecology of Freshwater Fish 2002;11(3) 158-167. DOI: 10.1034/j.1600-0633.2002.t01-2-00001.x

[53] Jensen H, Bøhn T, Amundsen P-A, Aspholm PE. Feeding ecology of piscivorous brown trout (*Salmo trutta* L.) in a subarctic watercourse. Annales Zoologici Fennici 2004;41(1) 319-328.

[54] Sánchez-Hernández J, Cobo F. Ontogenetic dietary shifts and food selection of endemic *Squalius carolitertii* (Actinopterygii: Cypriniformes: Cyprinidae) in River Tormes, Central Spain, in summer. Acta Ichthyologica et Piscatoria 2012b;42(2) 101-111. DOI: 10.3750/AIP2011.42.2.03

[55] Sagar PM, Glova GJ. Prey availability and diet of juvenile brown trout (*Salmo trutta*) in relation to riparian willows (*Salix* spp.) in three New Zealand streams. New Zealand Journal of Marine & Freshwater Research 1995;29(4) 527-537. DOI: 10.1080/00288330.1995.9516685

[56] Ochs G. The ecology and ethology of whirligig beetles. Archiv für Hydrobiologie 1969;37 375-404.

[57] Power G. Seasonal growth and diet of juvenile Chinook salmon (*Oncorhynchus tshawytscha*) in demonstration channels and the main channel of the Waitaki river, New Zealand, 1982–1983. Ecology of Freshwater Fish 1992;1(1) 12-25. DOI: 10.1111/j. 1600-0633.1992.tb00003.x

[58] Oscoz J, Escala MC, Campos F. La alimentación de la trucha común (*Salmo trutta* L., 1758) en un río de Navarra (N. España). Limnetica 2000;18(1) 29-35.

[59] Steingrímsson SÓ, Gíslason GM. Body size, diet and growth of landlocked brown trout, Salmo trutta, in the subarctic river Laxá, North-East Iceland. Environmental Biology of Fishes 2002;63(4) 417-426. DOI: 10.1023/A:1014976612970

[60] Blanco-Garrido F, Sánchez-Polaina FJ, Prenda J. Summer diet of the Iberian chub (*Squalius pyrenaicus*) in a Mediterranean stream in Sierra Morena (Yeguas Stream, Córdoba, Spain). Limnetica 2003;22(3-4) 99-106.

[61] Clapp DF, Clark RD, Diana JS. Range, activity, and habitat of large, free-ranging brown trout in a Michigan stream. Transactions of the American Fisheries Society 1990;119(6) 1022-1034. DOI: 10.1577/1548-8659(1990)119<1022:RAAHOL>2.3.CO;2

[62] Metcalfe NB, Fraser NHC, Burns MD. State-dependent shifts between nocturnal and diurnal activity in salmon. Proceedings of the Royal Society of London, Series B 1998;265(1405) 1503-1507. DOI: 10.1098/rspb.1998.0464

[63] Meyer CG, Holland KN. Movement patterns, home range size and habitat utilization of the bluespine unicornfish, *Naso unicornis* (Acanthuridae) in a Hawaiian marine reserve. Environmental Biology of Fishes 2005;73(2) 201-210. DOI: 10.1007/s10641-005-0559-7

[64] Ramos-Jiliberto R, Valdovinos FS, Arias J, Alcaraz C, García-Berthou E. A network-based approach to the analysis of ontogenetic diet shifts: An example with an endangered, small-sized fish. Ecological Complexity 2011;8(1) 123-129. DOI: 10.1016/j.ecocom.2010.11.005

[65] Haury J, Ombredane D, Baglinière JL. L'habitat de la truite commune (*Salmo trutta* L.) en cours d'eau. In: Baglinière JL, Maisse G. (eds.) La trutie: biologie et écologie. Paris: INRA éditions; 1991. p47-96.

[66] Parra I, Almodóvar A, Ayllón D, Nicola GG, Elvira B. Ontogenetic variation in density-dependent growth of brown trout through habitat competition. Freshwater Biology 2011;56(3) 530-540. DOI: 10.1111/j.1365-2427.2010.02520.x

[67] Roussel J, Bardonnet A. Ontogeny of diel pattern of stream-margin habitat use by emerging brown trout, *Salmo trutta*, in experimental channels: influence of food and predator presence. Environmental Biology of Fishes 1999;56(1-2) 253-262. DOI: 10.1023/A:1007504402613

[68] O'Connor RR, Rahel FJ. A patch perspective on summer habitat use by brown trout *Salmo trutta* in a high plains stream in Wyoming, USA. Ecology of Freshwater Fish 2009;18(3) 473-480. DOI: 10.1111/j.1600-0633.2009.00364.x

[69] Martínez-Capel F, García de Jalón D, Werenitzky D, Baeza D, Rodilla-Alamá M. Microhabitat use by three endemic Iberian cyprinids in Mediterranean rivers (Tagus River Basin, Spain). Fisheries Management and Ecology 2009;16(1) 52-60. DOI: 10.1111/j.1365-2400.2008.00645.x

[70] Bachman RA, Reynolds WW, Casterlin ME. Diel locomotor activity patterns of wild brown trout (*Salmo trutta* L.) in an electronic shuttlebox. Hydrobiologia 1979;66(1) 45-47.DOI: 10.1007/BF00019138

[71] Railsback SF, Harvey BC, Hayse JW, LaGory KE. Tests of theory for diel variation in salmonid feeding activity and habitat use. Ecology 2005;86(4) 947-959. DOI: 10.1890/04-1178

[72] Tachet H, Richoux P, Bournaud M, Usseglio-Polaterra P. Invertébrés d'eau douce. Paris: CNRS Éditions; 2002.

[73] Hegge O, Hesthagen T, Skurdal J. Vertical distribution and substrate preference of brown trout in a littoral zone. Environmental Biology of Fishes 1993;36(1) 17-24. DOI: 10.1007/BF00005975

[74] Heggenes J, Baglinière JL, Cunjak RA. Spatial niche variability for young Atlantic salmon (*Salmo salar*) and brown trout (*S. trutta*) in heterogeneous streams. Ecology of Freshwater Fish 1999;8(1) 1-21. DOI: 10.1111/j.1600-0633.1999.tb00048.x

[75] Deady D, Fives JM. The diet of corkwing wrasse, *Crenilabrus melops*, in Galway Bay, Ireland, and in Dinard, France. Journal of the Marine Biological Association of the United Kingdom 1995;75(3) 635-649. DOI: 10.1017/S0025315400039060

[76] Magalhães MF. Effects of season and body-size on the distribution and diet of the Iberian chub *Leuciscus pyrenaicus* in a lowland catchment. Journal of Fish Biology 1993;42(6) 875-888. DOI: 10.1111/j.1095-8649.1993.tb00397.x

[77] Oscoz J, Leunda PM, Miranda R, Escala MC. Summer feeding relationships of the co-occurring *Phoxinus phoxinus* and *Gobio lozanoi* (Cyprinidae) in an Iberian river. Folia Zoologica 2006;5(4) 418-432.

[78] Akpan BE. Ontogenetic and monthly feeding behaviour of *Liza falcipinnis* (Mugilidae) from Cross River Estuary, Nigeria. World Journal of Applied Science and Technology 2011;3(2) 48-56.

[79] Magnussen E. Food and feeding habits of cod (*Gadus morhua*) on the Faroe Bank. ICES Journal of Marine Science 2011;68(9) 1909-1917. DOI: 10.1093/icesjms/fsr104

[80] Sánchez-Hernández J, Cobo F, González MA. Biología y la alimentación del salvelino, *Salvelimls fontinalis* (Mitchill, 1814), en cinco lagunas glaciares de la Sierra de Gredos. Nova Acta Científica Compostelana 2007;16 129-144.

[81] Friberg N, Andersen TM, Hansen HO, Iversen TM, Jacobsen D, Krojgaard L, Larsen, E. The effect of brown trout (*Salmo trutta* L.) on stream invertebrate drift, with special reference to *Gammarus pulex* L. Hydrobiologia 1994;294(2) 105-110. DOI: 10.1007/BF00016850

[82] Grossman GD, Boulé V. An experimental study of competition for space between rainbow trout (*Oncorhynchus mykiss*) and rosyside dace (*Clinostomus funduloides*). Canadian Journal of Fisheries and Aquatic Sciences 1991;48(7) 1235-1243. DOI: 10.1139/f91-149

[83] Facey DE, Grossman GD. The relationship between water velocity, energetic costs and microhabitat use in four North American stream fishes. Hydrobiologia 1992;239(1) 1-6. DOI: 10.1007/BF00027524

[84] Hill J, Grossman GD. An energetic model of microhabitat use for rainbow trout and rosyside dace. Ecology 1993;74(3): 685-698. DOI: 10.2307/1940796

[85] Gabler H-M, Amundsen P-A. Feeding strategies, resource utilisation and potential mechanisms for competitive coexistence of Atlantic salmon and alpine bullhead in a sub-Arctic river. *Aquatic Ecology* 2010;44(2) 325-336. DOI: 10.1007/s10452-009-9243-x

[86] Encina L, Rodríguez-Ruiz A, Granado-Lorencio C. Trophic habits of the fish assemblage in an artificial freshwater ecosystem: the Joaquin Costa reservoir, Spain. *Folia Zoologica* 2004;53(4) 437-449.

[87] Novakowski GC, Hahn NS, Fugi R. Diet seasonality and food overlap of the fish assemblage in a pantanal pond. Neotropical Ichthyology 2008;6(4) 567-576. DOI: 10.1590/S1679-62252008000400004

[88] Grossman GD, de Sostoa A, Freeman MC, Lobón-Cerviá J. Microhabitat use in a Mediterranean riverine fish assemblage. Fishes of the lower Matarraña. Oecologia 1987a;73(4) 490-500. DOI: 10.1007/BF00379406

[89] Grossman GD, de Sostoa A, Freeman MC, Lobón-Cerviá J. Microhabitat use in a Mediterranean riverine fish assemblage. Fishes of the upper Matarraña. Oecologia 1987b;73(4) 501-512. DOI: 10.1007/BF00379407

[90] Jepsen DB, Winemiller KO, Taphorn DC. Temporal patterns of resource partitioning among Cichla species in a Venezuela blackwater River. Journal of Fish Biology 1997;51(6) 1085-1108. DOI: 10.1111/j.1095-8649.1997.tb01129.x

[91] Stevens M, Maes J, Ollevier F. Taking potluck: trophic guild structure and feeding strategy of an intertidal fish assemblage. In: Stevens M. (ed.) Intertidal and basin-wide habitat use of fishes in the Scheldt estuary. Heverlee (Leuven): Katholieke Universiteit Leuven; 2006. p37-59.

[92] Alanärä A, Burns MD, Metcalfe NB. Intraspecific resource partitioning in brown trout: the temporal distribution of foraging is determined by social rank. Journal of Animal Ecology 2001;70(6) 980-986. DOI: 10.1046/j.0021-8790.2001.00550.x

[93] David BO, Closs GP, Crow SK, Hansen EA. Is diel activity determined by social rank in a drift-feeding stream fish dominance hierarchy? Animal Behaviour 2007;74(2) 259-263. DOI: 10.1016/j.anbehav.2006.08.015

Advances and Applications of Tracer Measurements of Carbohydrate Metabolism in Fish

Ivan Viegas, Rui de Albuquerque Carvalho,
Miguel Ângelo Pardal and John Griffith Jones

Additional information is available at the end of the chapter

1. Introduction

1.1. Relevance of carbohydrate metabolism studies in fish

The zebrafish *Danio rerio* has long been used as an animal model for developmental studies due to a number of desirable characteristics for lab study including its short generation time, large number of offspring, transparent embryos, and *ex utero* development of the embryo. This setting proved to be particularly useful for research on vertebrate development [1] and modeling of human disease [2], namely the hematopoietic [3, 4] and the cardiac systems [5, 6]. However, the use of fish for scientific studies has evolved substantially in the last 30 years. The direction of metabolic fish studies changed drastically with the advent of modern aquaculture and a high demand for models to study nutrition, physiology and metabolism. Capture of fish products in the end of the 1990's represented almost 75% of total production but since 2001 that has leveled at around 90 million tons. Meanwhile, aquaculture production has been increasing at an average annual growth rate of 6.2% from 38.9 million tons in 2003 to 52.5 million tons in 2008 [7]. Aquaculture in 2009 already accounted for 38% of the 145 million tones of total fish products, 81% of which were for human consumption [8]. It is estimated that by 2030, half of the production for human consumption will be derived from aquaculture while the harvest of wild fish will not show any significant growth [9]. Farming of aquatic species has existed for many thousands of years: in ancient Asia, carp were left to grow in ponds and rice paddies and later harvested. Similar practices were thought to take place in ancient Egypt with tilapia and in southern Europe in a polyculture regime, including mullet *Mugil* spp, sole *Solea* spp, seabass *Dicentrarchus labrax* and gilthead seabream *Sparus aurata* [10]. These artisanal methods, with little or no active manipulation of the animals or diets, are still practiced in some parts of the globe. However, as it has evolved into a highly competitive and commercialized business, the technique of aquaculture has also significantly evolved both in targeted species and in farming methodologies. Aquaculture is highly dependent on capture fisheries to provide fishmeal and fish oil required to produce feeds [11, 12]. Thus, the development of well-suited and cost-effective substitute feeds based on carbohydrates has become a matter of extreme importance to the sustainability and profitability of the sector. Along with it, comes the need to improve our understanding of fish carbohydrate nutrition, its physiology, biochemistry and metabolism.

1.2. Carbohydrate metabolism in carnivorous fish

The central organization of metabolic pathways is highly conserved amongst vertebrates. While the metabolic machinery of fish is much the same as that of mammals, the main differences lie in the nutritional and endocrine control of the pathways via various feedback mechanisms. As a water-living organism, fish have specific adaptations in relation to their terrestrial vertebrate counter-parts. As consequence of living in water, their constant thermal equilibrium with the environment is a principal determinant of overall metabolic rates while excess dietary nitrogen can be cleared as ammonia, a highly toxic but rapidly diffusible molecule that is efficiently transferred from blood to the outside water via the gills [13]. Carbohydrates (CHO) are a basic nutritional source of energy and carbon, but carnivorous fish have a limited capability to digest and metabolize them since they are adapted to a diet high in protein [14]. CHO influence growth, feed utilization and deposition of nutrients according to species, quantity, origin and treatment of dietary CHO used [15, 16, 17]. Fish are considered to be glucose intolerant on the basis of sluggish clearance of a glucose load and their metabolism of glucose has often been compared to that of mammals with insulin-dependent diabetes *mellitus* (IDDM) [18]. However, unlike IDDM, the causes of this intolerance cannot be attributed to a simple deficiency of insulin but is instead a reflection of different enzymatic and hormonal control regimes for glucose regulation in fish compared to mammals. The liver utilizes, produces and stores CHO and is an important component of systemic glycemic control in vertebrates. The rate of nutrient absorption coupled with continuous monitoring of blood glucose levels by various organs including the brain and pancreas mediates a combination of nutritional, endocrine and nerve-mediated regulation of hepatic glucose metabolism. This results in net hepatic glucose uptake when CHO is abundant, such as after feeding and net hepatic glucose output when dietary CHO is unavailable, such as during starvation. Glucose is the major energy source for the central nervous system and the only energy source for erythrocytes and the functions of both are dependent on a threshold concentration of blood glucose. Blood glucose levels are maintained through a balance of several factors, including the rate of consumption and intestinal absorption of dietary CHO, the rate of utilization of glucose by peripheral tissues and the loss of glucose through the kidney and finally the rate of removal or release of glucose by the liver [19].

Figure 1 depicts the pathways by which the liver disposes glucose into the bloodstream during fasting and storage and utilization of imported glucose during feeding. The main enzymes involved in these pathways are also represented as well as possible final fates of glucose anabolism. As for mammals, many tissues such as skeletal muscle can oxidize alternative substrates to glucose for energy such as triglycerides and can therefore function with variable levels of blood glucose [20] but the brain is highly dependent on glucose as an energy substrate therefore a threshold level of plasma glucose is required for maintaining its function. Not surprisingly, the brain is regarded as key component of endogenous glucose sensing mechanisms that have recently been described in fish models [21]. The plasma glucose concentration also has considerable influence on the selection of myocardial substrates for oxidation in resting and active fish [22] and this adaptation is representative of the changes that occur in whole-body glucose utilization in response to changes in glucose availability [23]. During hypoxia, tissues such as the heart and brain upregulate anaerobic metabolic pathways and become more reliant on glycolysis for ATP production [24]. However under normoxic conditions, glucose uptake by muscle is surpassed by that of amino acids and lipids for growth and energy [25]; this also happens during the recovery of muscle glycogen following exhaustive exercise. Thus in general, the direct contribution of blood glucose to muscle energy metabolism seems to be minor [20, 26, 27, 28, 29]. The generally low reliance of fish muscle on glucose as an energy substrate is compatible with the limited availability of carbohydrate in fish diets [18].

1.3. Hepatic glucose storage and disposal

In mammals, glucose enters the hepatocyte through the GLUT-2 transporter, which is primarily expressed in the liver but is also present in kidney, intestine and β-cells of pancreatic islets [19]. Due to its low affinity and high capacity, GLUT-2 transports hexoses in a large range of physiological concentrations necessary for glucose homeostasis, displaying bidirectional fluxes in and out the cells, namely in hepatocytes [30]. Recently, the presence of GLUT-2 transporter was confirmed at both biochemical and molecular levels in species like rainbow trout *Oncorhynchus mykiss* [31, 32] and its expression in seabass *D. labrax* hepatocytes seems to be regulated by hypoxia.

Blood glucose concentration is the net result of the difference between rates of glucose appearance and disposal. Following uptake into the hepatocyte, the ATP-dependent conversion of glucose to glucose 6-phosphate (G6P) is catalyzed by hexokinase IV or glucokinase (GK). GK is the major hexokinase expressed in liver and due to its relatively high K_m, its role in glucose homeostasis, is to increase the hepatic capacity for glucose metabolism during hyperglycemia. Uniquely, GK is not inhibited by its G6P product so it is able to maintain high rates of glucose phosphorylation while the activities of other hexokinases become limited by increased G6P levels. Moreover, GK activity is hormonally controlled via the glucokinase regulatory protein (GKRP), such that its activity is increased by the high insulin levels of the fed state, but rapidly suppressed during the onset of fasting when both portal vein glucose and insulin levels are waning. As a result, when the liver becomes a net producer of glucose during the fasting state, GK is prevented from converting this back to G6P.

The G6P product can be utilized by the pentose phosphate pathway, be recruited for glycogen synthesis (glycogenesis) or be metabolized by the glycolytic pathway to pyruvate. Pyruvate in turn may be oxidized to acetyl coenzyme A (acetyl-CoA) or carboxylated to oxaloacetate or malate – a process known as anaplerosis. Acetyl-CoA may be oxidized to CO_2 by the tricarboxylic acid cycle (TCA cycle) or utilized for lipid production (lipogenesis) while anaplerotic products can be utilized for gluconeogenesis. G6P can also be hydrolyzed back to glucose, a reaction catalyzed by glucose 6-phosphatase (G6Pase). In fish, it has been shown that G6Pase is poorly regulated by dietary CHO in comparison to GK [33]. When G6Pase and GK are both active at the same time, glucose and G6P are interconverted in a "futile cycle" that results in the consumption of ATP but no net conversion of glucose to products. This activity has important implications for interpreting the metabolism of certain glucose tracers such as [2-^{2}H]- or [2-^{3}H]glucose since it results in disappearance of the label without any net glucose consumption.

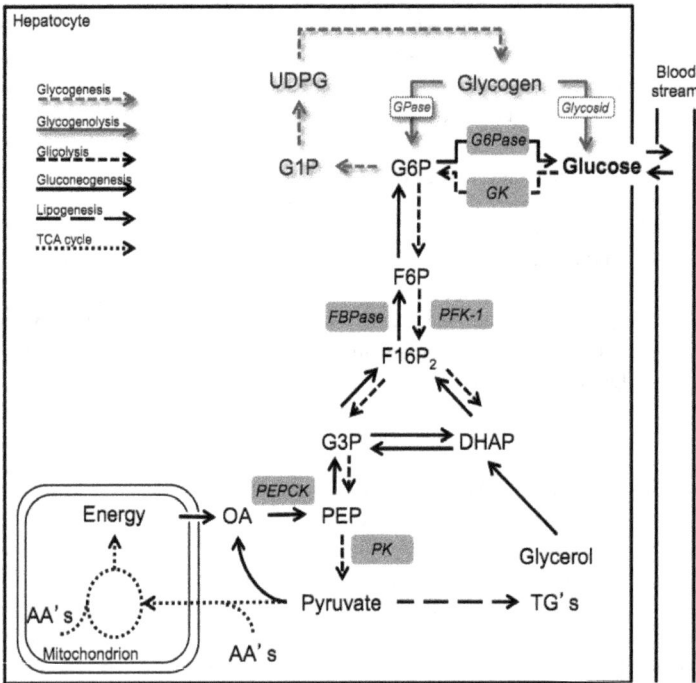

Figure 1. Metabolic model representing main pathways in the liver involving catabolism and anabolism of glucose. Gluconeogenic precursors are represented by pyruvate (and gluconeogenic amino acids, metabolized via the anaplerotic pathways of TCA cycle) as well as glycerol from lipolysis. Some metabolic intermediates were omitted for clarity. Abbreviations are as follows: G6P - glucose 6-phosphate; G1P - glucose 1-phosphate; UDPG - uridine diphosphoglucose; F6P - fructose 6-phosphate; F16P2 - fructose-1,6-bisphosphate; G3P - glyceraldehyde 3-phosphate; DHAP - dihydroxyacetone phosphate; PEP - phosphoenolpyruvate; TG's - triglycerides; OA - oxaloacetate; AA's - amino acids; GK - glucokinase; G6Pase - glucose-6-phosphatase; FBPase - fructose bisphosphatase; PFK-1 - 6-phosphofructo-1-kinase; PEPCK - phosphoenolpyruvate carboxykinase; PK - pyruvate kinase; GPase - glycogen phosphatase; Glycosid - glycosidic enzymes; TCA cycle - tricarboxylic acid cycle.

1.4. Glycogen synthesis and hydrolysis

The liver is the main storage site for glycogen, but tissues such as gills, kidney and brain also sustain relatively high rates of glycogen synthesis from glucose. Net glycogen synthesis results from the glycogen synthase (GSase) and glycogen phosphorylase (GPase) activities that are reciprocally regulated both allosterically and by phosphorylation [34].

GSase exists in a phosphorylated inactive form in the cytosol and it is activated by binding to G6P and this causes a conformational change that makes it a better substrate for protein phosphatases, which then convert the enzyme to the active dephosphorylated isoform. In glycogenesis, G6P is first converted to glucose 1-phosphate (G1P) by phosphoglucomutase and subsequently to uridine diphosphate glucose (UDPG) by UDPG pyrophosphorylase. Glycogen synthase then adds glucose residues from the UDPG donor to the growing glycogen molecule via α-1,4 glycosidic linkages. Hepatic glycogen can be synthesized by the incorporation of glucose into glycogen via a sequence of reactions that was initially characterized in muscle. In this so-called "direct pathway", the glucosyl moiety is transferred into glycogen as an intact entity. It was subsequently discovered that in the liver, glycogen can also be synthesized independently from glucose via gluconeogenic precursors [19]. These contributions are collectively referred to as the "indirect pathway. Since the indirect pathway bypasses GK, and uses precursors that are independent of blood glucose, its contribution to glycogen synthesis may respond to nutritional and hormonal states in a different way to that of the direct pathway. Quantifying the direct and indirect pathway flux components of hepatic glycogen synthesis can therefore further inform adaptations of hepatic carbohydrate metabolism to different diets and hormonal states.

Under fasting conditions, the liver becomes a net producer of glucose and the hydrolysis of glycogen is an important source of glucose production at least in the early stages of fasting. To avoid futile cycling between G6P and glycogen, the activities of GSase and GPase are regulated in a reciprocal manner such that only one or the other is active at a given time. Glycogen may be also hydrolyzed to glucose by the action of glucosidases enzymes that hydrolyze the non-reducing end of polysaccharides directly to glucose. There is evidence of glucosidase activity alongside that of GPase in living fish [35, 36, 37], but the extent of its contribution to glucose synthesis from glycogen and modulation of activity by feeding and fasting are little known.

Most fish species need to periodically cope with starvation as a result of seasonal food limitation or in some cases as natural consequence of their life cycle [38] and the liver plays an important role in the endogenous control of fuel storage and mobilization [39, 40, 41]. Glycogen represents an important source of glucose to be released in the fasted state or when responding to acute stressors. The transition from fasting to feeding states results in a comprehensive realignment of hepatic carbohydrate metabolic fluxes from minimal maintenance of glycemia and peripheral glucose demands by endogenous glucose production during the fasting phase to a high nutrient inflow and replenishment of liver glycogen stores during refeeding [39, 40, 42, 43]. The general consensus is that the liver of fish plays a role in glycemic control through the regulation of hepatic glucose storage and mobilization [33] but its actions are sluggish in comparison to mammalian liver.

1.5. Glycolysis and gluconeogenesis

Glycolysis involves the metabolism of glucose or glycogen to pyruvate. From glucose,the first step in the glycolytic pathway is the phosphorylation of glucose by GK into G6P, as described. From glycogen, G6P is generated following GPase and phosphoglucomutase activities. G6P is isomerized to F6P via G6P isomerase and this is followed by another phosphorylation of fructose 6-phosphate (F6P) to fructose 1,6-bisphosphate (F16P$_2$) catalyzed by 6-phosphofructo 1-kinase (PFK-1). These reactions commit the hexose carbon skeletons to pyruvate production. Both phosphorylations are highly regulated and along with conversion of phosphoenolpyruvate (PEP) to pyruvate catalyzed by pyruvate kinase (PK) constitute the irreversible steps of glycolysis. The energy charge of the cell, as well as allosteric and transcriptional processes, contributes to control fluxes through PFK-1 and PK. The free energy released in this process is coupled to substrate-level ATP and NADH formation. Pyruvate is the final product of glycolysis and can undergo further oxidation to acetyl-CoA via pyruvate dehydrogenase (PDH), carboxylation via pyruvate carboxylase (PC) or NADP-malic enzyme or reduction to lactate. In the liver, the relative activities of PDH and PC determine whether pyruvate is oxidized or utilized by anaplerotic pathways such as gluconeogensis hence the activities of these enzymes are highly regulated depending on the nutritional state. Gluconeogenesis is the principal metabolic pathway that generates glucose in fish. Its main precursors include lactate (converted into pyruvate by lactate dehydrogenase) and gluconeogenic amino acids (those that are metabolized to pyruvate or C4 and C5 TCA cycle intermediates), with minor contributions from glycerol, derived by lipolysis of triglycerides. While conversion of pyruvate to glucose occurs via the same intermediates as glycolysis, different enzymes are used to overcome the unfavorable free energy change of pyruvate to PEP conversion. The conversion of F16P$_2$ to F6P and G6P to glucose are also mediated by different enzymes to their glycolytic counter-parts (fructose 1,6-bisphosphatase and glucose 6-phosphatase, respectively). Pyruvate is converted to PEP at the expense of two ATP equivalents. First, pyruvate carboxylase generates oxaloacetate (OA) from pyruvate and second, PEP carboxykinase (PEPCK) converts OA to PEP. The free energy change of pyruvate conversion to glucose is positive, hence pyruvate gluconeogenesis requires energy in the form of ATP and reducing equivalents. This is not the case for all gluconeogenic precursors, notably glutamate and glutamine. These are metabolized to α-ketoglutarate, which during its oxidation to OA via the TCA cycle, generates ATP and reducing equivalents in excess of those consumed by PEPCK and glyceraldehyde 3-phosphate dehydrogenase. It is important to note that these amino acids are among the most abundant in dietary protein, therefore from a thermodynamic viewpoint, they facilitate gluconeogenesis from protein.

1.6. Enzymatic and hormonal regulation

Measurement of enzymatic activities involved directly or indirectly in CHO metabolism has been performed in various species and tissues and has proved to be of crucial importance in evaluating adaptation to changes in temperature [44, 45], salinity and osmoregulation [46, 47], feeding status [39, 41, 48, 49], rearing densities [50], diets [51, 52, 53, 54, 55, 56, 57] and to study the effects of hormones [58, 59, 60, 61].

Hormones in vertebrate organisms are a diverse but well conserved group of signaling molecules that regulate and modulate metabolic fluxes. The role of hormones in the regulation of CHO metabolism of fish are far from being completely known due to the variety of responses observed - many of which are dependent on whether studies are performed on intact fish, isolated organs or primary cell cultures. The mammalian paradigm, to which the hormonal regulation in fish is usually compared, has shown to be poorly representative of fish metabolism in a number of cases [62]. Glucose is weakly effective at stimulating insulin secretion in fish compared to mammalian species, although, as expected, an improved response is observed in CHO-tolerant species [63]. On the other hand, amino acids are potent stimulators of insulin secretion even though the intensity of response varies greatly between salmonids and other fish like carps and seabream. This bears some resemblance to mammals, where arginine is also a powerful insulin secretagogue.

In mammals, the actions of insulin are directly opposed by glucagon and adrenal hormones such as adrenaline and these counterregulatory actions are an important component of glucose homeostasis. However, it is unclear in fish to what extent the regulatory and counterregulatory processes are coupled [62]. For example, in response to stress or hypoglycemia, the adrenaline-induced hyperglycemic response of rainbow trout is caused by the stimulation of hepatic glucose production that happens in a dose-dependent fashion [64]. However, the accompanying suppression of glucose clearance described in mammals is not observed in fish. Cortisol is a major regulator of intermediary metabolism and promotes hepatic glucose production and hyperglycemia, primarily as a result of increased hepatic gluconeogenesis fuelled in part by amino acid products of peripheral proteolysis [65].

2. Tracer studies of carbohydrate metabolism in fish

2.1. Overview

A metabolic tracer is, by definition, a substance used to follow the biological transformation of an endogenous substrate (tracee). The tracer must have a unique property that allows its detection but at the same time be chemically identical to the tracee. In most cases, tracers consist of synthetic substrate molecules where one or more atoms in the molecule are substituted for an atom of the same chemical element, but of a different isotope. Isotopic tracers may be identified by radioactivity or differences in mass and/or nuclear spin from the tracee. In order that the tracee pool of interest be not altered by administration of the tracer (thereby potentially perturbing the metabolic activity under observation), its chemical concentration should be insignificant in comparison to that of the tracee. At the same time, this amount of material must possess a high abundance of label relative to the tracee since metabolic flux parameters are typically derived by administering a high density of label and monitoring its dilution by unlabeled tracee molecules. With radioisotopes, the label concentration is measured in terms of specific activity, defined as the number of decays per minute measured per mol of glucose. With stable isotopes, tracer concentration is defined in terms of percent excess enrichment per mol of glucose. Since most stable isotopes are typically present at background levels (for example, 1.1% of carbon in nature is ^{13}C), this must be accounted for when attributing metabolite enrichment to inflow from a tracer.

Since blood glucose and other carbohydrates can be synthesized endogenously as well as absorbed from the diet, several tracer strategies may be applied to inform carbohydrate metabolism. The first and most easily understood is to incorporate the tracer into dietary carbohydrate and monitor its rate of appearance in endogenous carbohydrate pools (i.e. plasma glucose and liver glycogen) as well as metabolic endproducts (i.e. respired CO_2). This approach provides a direct measure of how a specific dietary carbohydrate source is utilized and disposed into endogenous oxidative and nonoxidative pathways. Oral administration of labeled food [66], either by pellets with ^{13}C-labeled starch and ^{15}N-labeled protein [67] or ^{13}C-labeled rotifers [68], and measurement of isotope appearance in endogenous carbohydrate and protein have also been used to measure substrate utilization in fish. In the case of carbohydrate studies, while the dilution of ^{13}C-enrichment between the feed precursor and sampled product metabolite may qualitatively inform the extent of endogenous glucose production (i.e. high dilution equates to high rates of endogenous glucose production relative to absorption of the ^{13}C-enriched carbohydrate), quantitative data rely on knowing the rate of entry of the ^{13}C-enriched dietary carbohydrate into blood glucose. Since this is a function of absorption and cannot be directly measured, this approach provides only a limited insight into endogenous glucose production. In the second type of measurement, a glucose tracer is directly administered into the blood at a known rate or quantity, and by measuring its dilution by unlabeled endogenous glucose, the rate of appearance (R_a) of blood glucose may be quantitatively defined. Since both absorbed and endogenously-produced glucose contribute to the dilution of the labeled glucose, this measurement does not resolve the contributions of dietary and endogenous sources. Under fasting conditions where there is no input of dietary glucose, glucose R_a equates to endogenous glucose production. In mammals and humans, dietary and endogenous components of glucose R_a may be resolved by combining a feed tracer with a second glucose tracer administered intravenously [69]. The detection method must have the capability of resolving the labeling contribution of blood glucose from each tracer. For example, radioactivity from infused [3-^{3}H]glucose and ingested [U-^{14}C]glucose are resolved by scintillation counters on the basis of their different emission energies. Finally, as previously discussed, endogenous glucose can be derived from multiple sources including gluconeogenesis from a diverse range of precursors as well as glycogen hydrolysis. Neither the labeled dietary carbohydrate nor glucose isotope dilution measurements inform the sources of endogenous glucose production. This represents a significant limitation for understanding carbohydrate metabolism of carnivorous fish since their diet is low in carbohydrate therefore the plasma glucose rate of appearance (R_a) is likely to be dominated by synthesis from endogenous precursors. Moreover in aquaculture, weaning of carnivorous fish from pure fishmeal to diets supplemented with carbohydrates is an important objective for improved sustainability and reduced environmental impact [18]. Hence, there is continuing interest in understanding to what extent their metabolic phenotype can adapt to increased dietary carbohydrate availability and if gluconeogenic utilization of amino acids is spared under these conditions [16]. To effectively address such questions and to improve our overall understanding of piscine glycemic control, endogenous and exogenous contributions to glucose R_a need to be better defined than they are at present.

2.2. Radioactive versus stable isotope carbohydrate tracers

Historically, the use of radioactive isotopes such as ^{14}C and ^{3}H has provided valuable information characterizing the main pathways of hepatic carbohydrate metabolism. They are easily detected by scintillation counting and their background radioactivity is very low in comparison to metabolite specific activity. Due to the fact that they are weak α- and β-ray emitters, and are therefore easily contained and have relatively high maximum permissible dosages, these tracers are still used in some human and many animal studies. They are more widely used in cell cultures for substrate uptake studies and isotope-dilution measurements where the small scale of the experiments can be accommodated by reasonable radiation containment measures. For this same reason, radioactive isotopes have been particularly useful in the study of larval nutrition. A supply of good quality fish fry is essential for a successful production of juveniles however knowledge of their nutritional requirements is often qualitative rather than quantitative [70]. In this context the innovative methodologies of incorporation of ^{14}C-labeled amino acids [71] or fatty acids [72] in larvae live prey *Artemia* sp. nauplii has improved our understanding of larval nutrition. However, for conventional nutritional studies involving larger fish and tanks, containment of radioactive tracers is often not practically possible therefore in these settings, their stable isotope counterparts (^{13}C and ^{2}H) are typically deployed. Methods for obtaining glucose rates of appearance (R_a) that were originally developed using [3]H- and ^{14}C-labeled glucose [23, 73, 74, 75, 76] can be easily adapted for stable isotope studies. Moreover, stable isotope tracers can potentially yield much richer metabolic information than their radioactive counterparts due to advances in positional and isotopomer analysis of metabolite enrichment from ^{13}C and ^{2}H that go above and beyond those which are possible for radiolabeled tracers.

2.3. Tracer administration and measurements in fish compared to mammals

Tracer measurements of whole-body carbohydrate metabolism were originally developed in humans and experimental animal models such as rodents, dogs and primates. Translating these methodologies to fish presents some additional challenges, the most important being the following: firstly, most mammals have accessible peripheral veins that can be catheterized for tracer administration and blood sampling. Catheters may be installed temporarily under anesthesia for the duration of the tracer study, or as indwelling entities for longitudinal studies of conscious free-moving animals. In fish, the dorsal aorta proximal to the gill arches is the main accessible vessel for tracer studies. It has been successfully cannulated for tracer infusion and blood sampling in undisturbed, non-anaesthetized rainbow trout of 800-1200 grams in weight [77, 78, 79]. However, it is unclear if this method can be easily applied to smaller sized fish, or to fish with a different gill-body configuration to rainbow trout, for example flatfish. The second major impediment for fish carbohydrate tracer studies is the fact that glucose appearance and turnover rates are much slower in comparison to mammals, necessitating far longer infusion periods for achieving isotopic steady-state conditions – a critical requirement for investigating precursor-product relationships of biosynthetic pathways. This is particularly problematic for quantitative studies of gluconeogenesis based on the delivery of a labeled precursor substrate such as ^{14}C-lactate or alanine, since both the tracee precursor and glucose product pools must be at isotopic equilibrium during the sampling period.

2.4. Measurement of glucose R_a by bolus injection and by primed infusion

There are two approaches for quantifying plasma glucose R_a by isotope dilution. The first involves an intravenous injection of a tracer bolus and monitoring the decrease in specific activity or enrichment of plasma glucose as the tracer is being diluted by unlabeled glucose [75]. The dilution kinetics are best represented by more than one exponential decay function indicating the presence of separate pools of glucose in the body with different clearance characteristics. Because of the relative simplicity of a single injection tracer delivery, this method has been applied in many fish species including kelpbass *Paralabrax clathratus* [73], seabass *D. labrax* [74], *Hoplias malabaricus* [76], common carp *C. carpio* [23]. However, this approach requires frequent blood sampling over a sustained period to adequately describe the complex tracer clearance kinetics. Limited by the number of blood samples that could be drawn from the fish, the study described in [73] extrapolated the clearance kinetics from a smaller set of initial measurements, but the uncertainties of this approach were acknowledged. The alternative primed-infusion method establishes a constant level of tracer in the bloodstream following an appropriate priming dose. Under these conditions, the dilution of the tracer, measured from the ratio of specific activity or enrichment of infused label to that of blood glucose, is equal to the ratio of endogenous glucose appearance and tracer infusion rates. If glucose R_a is variable (for example during meal ingestion) then the rate of infusion needs to be adjusted accordingly in order to maintain a constant ratio of infused glucose to blood glucose specific activity or enrichment. Compared to the bolus injection method, primed infusion measurements require fewer blood samplings; in fact the sole rationale for multiple blood samplings is to verify a constant ratio of infused to blood glucose specific activity or enrichment. Calculation of glucose R_a from steady-state isotope data is more robust compared to single injection since it is independent of the complex and often poorly defined clearance kinetics. However, a primed infusion requires catheterization of a vein to deliver the tracer over an extended period and is technically more difficult than a single bolus injection. While these procedures now allowed glucose R_a to be well determined, and by combination with dietary tracers can determine the contributions of absorbed and endogenous glucose production to glucose R_a, they do not inform the sources of endogenous glucose production. Novel methodologies of resolving the sources of glucose R_a using deuterated water (^2H_2O) have been developed and can be integrated with primed-infusion glucose R_a measurements [80, 81, 82]. ^2H_2O is ideally suited for fish metabolic studies since it can be incorporated into aquarium water for an indefinite period and is rapidly incorporated into the fish body water such that the ^2H-enrichment level of the fish tissue water is also fixed for the duration in the ^2H-enriched tank water. As a result of enzyme-catalyzed exchange reactions between bulk water and metabolite hydrogens, many important precursor metabolites of glucose and glycogen biosynthesis, such as lactate, pyruvate and alanine also become rapidly enriched to the same level as body water for the study duration. Thus, a constant precursor enrichment that can be indefinitely maintained is possible. ^2H_2O has recently been used to study protein synthesis kinetics in the catfish *Ictalarus punctatus* [83] and gilthead seabream *S. aurata* [84] as well as blood glucose and hepatic glycogen turnover in seabass *D. labrax* [43, 85]. Because of its enormous potential for informing endogenous carbohydrate metabolism in fish, we will discuss its use in more detail.

3. Deuterated water as a tracer of endogenous glucose and glycogen synthesis

3.1. Overview

Fish and mammals share common pathways for glucose production and consumption [86] hence the underlying principles of plasma glucose ^2H-enrichment from ^2H_2O that are well described and validated for mammals [80, 87, 88] can be applied to fish. Deuterated water (^2H_2O) is a relatively inexpensive non-radioactive tracer that can be incorporated in drinking water, or in the case of fish studies, in the tank water. It has been successfully used in humans and other mammals for the study of hepatic intermediary metabolism in both normal and pathological conditions. It rapidly equilibrates with total body water and is distributed evenly into all tissues. It is a practical tracer for both short and long-term metabolic studies. ^2H_2O is ideally suited for studying fish metabolism since it can be added to the tank water for an indefinite period, during which time it is incorporated into hepatic metabolites such as glycogen and glucose by specific enzymatic reactions in their biosynthetic pathways, as previously described for mammals. Applying these principles to free-swimming fish provides an authentic metabolic profile that is unadulterated by anesthesia or infusion procedures that characterize the administration of classical carbon tracers.

3.2. Basic principles

Deuterium (^2H) is a stable isotope of hydrogen with a nucleus containing one proton and one neutron (the ^1H nucleus contains no neutron). In the NMR experiment, ^2H resonates at a different frequency compared to its ^1H counterpart, allowing tracer levels of ^2H to be observed in the presence of the tracee ^1H. The inherent sensitivity of ^2H (at constant field and with an equivalent number of nuclei) is about 0.9% that of ^1H. Metabolism of ^2H is not exactly equivalent to that of ^1H because of kinetic isotope effects. The strength of a chemical bond between two atoms is dependent in part on their relative masses, hence a $C-^2H$ bond is stronger than a $C-^1H$ for any compound. Since metabolite transformation is governed in part by breaking and formation of C-H bonds, the presence of ^2H makes the bonds harder to break thereby potentially slowing the rate of $C-^2H$ *vs.* $C-^1H$ transformation. This can discriminate the transformation of ^2H-enriched metabolites compared to their tracees resulting in apparently slower rates of transformation. Moreover, with bulk levels of ^2H tracers, notably

^2H_2O, the aggregate isotope effects are toxic and indeed lethal to most living organisms. With tracer studies that utilize ^2H_2O, toxicity from isotope effects is minimized by substituting a relatively low proportion of ^1H by ^2H (<10%). Furthermore, discrimination against ^2H incorporation into metabolites via enzymatic reactions is minimized when the reaction that transfers ^2H from water to the metabolite hydrogen is reversible. Most of the enzymatic steps of intermediary metabolism are reversible with extensive exchange of precursor and product, and under these conditions, discrimination of ^2H incorporation via kinetic isotope effects is not significant. Indeed, when ^2H-discrimination is observed, it informs the unidirectionality of a particular enzymatic step.

These caveats notwithstanding, ^2H_2O is an inexpensive tracer that is easily delivered into body water by immersion of fish in ^2H-enriched water. It has been successfully used in humans [89, 90, 91] and other mammals [82, 92, 93], for study of hepatic carbohydrate metabolism in physiological and pathophysiological conditions. The methodology was pioneered in humans [80, 87, 94] and was rapidly adopted by others in tissues [81, 88, 95]. As previously discussed, G6P is a common precursor to both glycogen and glucose and in the presence of ^2H_2O, G6P is labeled with ^2H in several positions due to the incorporation of that isotope via exchange with body water. It rapidly equilibrates with total body water and distributes homogeneously within tissues and the body water ^2H-enrichment level can be maintained indefinitely [96]. In fish, incorporation of ^2H from a 5%-enriched saltwater tank into plasma water is rapid, reaching more than 1% in 15 minutes, half of the enrichment of the tank water within 1 hour and approaching that of the tank water after 6h [85].

For studies of carbohydrate metabolism, the ^2H-enrichment distribution of plasma glucose from ^2H_2O is established according to origin of the G6P precursor (Figure 2.).

If produced from gluconeogenic substrates (gluconeogenic amino acids, pyruvate or glycerol) enrichment in position 2 (H2) is obligatory since conversion of F6P to G6P (facilitated by G6P-isomerase) is part of the gluconeogenic pathway. In mammals, G6P-F6P exchange is extensive and essentially complete hence the hepatic G6P pool is quantitatively enriched in H2 regardless of its origin. This means that newly-synthesized glycogen from G6P is also enriched in this position, regardless of whether the G6P was derived via the direct or indirect pathway [93, 97]. There is some evidence that in fish, hepatic G6P-isomerase activity is sensitive to the nutritional state [43]. In common carp *C. carpio* hepatic G6P-isomerase activity decreased in direct relation to feeding rates [98, 99] but there is no information on how G6P-isomerase activity is modified during the fasting to feeding transition. One possible explanation is that induction of G6P-isomerase activity by feeding is slow compared to activation of glycogen synthesis fluxes. Under these conditions, G6P-isomerase activity could be a rate-liming step for indirect pathway synthesis of glycogen, at least in the initial stages of refeeding. In principle, sub-maximal G6P-isomerase activity could limit the glycolytic metabolism of G6P derived from glucose and favor its conversion to glycogen via the direct pathway or its utilization by the pentose phosphate pathway.

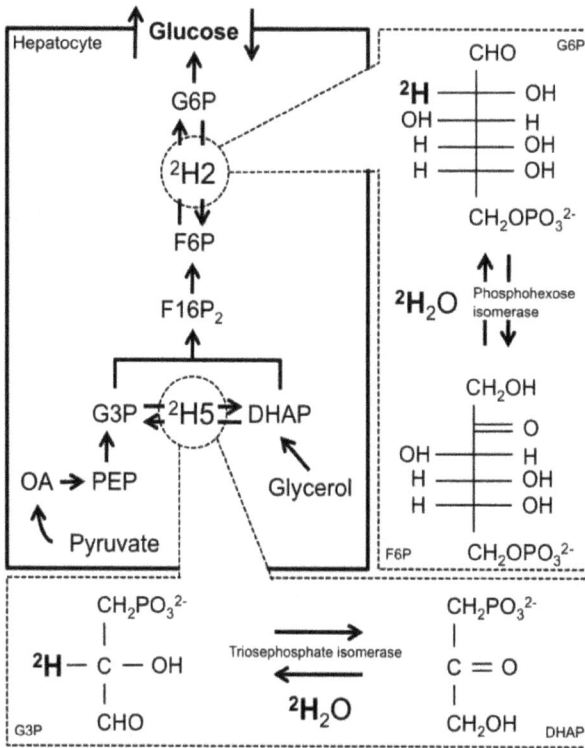

Figure 2. Metabolic model representing gluconeogenesis in the liver with special detail to the labeling with deuterium in position 2 ($^2H2$) and 5 ($^2H5$) of the glucose molecule. Some metabolic intermediates were omitted for clarity. Abbreviations are as follows: G6P - glucose 6-phosphate; F6P - fructose 6-phosphate; F16P$_2$ - fructose 1,6-bisphosphate; G3P - glyceraldehyde 3-phosphate; DHAP - dihydroxyacetone phosphate; PEP - phosphoenolpyruvate; OA - oxaloacetate; ^2H_2O - deuterated water.

Besides, ^2H-Enrichment in position 5 of G6P occurs at the level of triose phosphates as shown in Figure 2. Therefore, enrichment of plasma glucose or hepatic glycogen in position 5 reflects the contribution of gluconeogenic fluxes to endogenous glucose production or indirect pathway contributions to hepatic glycogen synthesis.

3.3. NMR and MS methods for ^2H enrichment analysis

Many improvements on the measurement of stable isotope tracer enrichment and analysis of positional labeling information have been largely driven by the development of nuclear magnetic resonance (NMR) and mass spectrometry (MS) technologies. Choosing between these methods depends on the kind of enrichment information that is required from the experiment, the available sample size and access to instrumentation. MS techniques quantify metabolite enrichment by resolving heavier labeled molecules from lighter unlabeled ones. For most MS instruments, the presence of two isotopes with similar increase in the molecular mass, (i.e. ^{2}H and ^{13}C) cannot be resolved, placing limitations on multiple isotope studies. Positional enrichment can be inferred from fragmentation and analysis of the mass of the daughter fragments (MS-MS). Nevertheless, as fragmentation is dependent on the molecule's chemical structure, the label of interest may or may not be isolated. Chemical derivatization of metabolites is often used to facilitate fragmentation and positional enrichment analysis [87]. MS is highly sensitive and can quantify enrichments from submicromole to picomole amounts of analyte and with appropriate signal calibration and sample purification safeguards, it can be configured for high throughput measurements.

Following a simple method based on a LC-MS/MS procedure for quantifying plasma [6,6-^{2}H$_2$]glucose enrichment [100] which does not require glucose derivatization, analysis of glucose can be performed on a few microliters of blood, either whole or as a dried spot on filter paper. This means it can be applied to any size fish and can also be used for repeated sampling of the same fish [85]. This LC-MS/MS measurement provides the mole percent enrichment (MPE) of the glucose molecule, equivalent to the sum of all seven positional enrichments. The principal uncertainties of utilizing plasma MPE levels as a marker of gluconeogenic contribution include the incomplete incorporation of ^{2}H into sites other than position 5, as seen by the tendency for lower enrichments in positions 1, 3, 4, 6_R and 6_S compared to position 5 by ^{2}H-NMR analysis [85].

Analysis of ^{2}H enrichment by ^{2}H NMR spectroscopy is a method with much lower sensitivity compared to MS, requiring 5-50 μmol of analyte in the typical experimental setting for ^{2}H$_2$O studies (0.5-5.0 % body water enrichment). However, in addition to being nondestructive to the sample, NMR provides a much higher level of positional enrichment information, allows enrichment from multiple stable isotope tracers to be selectively observed, and can provide a global analysis of metabolite enrichments from a complex mixture of metabolites, such as cellular extracts, biological fluids, and intact tissues [101]. This technique relies on the ability of atomic nuclei with odd mass and/or atomic number to align if subjected to an external magnetic field. When irradiated with a certain frequency signal the nuclei in a molecule can change their alignment and the energy frequency at which this occurs can be measured and displayed as an NMR spectrum. Common biologically relevant nuclei that are present at ~100% natural abundance and are observed by NMR include ^{1}H and ^{31}P and ^{23}Na. Isotopes that are more rare in nature such as ^{2}H (0.015% of hydrogen) and ^{13}C (1.11% of carbon) can also be observed at natural abundance levels, but molecules that are enriched to higher levels from ^{2}H- or ^{13}C-enriched precursors can be measured against this background. Since isotopes resonate at a specific frequency, its signals can be uniquely isolated from any other isotope that may be present. Derivatization of the target molecule can be used to provide a more heterogeneous chemical environment therefore improving signal dispersion [102, 103, 104]. This is particularly important for analysis of carbohydrate ^{2}H enrichment, which feature highly crowded hydrogen signals that are poorly resolved by the inherently small dispersion of ^{2}H signals (~15% of ^{1}H signals).

4. Conclusions

There is a compelling need to better understand the metabolism of carbohydrates by fish in general and aquaculture species in particular. Stable-isotope tracer methodologies have evolved such that safe, inexpensive and practical measurements of carbohydrate metabolism may be directly performed on naturally feeding fish in the aquaculture setting. These studies have great potential for informing the efficacy of novel dietary supplements in sparing the conversion of feed protein to carbohydrate as well as improving our general understanding of fish nutritional physiology.

Acknowledgements

We would like to thank Fundação para a Ciência e Tecnologia (FCT) for funding the project "Optimizing carbohydrate utilization in farmed sea-bass through metabolic profiling" (PTDC/EBB-BIO/098111/2008) through the programme COMPETE - Operational Competitiveness Programme.

Author details

Ivan Viegas[1,2,3], Rui de Albuquerque Carvalho[1,2], Miguel Ângelo Pardal[1,3] and John Griffith Jones[2]

*Address all correspondence to: iviegas@ci.uc.pt

1 Department of Life Sciences, Faculty of Sciences and Technology, University of Coimbra, Portugal

2 CNC - Center for Neuroscience and Cell Biology, University of Coimbra, Portugal

3 CFE - Centre for Functional Ecology, University of Coimbra, Portugal

References

[1] Roush W. Zebrafish Embryology Builds Better Model Vertebrate. Science 1996;272: 1103.

[2] Lieschke GJ, Currie PD. Animal models of human disease: zebrafish swim into view. Nature Reviews Genetics 2007;8: 353-367.

[3] Berman JN, Kanki JP, Look AT. Zebrafish as a model for myelopoiesis during embryogenesis. Experimental Hematology 2005;33: 997-1006.

[4] Ellett F, Lieschke GJ. Zebrafish as a model for vertebrate hematopoiesis. Current Opinion in Pharmacology 2010;10: 563-570.

[5] Bakkers J. Zebrafish as a model to study cardiac development and human cardiac disease. Cardiovascular Research 2011:

[6] Feitsma H, Cuppen E. Zebrafish as a Cancer Model. Molecular Cancer Research 2008;6: 685-694.

[7] FAO Fisheries and Aquaculture Department. The State of World Fisheries and Aquaculture, 2010. Rome: FAO; 2010.

[8] FAO Statistics and Information Service of the Fisheries and Aquaculture. FAO yearbook. Fishery and Aquaculture Statistics, 2009. Rome: FAO; 2011.

[9] Failler P. Future prospects for fish and fishery products. 4. Fish consumption in the European Union in 2015 and 2030. Part 1. European overview. FAO Fisheries Circular No 972/4, Part 1 Rome, FAO 2007 204p 2007:

[10] Pillay TVR, Kutty MN, Aquaculture: Principles and Practices. Wiley-Blackwell; 2005.

[11] Kristofersson D, Anderson JL. Is there a relationship between fisheries and farming? Interdependence of fisheries, animal production and aquaculture. Marine Policy 2006;30: 721-725.

[12] Tacon AGJ, Metian M. Global overview on the use of fish meal and fish oil in industrially compounded aquafeeds: Trends and future prospects. Aquaculture 2008;285: 146-158.

[13] Ip YK, Chew SF. Ammonia Production, Excretion, Toxicity, and Defense in Fish: A Review. Frontiers in Physiology 2010;1: 134.

[14] Oliva-Teles A. Nutrition and health of aquaculture fish. Journal of Fish Diseases 2012;35: 83-108.

[15] Enes P, Panserat S, Kaushik S, Oliva-Teles A. Dietary Carbohydrate Utilization by European Sea Bass (*Dicentrarchus labrax* L.) and Gilthead Sea Bream (*Sparus aurata* L.) Juveniles. Reviews in Fisheries Science 2011;19: 201-215.

[16] Hemre GI, Mommsen TP, Krogdahl Ã. Carbohydrates in fish nutrition: effects on growth, glucose metabolism and hepatic enzymes. Aquaculture Nutrition 2002;8: 175-194.

[17] Stone DAJ. Dietary Carbohydrate Utilization by Fish. Reviews in Fisheries Science 2003;11: 337 - 369.

[18] Moon TW. Glucose intolerance in teleost fish: fact or fiction? Comparative Biochemistry and Physiology - Part B: Biochemistry & Molecular Biology 2001;129: 243-249.

[19] Nordlie RC, Foster JD. Regulation of glucose production by the liver. Annual Review of Nutrition 1999;19: 379–406.

[20] Blasco J, Marimón I, Viaplana I, Fernández-Borrás J. Fate of plasma glucose in tissues of brown trout *in vivo*: effects of fasting and glucose loading. Fish Physiology and Biochemistry 2001;24: 247–258.

[21] Polakof S, Mommsen TP, Soengas J. Glucosensing and glucose homeostasis: From fish to mammals. Comparative Biochemistry and Physiology - Part B: Biochemistry & Molecular Biology 2011;160: 123-149.

[22] West T, Schulte P, Hochachka P. Implications of hyperglycemia for post-exercise resynthesis of glycogen in trour skeletal muscle. Journal of Experimental Biology 1994;189: 69-84.

[23] West TG, Brauner CJ, Hochachka PW. Muscle glucose utilization during sustained swimming in the carp (*Cyprinus carpio*). American Journal of Physiology - Regulatory, Integrative and Comparative Physiology 1994;267: R1226-R1234.

[24] MacCormack TJ, Driedzic WR. The impact of hypoxia on *in vivo* glucose uptake in a hypoglycemic fish, *Myoxocephalus scorpius*. American Journal of Physiology - Regulatory, Integrative and Comparative Physiology 2007;292: R1033-1042.

[25] Thillart G. Energy metabolism of swimming trout (*Salmo gairdneri*). Journal of Comparative Physiology B: Biochemical, Systemic, and Environmental Physiology 1986;156: 511-520.

[26] Blasco J, Fernàndez-Borràs J, Marimon I, Requena A. Plasma glucose kinetics and tissue uptake in brown trout *in vivo*: effect of an intravascular glucose load. Journal of Comparative Physiology B: Biochemical, Systemic, and Environmental Physiology 1996;165: 534-541.

[27] Milligan CL. Metabolic recovery from exhaustive exercise in rainbow trout. Comparative Biochemistry and Physiology - Part A: Molecular & Integrative Physiology 1996;113: 51-60.

[28] Pagnotta A, Milligan CL. The role of blood glucose in the restoration of muscle glycogen during recovery from exhaustive exercise in rainbow trout (*Oncorhynchus mykiss*) and winter flounder (*Pseudopleuronectes americanus*). Journal of Experimental Biology 1991;161: 489-508.

[29] West TG, Arthur PG, Suarez RK, Doll CJ, Hochachka PW. *In vivo* utilization of glucose by heart and locomotory muscles of exercising rainbow trout (*Oncorhynchus mykiss*). Journal of Experimental Biology 1993;177: 63-79.

[30] Leturque A, Brot-Laroche E, Le Gall M, Stolarczyk E, Tobin V. The role of GLUT2 in dietary sugar handling. Journal of Physiology and Biochemistry 2005;61: 529-537.

[31] Polakof S, Míguez JM, Moon TW, Soengas JL. Evidence for the presence of a glucosensor in hypothalamus, hindbrain, and Brockmann bodies of rainbow trout. American Journal of Physiology - Regulatory, Integrative and Comparative Physiology 2007;292: R1657-R1666.

[32] Polakof S, Míguez JM, Soengas JL. *In vitro* evidences for glucosensing capacity and mechanisms in hypothalamus, hindbrain, and Brockmann bodies of rainbow trout. American Journal of Physiology - Regulatory, Integrative and Comparative Physiology 2007;293: R1410-R1420.

[33] Enes P, Panserat S, Kaushik S, Oliva-Teles A. Nutritional regulation of hepatic glucose metabolism in fish. Fish Physiology and Biochemistry 2009;35: 519-539.

[34] Ferrer JC, Favre C, Gomis RR, Fernandez-Novell JM, Garcia-Rocha M, de la Iglesia N, Cid E, Guinovart JJ. Control of glycogen deposition. FEBS Lett 2003;546: 127-132.

[35] Mehrani H, Storey KB. Characterization of α-Glucosidases from Rainbow Trout Liver. Archives of Biochemistry and Biophysics 1993;306: 188-194.

[36] Murat J-C. Studies on glycogenolysis in carp liver: Evidence for an amylase pathway for glycogen breakdown. Comparative Biochemistry and Physiology Part B: Comparative Biochemistry 1976;55: 461-465.

[37] Picukans I, Umminger BL. Comparative activities of glycogen phosphorylase and α-amylase in livers of carp (*Cyprinus carpio*) and goldfish (*Carassius auratus*). Comparative Biochemistry and Physiology Part B: Comparative Biochemistry 1979;62: 455-457.

[38] Navarro I, Gutiérrez J. Fasting and starvation. In Biochemistry and Molecular Biology of Fishes. Elsevier; 1995. 393-434

[39] Metón I, Fernández F, Baanante IV. Short- and long-term effects of refeeding on key enzyme activities in glycolysis-gluconeogenesis in the liver of gilthead seabream (*Sparus aurata*). Aquaculture 2003;225: 99-107.

[40] Pérez-Jiménez A, Guedes MJ, Morales AE, Oliva-Teles A. Metabolic responses to short starvation and refeeding in *Dicentrarchus labrax*. Effect of dietary composition. Aquaculture 2007;265: 325-335.

[41] Soengas JL, Strong EF, Fuentes J, Veira JAR, Andrés MD. Food deprivation and refeeding in Atlantic salmon, *Salmo salar*: effects on brain and liver carbohydrate and ketone bodies metabolism. Fish Physiology and Biochemistry 1996;15: 491-511.

[42] Pérez-Jiménez A, Cardenete G, Hidalgo M, García-Alcázar A, Abellán E, Morales A. Metabolic adjustments of *Dentex dentex* to prolonged starvation and refeeding. Fish Physiology and Biochemistry 2012: 1-13.

[43] Viegas I, Rito J, Jarak I, Leston S, Carvalho RA, Metón I, Pardal MA, Baanante IV, Jones JG. Hepatic glycogen synthesis in farmed European seabass (*Dicentrarchus labrax* L.) is dominated by indirect pathway fluxes. Comparative Biochemistry and Physiology - Part A: Molecular & Integrative Physiology 2012;163: 22-29.

[44] Couto A, Enes P, Peres H, Oliva-Teles A. Effect of water temperature and dietary starch on growth and metabolic utilization of diets in gilthead sea bream (*Sparus aur-*

ata) juveniles. Comparative Biochemistry and Physiology - Part A: Molecular & Integrative Physiology 2008;151: 45-50.

[45] Enes P, Panserat S, Kaushik S, Oliva-Teles A. Rearing temperature enhances hepatic glucokinase but not glucose-6-phosphatase activities in European sea bass (*Dicentrarchus labrax*) and gilthead sea bream (*Sparus aurata*) juveniles fed with the same level of glucose Comparative Biochemistry and Physiology - Part A: Molecular & Integrative Physiology 2008;150: 355-358.

[46] Laiz-Carrión R, Sangiao-Alvarellos S, Guzman JM, Martin del Rio MP, Soengas JL, Mancera JM. Growth performance of gilthead sea bream *Sparus aurata* in different osmotic conditions: Implications for osmoregulation and energy metabolism. Aquaculture 2005;250: 849-861.

[47] Sangiao-Alvarellos S, Laiz-Carrion R, Guzman JM, Martin del Rio MP, Miguez JM, Mancera JM, Soengas JL. Acclimation of *S. aurata* to various salinities alters energy metabolism of osmoregulatory and nonosmoregulatory organs. American Journal of Physiology - Regulatory, Integrative and Comparative Physiology 2003;285: R897-907.

[48] Caseras A, Metón I, Fernández F, Baanante IV. Glucokinase gene expression is nutritionally regulated in liver of gilthead sea bream (*Sparus aurata*). Biochimica et Biophysica Acta (BBA) - Gene Structure and Expression 2000;1493: 135-141.

[49] Soengas JL, Aldegunde M. Energy metabolism of fish brain. Comparative Biochemistry and Physiology - Part B: Biochemistry & Molecular Biology 2002;131: 271-296.

[50] Sangiao-Alvarellos S, Guzmán JM, Láiz-Carrión R, Míguez JM, Martín Del Río MP, Mancera JM, Soengas JL. Interactive effects of high stocking density and food deprivation on carbohydrate metabolism in several tissues of gilthead sea bream *Sparus auratus*. Journal of Experimental Zoology Part A: Comparative Experimental Biology 2005;303A: 761-775.

[51] Fernández F, Miquel AG, Cordoba M, Varas M, Metón I, Caseras A, Baanante IV. Effects of diets with distinct protein-to-carbohydrate ratios on nutrient digestibility, growth performance, body composition and liver intermediary enzyme activities in gilthead sea bream (*Sparus aurata*, L.) fingerlings. Journal of Experimental Marine Biology and Ecology 2007;343: 1-10.

[52] Melo JFB, Lundstedt LM, Metón I, Baanante IV, Moraes G. Effects of dietary levels of protein on nitrogenous metabolism of *Rhamdia quelen* (Teleostei: Pimelodidae). Comparative Biochemistry and Physiology - Part A: Molecular & Integrative Physiology 2006;145: 181-187.

[53] Metón I, Mediavilla D, Caseras A, Cantó M, Fernández F, Baanante IV. Effect of diet composition and ration size on key enzyme activities of glycolysis-gluconeogenesis, the pentose phosphate pathway and amino acid metabolism in liver of gilthead sea bream (*Sparus aurata*). British Journal of Nutrition 1999;82: 223-232.

[54] Panserat S, Medale F, Blin C, Breque J, Vachot C, Plagnes-Juan E, Gomes E, Krishna-moorthy R, Kaushik S. Hepatic glucokinase is induced by dietary carbohydrates in rainbow trout, gilthead seabream, and common carp. American Journal of Physiology - Regulatory, Integrative and Comparative Physiology 2000;278: R1164-1170.

[55] Panserat S, Plagnes-Juan E, Breque J, Kaushik S. Hepatic phosphoenolpyruvate car-boxykinase gene expression is not repressed by dietary carbohydrates in rainbow trout (Oncorhynchus mykiss). Journal of Experimental Biology 2001;204: 359-365.

[56] Panserat S, Plagnes-Juan E, Kaushik S. Nutritional regulation and tissue specificity of gene expression for proteins involved in hepatic glucose metabolism in rainbow trout (Oncorhynchus mykiss). Journal of Experimental Biology 2001;204: 2351-2360.

[57] Pérez-Jiménez A, Hidalgo MC, Morales AE, Arizcun M, Abellón E, Cardenete G. Use of different combinations of macronutrients in diets for dentex (Dentex dentex): Ef-fects on intermediary metabolism. Comparative Biochemistry and Physiology - Part A: Molecular & Integrative Physiology 2009;152: 314-321.

[58] Enes P, Peres H, Sanchez-Gurmaches J, Navarro I, Gutiérrez J, Oliva-Teles A. Insulin and IGF-I response to a glucose load in European sea bass (Dicentrarchus labrax) juve-niles. Aquaculture 2011;315: 321-326.

[59] Enes P, Sanchez-Gurmaches J, Navarro I, Gutiérrez J, Oliva-Teles A. Role of insulin and IGF-I on the regulation of glucose metabolism in European sea bass (Dicen-trarchus labrax) fed with different dietary carbohydrate levels. Comparative Biochem-istry and Physiology - Part A: Molecular & Integrative Physiology 2010;157: 346-353.

[60] Laiz-Carrión R, Martín del Río M, Míguez J, Mancera J, Soengas J. Influence of Corti-sol on Osmoregulation and Energy Metabolism in Gilthead Seabream Sparus aurata. Journal of Experimental Zoology 2003;298A: 105-118.

[61] Vijayan MM, J. MA, B. SC, Moon TW. Hormonal Control of Hepatic Glycogen Me-tabolism in Food-deprived, Continuously Swimming Coho Salmon (Oncorhynchus ki-sutch) Canadian Journal of Fisheries and Aquatic Sciences 1993;50: 1676–1682.

[62] Moon TW. Hormones and fish hepatocyte metabolism: "the good, the bad and the ugly!". Comparative Biochemistry and Physiology - Part B: Biochemistry & Molecu-lar Biology 2004;139: 335-345.

[63] Navarro I, Rojas P, Capilla E, Albalat A, Castillo J, Montserrat N, Codina M, Gutiér-rez J. Insights into Insulin and Glucagon Responses in Fish. Fish Physiology and Bio-chemistry 2002;27: 205-216.

[64] Weber J-M, Shanghavi DS. Regulation of glucose production in rainbow trout: role of epinephrine in vivo and in isolated hepatocytes. American Journal of Physiology - Regulatory, Integrative and Comparative Physiology 2000;278: R956-963.

[65] Mommsen TP, Vijayan MM, Moon TW. Cortisol in teleosts: dynamics, mechanisms of action, and metabolic regulation. Reviews in Fish Biology and Fisheries 1999;9: 211-268.

[66] Hemre GI, Storebakken T. Tissue and organ distribution of ^{14}C-activity in dextrin-adapted Atlantic salmon after oral administration of radiolabelled [14]C1-glucose. Aquaculture Nutrition 2000;6: 229-234.

[67] Felip O, Ibarz A, Fernández-Borràs J, Beltrán M, Martín-Pérez M, Planas JV, Blasco J. Tracing metabolic routes of dietary carbohydrate and protein in rainbow trout (*Oncorhynchus mykiss*) using stable isotopes ([^{13}C]starch and [15N]protein): effects of gelatinisation of starches and sustained swimming. British Journal of Nutrition 2012;107: 834-844.

[68] Conceição LEC, Grasdalen H, Dinis MT. A new method to estimate the relative bioavailability of individual amino acids in fish larvae using ^{13}C-NMR spectroscopy. Comparative Biochemistry and Physiology - Part B: Biochemistry and Molecular Biology 2003;134: 103-109.

[69] Mari A, Stojanovska L, Proietto J, Thorburn AW. A circulatory model for calculating non-steady-state glucose fluxes. Validation and comparison with compartmental models. Computer methods and programs in biomedicine 2003;71: 269-281.

[70] Conceição LEC, Morais S, Ronnestad I. Tracers in fish larvae nutrition: A review of methods and applications. Aquaculture 2007;267: 62-75.

[71] Morais S, Conceição LE, Dinis MT, Ronnestad I. A method for radiolabeling Artemia with applications in studies of food intake, digestibility, protein and amino acid metabolism in larval fish. Aquaculture 2004;231: 469-487.

[72] Morais S, Torten M, Nixon O, Lutzky S, Conceição LEC, Dinis MT, Tandler A, Koven W. Food intake and absorption are affected by dietary lipid level and lipid source in seabream (*Sparus aurata* L.) larvae. Journal of Experimental Marine Biology and Ecology 2006;331: 51-63.

[73] Bever K, Chenoweth M, Dunn A. Glucose turnover in kelp bass (*Paralabrax* sp.): *in vivo* studies with [6-^{3}H,6-^{14}C]glucose. American Journal of Physiology - Regulatory, Integrative and Comparative Physiology 1977;232: R66-72.

[74] Garin D, Rombaut A, Fréminet A. Determination of glucose turnover in sea bass *Dicentrarchus labrax*. Comparative aspects of glucose utilization. Comparative Biochemistry and Physiology - Part B: Biochemistry & Molecular Biology 1987;87: 981-988.

[75] Katz J, Rostami H, Dunn A. Evaluation of glucose turnover, body mass and recycling with reversible and irreversible tracers. The Biochemical Journal 1974;142: 161-170.

[76] Machado CR, Garofalo MA, Roselino JE, Kettelhut IC, Migliorini RH. Effect of fasting on glucose turnover in a carnivorous fish (*Hoplias* sp). American Journal of Physiology - Regulatory, Integrative and Comparative Physiology 1989;256: R612-615.

[77] Haman F, Powell M, Weber J. Reliability of continuous tracer infusion for measuring glucose turnover rate in rainbow trout. Journal of Experimental Biology 1997;200: 2557-2563.

[78] Haman F, Weber JM. Continuous tracer infusion to measure *in vivo* metabolite turn-over rates in trout. Journal of Experimental Biology 1996;199: 1157-1162.

[79] Soivio A, Nynolm K, Westman K. A technique for repeated sampling of the blood of individual resting fish. Journal of Experimental Biology 1975;63: 207-217.

[80] Chandramouli V, Ekberg K, Schumann WC, Kalhan SC, Wahren J, Landau BR. Quantifying gluconeogenesis during fasting. American Journal of Physiology - Endocrinology And Metabolism 1997;273: E1209-1215.

[81] Jones JG, Solomon MA, Cole SM, Sherry AD, Malloy CR. An integrated ^{2}H and ^{13}C NMR study of gluconeogenesis and TCA cycle flux in humans. American Journal of Physiology - Endocrinology And Metabolism 2001;281: E848-E856.

[82] Nunes PM, Jones JG. Quantifying endogenous glucose production and contributing source fluxes from a single ^{2}H NMR spectrum. Magnetic Resonance in Medicine 2009;62: 802-807.

[83] Gasier HG, Previs SF, Pohlenz C, Fluckey JD, Gatlin III DM, Buentello JA. A novel approach for assessing protein synthesis in channel catfish, *Ictalurus punctatus*. Comparative Biochemistry and Physiology - Part B: Biochemistry & Molecular Biology 2009;154: 235-238.

[84] González JD, Caballero A, Viegas I, Metón I, Jones JG, Barra J, Fernández F, Baanante IV. Effects of alanine aminotransferase inhibition on the intermediary metabolism in *Sparus aurata* through dietary amino-oxyacetate supplementation. British Journal of Nutrition 2012;107: 1747-1756.

[85] Viegas I, Mendes VM, Leston S, Jarak I, Carvalho RA, Pardal MA, Manadas B, Jones JG. Analysis of glucose metabolism in farmed European sea bass (*Dicentrarchus labrax* L.) using deuterated water. Comparative Biochemistry and Physiology - Part A: Molecular & Integrative Physiology 2011;160: 341-347.

[86] Dabrowski K, Guderley H. Intermediary Metabolism. In Fish Nutrition (Third Edition). San Diego: Academic Press; 2003. 309-365

[87] Landau BR, Wahren J, Chandramouli V, Schumann WC, Ekberg K, Kalhan SC. Contributions of gluconeogenesis to glucose production in the fasted state. The Journal of Clinical Investigation 1996;98: 378-385.

[88] Saadatian M, Peroni O, Diraison F, Beylot M. *In vivo* measurement of gluconeogenesis in animals and humans with deuterated water: a simplified method. Diabetes Metab 2000;26: 202-209.

[89] Barosa C, Silva C, Fagulha A, Barros Ls, Caldeira MM, Carvalheiro M, Jones JG. Sources of hepatic glycogen synthesis following a milk-containing breakfast meal in healthy subjects. Metabolism 2012;61: 250-254.

[90] Chacko SK, Sunehag AL, Sharma S, Sauer PJJ, Haymond MW. Measurement of gluconeogenesis using glucose fragments and mass spectrometry after ingestion of deuterium oxide. Journal of Applied Physiology 2008;104: 944-951.

[91] Jones JG, Garcia P, Barosa C, Delgado TC, Diogo L. Hepatic anaplerotic outflow fluxes are redirected from gluconeogenesis to lactate synthesis in patients with Type 1a glycogen storage disease. Metabolic Engineering 2009;11: 155-162.

[92] Sena CM, Barosa C, Nunes E, Seiça R, Jones JG. Sources of endogenous glucose production in the Goto–Kakizaki diabetic rat. Diabetes & Metabolism 2007;33: 296-302.

[93] Soares AF, Viega FJ, Carvalho RA, Jones JG. Quantifying hepatic glycogen synthesis by direct and indirect pathways in rats under normal ad libitum feeding conditions. Magnetic Resonance in Medicine 2009;61: 1-5.

[94] Landau BR. Methods for measuring glycogen cycling. American Journal of Physiology - Endocrinology And Metabolism 2001;281: E413-E419.

[95] Gastaldelli A, Baldi S, Pettiti M, Toschi E, Camastra S, Natali A, Landau BR, Ferrannini E. Influence of obesity and type 2 diabetes on gluconeogenesis and glucose output in humans: a quantitative study. Diabetes 2000;49: 1367-1373.

[96] Jones JG. Tracing Hepatic Glucose and Glycogen Fluxes with ^2H_2O. In Clinical Diabetes Research. London: John Wiley, Sons, Ltd; 2007. pp. 139-149.

[97] Jones JG, Fagulha A, Barosa C, Bastos M, Barros L, Baptista C, Caldeira MM, Carvalheiro M. Noninvasive Analysis of Hepatic Glycogen Kinetics Before and After Breakfast with Deuterated Water and Acetaminophen. Diabetes 2006;55: 2294-2300.

[98] Shimeno S, Shikata T. Effects of Acclimation Temperature and Feeding Rate on Carbohydrate-Metabolizing Enzyme Activity and Lipid Content of Common Carp. Nippon Suisan Gakk 1993;59: 661-666.

[99] Shimeno S, Shikata T, Hosokawa H, Masumoto T, Kheyyali D. Metabolic response to feeding rates in common carp, *Cyprinus carpio*. Aquaculture 1997;151: 371-377.

[100] Ullah S, Wahren J, Beck O. Precise determination of glucose-d2/glucose ratio in human serum and plasma by APCI LC-MS/MS. Scandinavian Journal of Clinical & Laboratory Investigation 2009;69: 837-842.

[101] Wu H, Southam AD, Hines A, Viant MR. High-throughput tissue extraction protocol for NMR- and MS-based metabolomics. Analytical Biochemistry 2008;372: 204-212.

[102] Jones JG, Perdigoto R, Rodrigues TB, Geraldes CFGC. Quantitation of absolute ^2H enrichment of plasma glucose by ^2H NMR analysis of its monoacetone derivative. Magnetic Resonance in Medicine 2002;48: 535-539.

[103] Kunert O, Stingl H, Rosian E, Krssak M, Bernroider E, Seebacher W, Zangger K, Staehr P, Chandramouli V, Landau BR, Nowotny P, Waldhausl W, Haslinger E, Roden M. Measurement of fractional whole-body gluconeogenesis in humans from

blood samples using ^{2}H nuclear magnetic resonance spectroscopy. Diabetes 2003;52: 2475-2482.

[104] Schleucher J, Vanderveer P, Markley JL, Sharkey TD. Intramolecular deuterium distributions reveal disequilibrium of chloroplast phosphoglucose isomerase. Plant, Cell & Environment 1999;22: 525-533.

Permissions

The contributors of this book come from diverse backgrounds, making this book a truly international effort. This book will bring forth new frontiers with its revolutionizing research information and detailed analysis of the nascent developments around the world.

We would like to thank Hakan Turker, Ph.D., for lending his expertise to make the book truly unique. He has played a crucial role in the development of this book. Without his invaluable contribution this book wouldn't have been possible. He has made vital efforts to compile up to date information on the varied aspects of this subject to make this book a valuable addition to the collection of many professionals and students.

This book was conceptualized with the vision of imparting up-to-date information and advanced data in this field. To ensure the same, a matchless editorial board was set up. Every individual on the board went through rigorous rounds of assessment to prove their worth. After which they invested a large part of their time researching and compiling the most relevant data for our readers. Conferences and sessions were held from time to time between the editorial board and the contributing authors to present the data in the most comprehensible form. The editorial team has worked tirelessly to provide valuable and valid information to help people across the globe.

Every chapter published in this book has been scrutinized by our experts. Their significance has been extensively debated. The topics covered herein carry significant findings which will fuel the growth of the discipline. They may even be implemented as practical applications or may be referred to as a beginning point for another development. Chapters in this book were first published by InTech; hereby published with permission under the Creative Commons Attribution License or equivalent.

The editorial board has been involved in producing this book since its inception. They have spent rigorous hours researching and exploring the diverse topics which have resulted in the successful publishing of this book. They have passed on their knowledge of decades through this book. To expedite this challenging task, the publisher supported the team at every step. A small team of assistant editors was also appointed to further simplify the editing procedure and attain best results for the readers.

Our editorial team has been hand-picked from every corner of the world. Their multi-ethnicity adds dynamic inputs to the discussions which result in innovative outcomes. These outcomes are then further discussed with the researchers and contributors who give their valuable feedback and opinion regarding the same. The feedback is then collaborated with the researches and they are edited in a comprehensive manner to aid the understanding of the subject.

Apart from the editorial board, the designing team has also invested a significant amount of their time in understanding the subject and creating the most relevant covers. They scrutinized every image to scout for the most suitable representation of the subject and create an appropriate cover for the book.

The publishing team has been involved in this book since its early stages. They were actively engaged in every process, be it collecting the data, connecting with the contributors or procuring relevant information. The team has been an ardent support to the editorial, designing and production team. Their endless efforts to recruit the best for this project, has resulted in the accomplishment of this book. They are a veteran in the field of academics and their pool of knowledge is as vast as their experience in printing. Their expertise and guidance has proved useful at every step. Their uncompromising quality standards have made this book an exceptional effort. Their encouragement from time to time has been an inspiration for everyone.

The publisher and the editorial board hope that this book will prove to be a valuable piece of knowledge for researchers, students, practitioners and scholars across the globe.

List of Contributors

Barbara A. Katzenback and Fumihiko Katakura
Department of Biological Sciences, University of Alberta, Edmonton, Alberta, Canada

Miodrag Belosevic
Department of Biological Sciences, University of Alberta, Edmonton, Alberta, Canada
Department of Biological Sciences, School of Public Health, University of Alberta, Edmonton, Alberta, Canada

Sebastián Reyes-Cerpa and Daniela Toro-Ascuy
Dept. Biology, Laboratory of Virology, University of Santiago of Chile, Santiago, Chile

Kevin Maisey and Mónica Imarai
Dept. Biology, Laboratory of Immunology, University of Santiago of Chile, Santiago, Chile

Felipe Reyes-López
Dept. Cell Biology, Physiology & Immunology, Unit of Animal Physiology, Autonomous University of Barcelona, Barcelona, Spain

Ana María Sandino
Dept. Biology, Laboratory of Virology, University of Santiago of Chile, Santiago, Chile; Activaq S.A., Santiago, Chile

Leon Grayfer
Department of Biological Sciences, University of Alberta, Canada

Jacinto Elías Sedeño-Díaz
Polytechnic Coordination for Sustainability, National Polytechnic Institute, Mexico, Distrito
Federal, México

Eugenia López-López
Ichtiology and Limnology Laboratory, National School of Biological Sciences, México, Distrito Federal, México

Cláudia Turra Pimpão
Laboratory of Pharmacology and Toxicology/PUCPR, College of Agricultural, Environmental Sciences and Veterinary Medicine, Pontifícia Universidade Católica do Paraná, Curitiba, Brazil

Ênio Moura
Service of Medical Genetics/PUCPR, College of Agricultural, Environmental Sciences
and Veterinary Medicine, Pontifícia Universidade Católica do Paraná, Curitiba, Brazil

Ana Carolina Fredianelli and Luciana G. Galeb
College of Agricultural, Environmental Sciences and Veterinary Medicine, Pontifícia
Universidade Católica do Paraná, Curitiba, Brazil

Rita Maria V. Mangrich Rocha
Service of Veterinary Clinical Pathology/PUCPR, College of Agricultural, Environmental
Sciences and Veterinary Medicine, Pontifícia Universidade Católica do Paraná, Curitiba,
Brazil

Francisco P. Montanha
Laboratory of Pharmacology and Toxicology/FAMED, College of Agricultural,
Environmental Sciences and Veterinary Medicine,Pontifícia Universidade Católica do
Paraná, Curitiba, Brazil

Carlos Edwar de Carvalho Freitas, Alexandre A. F. Rivas and James Randall Kahn
Federal University of Amazonas, Manaus, Amazonas, Brazil
Washington and Lee University, Manaus, Amazonas, Brazil

Caroline Pereira Campos and Michel Fabiano Catarino
National Institute for Amazonian Research, Manaus, Amazonas, Brazil

Igor Sant'Ana and Maria Angélica de Almeida Correa
Federal University of Amazonas, Manaus, Amazonas, Brazil

Christine Genge
Molecular Cardiac Physiology Group, Simon Fraser University, Burnaby, Canada

Leif Hove-Madsen
Cardiovascular Research Centre CSIC-ICCC, Hospital de Sant Pau, Barcelona, Spain

Glen F. Tibbits
Molecular Cardiac Physiology Group, Simon Fraser University, Burnaby, Canada
Cardiovascular Sciences, Child and Family Research Institute, Vancouver, Canada

Javier Sánchez-Hernández and Fernando Cobo
Department of Zoology and Physical Anthropology, Faculty of Biology, University of
Santiago de Compostela, Spain
Station of Hydrobiology "Encoro do Con", Castroagudín s/n, Vilagarcía de Arousa,
Pontevedra, Spain

María J. Servia
Department of Animal Biology, Vegetal Biology and Ecology, Faculty of Science,
University of A Coruña, Spain

Rufino Vieira-Lanero
Station of Hydrobiology "Encoro do Con", Castroagudín s/n, Vilagarcía de Arousa, Pontevedra, Spain

Ivan Viegas
Department of Life Sciences, Faculty of Sciences and Technology, University of Coimbra, Portugal
CNC - Center for Neuroscience and Cell Biology, University of Coimbra, Portugal
CFE - Centre for Functional Ecology, University of Coimbra, Portugal

Rui de Albuquerque Carvalho
Department of Life Sciences, Faculty of Sciences and Technology, University of Coimbra, Portugal
CNC - Center for Neuroscience and Cell Biology, University of Coimbra, Portugal

Miguel Ângelo Pardal
Department of Life Sciences, Faculty of Sciences and Technology, University of Coimbra, Portugal
CFE - Centre for Functional Ecology, University of Coimbra, Portugal

John Griffith Jones
CNC - Center for Neuroscience and Cell Biology, University of Coimbra, Portugal